Water Sustainability and Climate Change

Water Sustainability and Climate Change

Edited by **Bruce Horak**

SYRAWOOD
PUBLISHING HOUSE
New York

Published by Syrawood Publishing House,
750 Third Avenue, 9th Floor,
New York, NY 10017, USA
www.syrawoodpublishinghouse.com

Water Sustainability and Climate Change
Edited by Bruce Horak

International Standard Book Number: 978-1-68286-224-7 (Hardback)

Printed in China.

Contents

Preface

This book has been a concerted effort by a group of academicians, researchers and scientists, who have contributed their research works for the realization of the book. This book has materialized in the wake of emerging advancements and innovations in this field. Therefore, the need of the hour was to compile all the required researches and disseminate the knowledge to a broad spectrum of people comprising of students, researchers and specialists of the field.

The greenhouse effect and overall rise in the average temperatures across the globe have had a major impact on water bodies and their sustainability. This has made the integrated study of climate change and its impact on natural resources very important. Impacts of climate change on water resource systems, sustainable use and planning of water resources, storm water management under changing climate, etc. are some of the topics that have been elaborated in detail. This book includes various researches and studies conducted by experts and will prove to be highly beneficial for students and research scholars.

At the end of the preface, I would like to thank the authors for their brilliant chapters and the publisher for guiding us all-through the making of the book till its final stage. Also, I would like to thank my family for providing the support and encouragement throughout my academic career and research projects.

Editor

1

Impact of Extreme Floods on Groundwater Quality (in Pakistan)

Tariq Usman Saeed[1*] and Haleema Attaullah[2]

[1]Department of Transport and Mass Transit, Government of Khyber Pakhtunkhwa, Peshawar, Pakistan.
[2]Department of Civil Engineering, University of Engineering and Technology, Peshawar, Pakistan.

Authors' contributions

This work was carried out in collaboration between both authors. Both authors read and approved the final manuscript.

ABSTRACT

Study of long and short-term impact of hydro-meteorologically induced extreme flood on groundwater from well is a baby science, yet to grow and groom. This article focuses on the environmental impacts of the worst Pakistani floods on water quality of affected areas, Charsadda and Nowshera districts in Khyber Pakhtunkhwa province which experienced a disastrous flood in its record due to torrential monsoon rains in late July 2010. For this purpose, consuming water products from 10 main sources (tube wells), 10 intermediate points in water supply distribution system and 10 consumers' ends in 30 selected sites of flood affected areas were collected and analyzed for 12 key factors. Most of the parameters with respect to the standard limits of WHO guidelines indicated contamination in all samples that are directly available from tube wells as well as the one supplied through damaged pipe distribution system. This result becomes more fatal in the presence of microbial contamination and makes water risky for domestic consumption. A concrete policy addressing post-flood environmental effects on life and human health should be devised and strictly followed. Individual cases must be assessed from a basin-wide perspective in order to make sure that environmental concerns are judiciously and properly represented in flood planning and risk management decisions.

Keywords: KP (Khyber Pakhtunkhwa); WHO; Nowshera; Charsadda; climatic; groundwater; tubewells; water sample.

**Corresponding author: Email: tariqusaeed@gmail.com*

1. INTRODUCTION

Water quality is of key importance to man and nature [1] and one can't deny the fact that it is one of the basic factors for the safe and healthy survival of humanity at large. A healthy society can't evolve in the absence of good quality water. It is quite clear from the previous examples that poor quality or contaminated water has led to the destruction of a whole community as the most tragic example is the 1972 flood in Buffalo Creek, West Virginia. A dam, layered with mine tailings over a period of time to raise its height from 20 feet and increase storage capacity, broke on February 26th, 1972 and discharged a violently fast stream of water contaminated with mine wastes on communities living downstream. More than 125 people were killed as the polluted water in the form of black wave of fluid containing more than two million tons of waste products coursed through streamside communities [2]. The community of Saunders met a dreadful destruction in a minute and the valley lay silent in shock, covered by a layer of sludge [1]. It is obvious that flooding causes a considerable reduction in water quality which then contributes to the adverse climatic impacts. The poor quality of water may be due to huge amount of man-made and natural contaminants and erosion. Man has altered low lying areas to farmlands by considerable removal of floodplain vegetation and wetland regions which serve as sediment intercept, hydraulic sponges, natural stilling ponds and erosion protection. Besides all these, overland flows have been a source of unwanted chemicals including heavy metals in the water on surface. Submerged municipal and industrial sites like wastewater treatment plants, chemical processing and manufacturing centers, and disposal or holding areas serve as point sources for chemical contamination of water [3] while runoff from agricultural land is the main non-point pollution origin. Large amounts of herbicides and nitrate were transported from urban and agricultural areas during the ruinous Mississippi River basin flood of 1993 [3]. 65 percent of the cropland in America is contained in the Mississippi River basin where an approximated one hundred thousand metric tons of pesticides and 6,300,000 metric tons of nitrogen fertilizers are utilized on annual basis. Goolsby et al. [3] compared 1992 and 1993 loads at the Mississippi mouth and noticed a 235 percent increase in the load of atrazine herbicide used in corn production and a 112 percent increase in the load of nitrate. Short- and long-term adverse climatic impacts can result from chemical loading and poor quality water. Goolsby et al. [3] cited Michael Dowgiallo of the National Oceanic and Atmospheric Administration in saying that the chemical loads were promoting primary production which caused an elevation in the levels of marine phytoplankton in the Gulf of Mexico. Zooplanktons, being a potent part of aquatic food chains, consume phytoplanktonic algae and provide a forage base for higher predators but this may cause adverse algal peaks.

Study of long and short-term environmental impacts of extreme floods is a baby science, yet to grow and groom. The quantitative impact analysis of these violent and short-lived environmental events requires reliable field data. The environmental impacts are complicated and unused in policy making. The focus of water management has recently changed from the need to dominate water resources to a more harmonious philosophy that seeks a balance between the structural flow control required to support and protect growing populations and environmental well-being [4-8]. Hickey et al. [1] identified status quo policy, the need for constant flood control and an incomplete understanding of environmental flood dynamics as hurdle for the implementation of this philosophy.

The nature and ingredients of flood flow are one of the important factors which define the quality of water of a specific area. Floods show diverse environmental and climatic impacts on water quality in case of different living beings; some are favored by these impacts while others have to face a non-compensatory and irrecoverable loss. The adverse climatic impacts of floods are found in the form of widespread chronic epidemics in humans.

1.1 Case Study of Pakistan

A flood is a natural disaster and integral component of natural cycle which often happens in multitude of ways and inundates area and communities as well. This natural phenomenon as a part of the earth's bio-physical processes becomes devastating due to human activity such as clearing of vegetation, intervention in natural drainage lines, deforestation and permanent occupancy of wetlands in the buffer zones of riverbank ecosystem.

In late July 2010, hard monsoon rainstorms led massive deluge in Pakistan which left its disastrous widespread footprints in seven districts of Pakistan. The heavy rainfall due to monsoon patterns and excessive heat due to seasonal weather conditions melted the glaciers which were responsible for the flash floods in Pakistan [9]. Unwanted legal or illegal excessive deforestation and lack of storage dams in Pakistan also played an additional significant role in increasing the chances of intense and huge damages caused by such floods. The severe flood, resulting from heavy monsoon rains during the four day wet spell from 27 to 30 July 2010, broke a long standing, 100 years record flood in early 1900s and rapidly became devastating for the province of Khyber Pakhtunkhwa. These rains inflicted heavy damage in districts of Charsadda and Nowshera. Heavy rainfall of more than 200 millimeters (7.9 in) was recorded in Khyber Pakhtunkhwa based on data from the Pakistan Meteorological Department. Gulf News [10] reported that the flood affected more than 20 million people exceeding the combined total of individuals affected by the 2004 Indian Ocean Tsunami, the 2005 Kashmir Earthquake and the 2010 Haiti Earthquake. UN Secretary-General Ban Ki-moon noted that the flood was the worst disaster he had ever seen. The World Health Organization [11] reported that ten million people were forced to drink unsafe water. It was reported in Dawn News [12] that the Pakistani economy was harmed by extensive damage to infrastructure and crops. The United Nations rated the floods in Pakistan as the greatest humanitarian crisis in recent history with more people affected than the South-East Asian tsunami and the recent earthquakes in Kashmir and Haiti combined.

If one portrays the trail of devastations that took place due to 2010 flash floods in all over the country, the noticeable and serious damages can be seen in Khyber Pakhtunkhwa (KP) province. The people's lives, economy, agriculture, infrastructure, buildings and communication networks were badly affected in KP. Land contaminants, toxic chemicals from septic tanks and industrial effluents were transferred and propagated during surging and caused serious health issues to every age group of people. Serious flood casualties also occurred to the population residing along and near River Kabul and its tributaries.

The two districts Charsadda and Nowshera were the two main flood affected zones of KP as almost all the villages of these districts were horribly hit and damaged by the floods on a large scale (Figs. 1, 2, 7).

After the flood, many organizations had an immediate focus on human losses, damages to livelihoods, properties and infrastructures in response to the flood damages assessment, emergency rescue and rehabilitation but the environmental impacts on human life and nature were neglected. This worst flood triggered damages not only to the life of individuals and destroyed infrastructure but also most of the drinking and irrigation water sources i.e. tube wells got submerged that made it difficult for the flood affected population to find potable and safe water from tube wells for drinking and other domestic purposes. As majority of the population of these areas is reliant on the consumption of ground water from the wells for drinking, cooking, preparation of food, washing cloths and utensils, hygiene and sanitation purposes, therefore, the need of safe and clean water in these areas was more and very immediate. The important environmental issue that people are experiencing nowadays due to hefty flooding in these areas of KP is the contamination of groundwater which is needed to be explored to provide safe and healthy water. The surface runoff created by heavy floods in these areas, when flowed over the ground, picked up and transmitted the natural and man-made contaminants with unsafe concentrations [13]. It also transmitted toxic materials and chemicals from the industrial effluents, pesticides and fertilizers to be used for the crop production. These contaminants finally deposited into water sheds and also subterranean rivers and polluted the whole aquifer making it unsuitable for use in daily routine.

The quality of water from tube wells in these areas was tested at three different points which are main source, intermediate point of the distribution line and the consumer's end. Selecting three significant points to determine the water quality from a single source was only because of the variation in the conditions of tube wells and distributaries in these areas. It was observed that few of the wells in the study areas had concrete lining. Some tube wells due to their substandard designs especially the private and the older ones which were not designed and maintained according to the design requirements increased the possibilities of polluting the subterranean water. It was also observed that 90 percent of the pipe lines were running through the open drains and were leaked from joints at many places ensured the contamination of water. Most of the inhabitants in these areas had installed small and cheap filter units at their homes which facilitate them in the availability of clean water. However, for this study the consumer's ends were selected at such areas where the people did not have the facility of filter units in order to get the clear picture of the quality of available water used by the majority of the population.

Fig. 1. Fully or partially submerged buildings in Nowshera under 15 feet of water, 2010

Fig. 2. In a flood-hit village near Nowshera, the rescue operation is in progress, 2010

The drinking water due to its sub-optimal quality may have serious adverse impacts on the health of population of all age groups causing high occurrence of water-borne diseases like diarrhea, liver disorder, cholera, giardiasis and abdominal viruses [14]. This polluted water also results in high chances of skin diseases during bathing and washing. Presence of some of the non-biodegradable heavy metals like lead, zinc, copper, manganese etc. in poor quality drinking water also causes chronic health issues such as weakness of neurological system, hair loss, liver cirrhosis and failure of renal system [13]. As most of the population in flood affected areas is reliant on ground water from the wells, the lack of the safe drinking water management practices in these areas specified the urgent need to ascertain if the water supply is suitable for human drinking purpose and for other domestic uses and whether it satisfies WHO guidelines or not.

In this study, water samples from 10 tube wells in the flood affected areas were collected and tested by routine techniques for 12 key parameters of water quality. The article evaluates the water quality of flood hit areas of Pakistan for human use and its climatic implications.

1.2 Study area

Charsadda, 34°09′N 71°44′E, at an altitude of 276 m and 29 km from the provincial capital Peshawar and Nowshera, 34°0'55N 71°58'29E, 27 miles due east of Peshawar are the focus of this research; shown in Fig. 3. Three rivers flowing in Charsadda: the River Jindi, the River Kabul and the River Swatmerge and join the River Indus.

Fig. 3. Map showing location of study areas in Pakistan (source: Wikipedia)

The local population in Charsadda and Nowshera faced the accumulated flood waters from Kabul and Swat rivers and from numerous flood channels flowing from the adjoining Malakand Division and FATA regions as can be seen from Fig. 4 and Fig. 5. River Kabul continued to flow much beyond its normal span (Fig. 4 and Fig. 5). Flood Monitoring Map in Fig. 6 identifies the areas hit by the river flood flow. Indus River had caused a massive flood water back flow resulting in extensive flooding in the southern parts of Nowshera district. The Kabul River is a 700 km long river that starts in the Sanglakh Range of the Hindu Kush Mountains in Afghanistan and ends in the Indus River near Attock, Pakistan. An extensive loss of standing sugarcane and other crops in particular along the eastern bank of River

Kabul was observed. Standing crops of maize, sugarcane and rice besides crops of various vegetables were destroyed and the lands, pastures and grazing areas got submerged under water. Nowshera was the worst hit out of the two districts as massive flooding along either banks of the Kabul River caused immense losses in habitats, standing crops and in livelihoods. The stagnant flood waters caused a major threat of water borne diseases in both the districts: diarrhea, AWT, malaria etc.

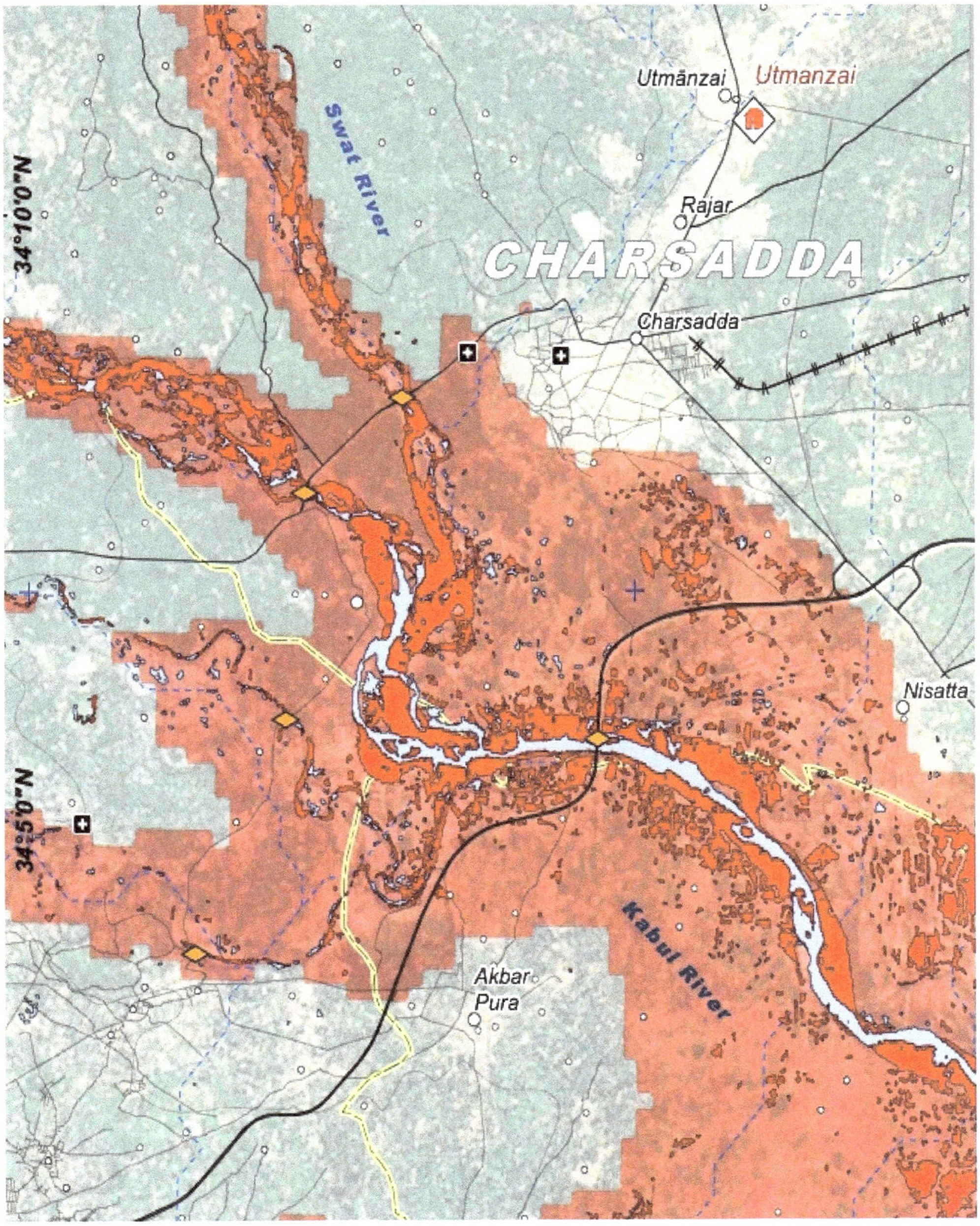

Fig. 4. Flooded River Kabul spreading and flooding Charsadda

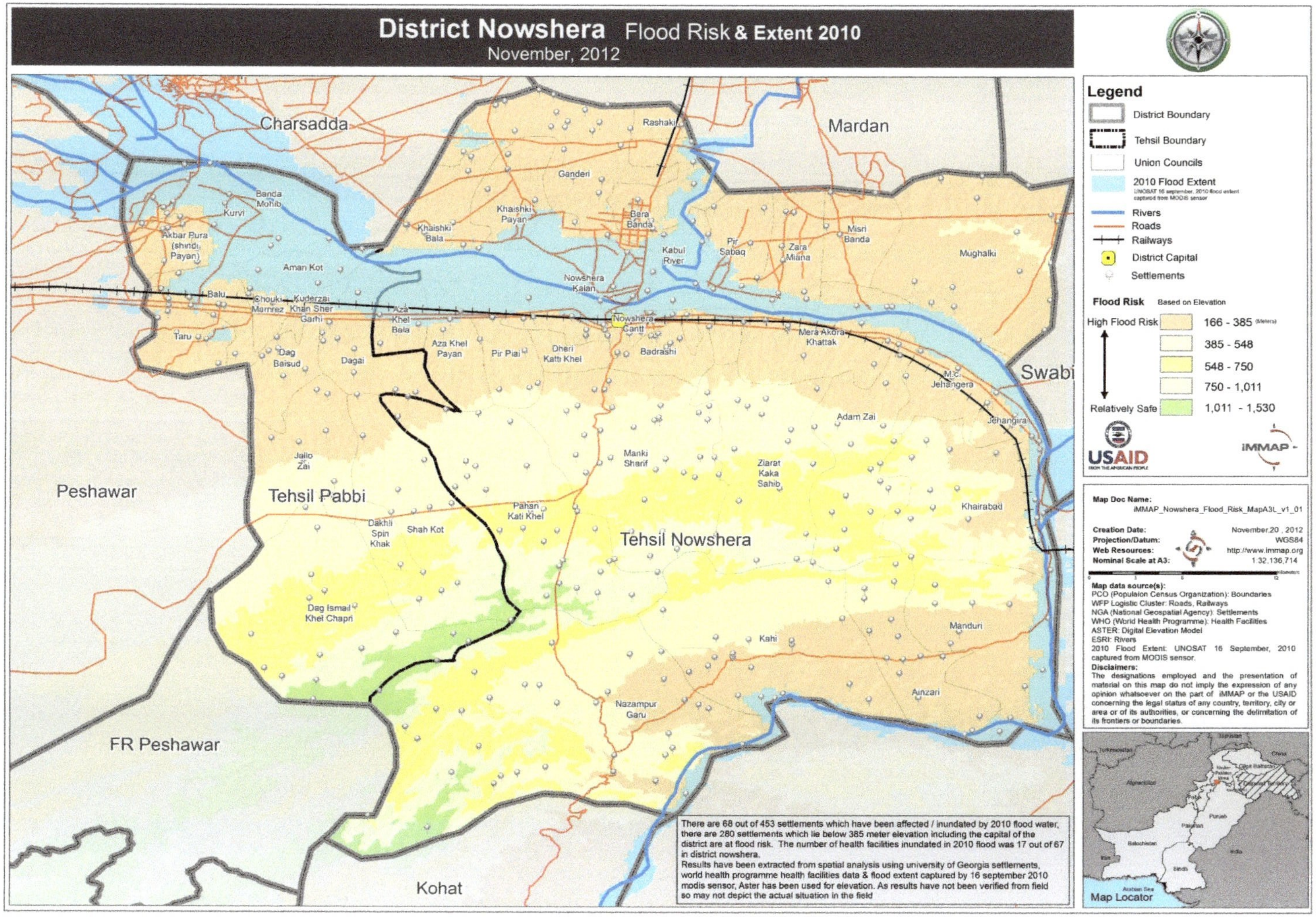

Fig. 5. Map developed by USAID to show the extent of Floods 2010 in District Nowshera

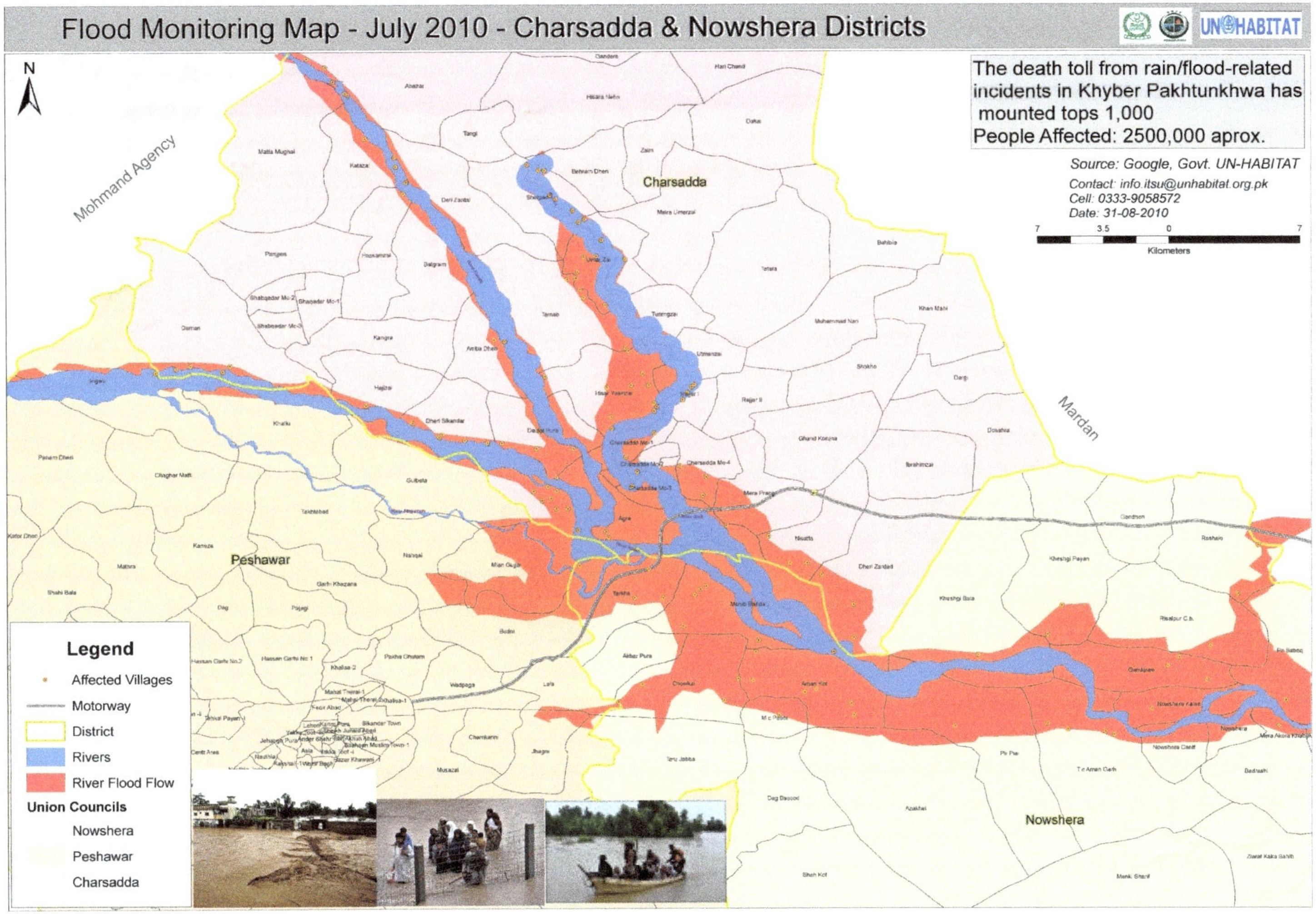

Fig. 6. Flood monitoring Map

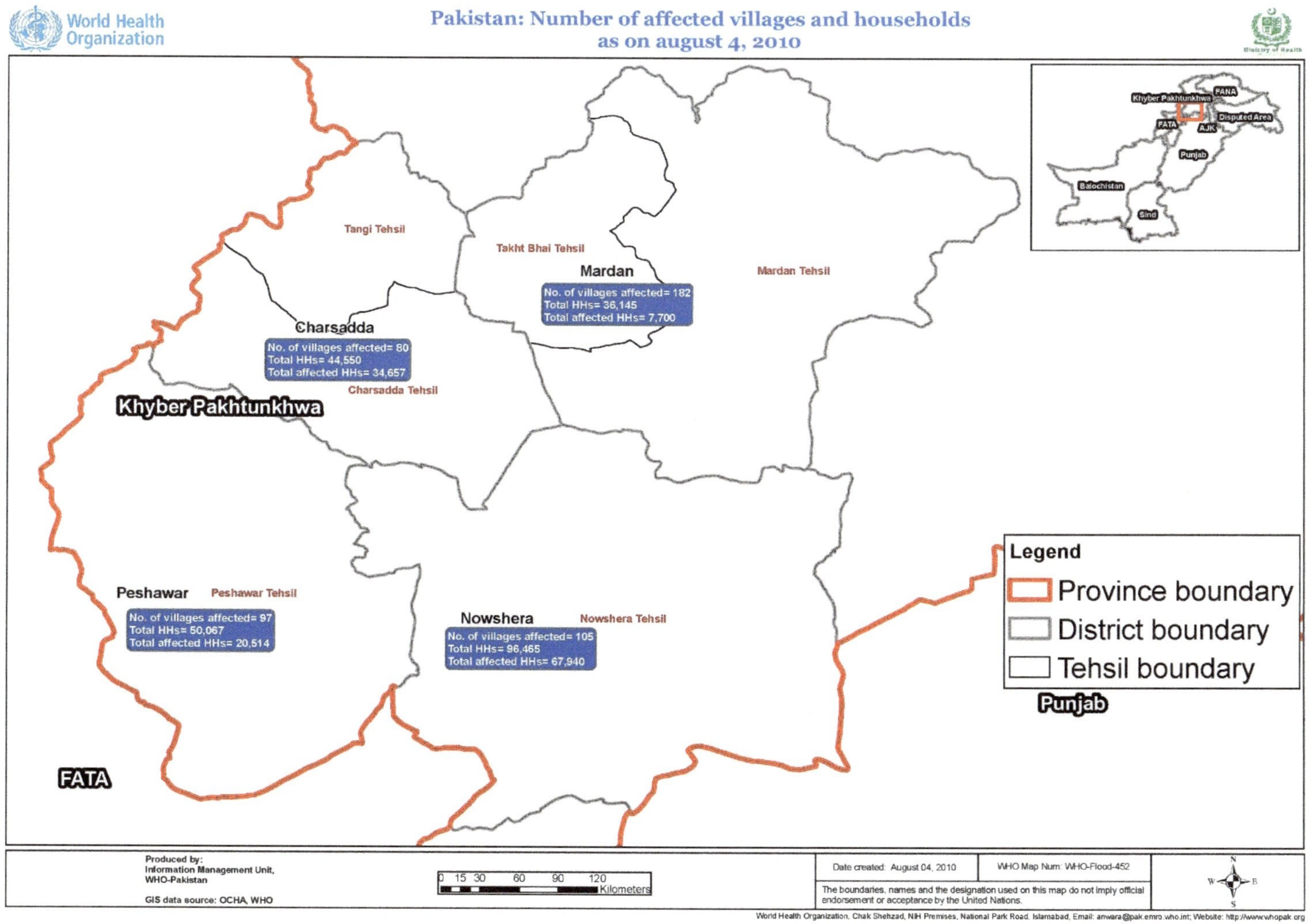

Fig. 7. Map developed by WHO to show the number of affected villages and households

The development of livestock and poultry sectors in Pakistan is dependent on promotion of maize crop. The crop, therefore, occupies high priority in agricultural development planning. The Kharif fodders include a number of crops as sorghum, millets and a variety of others. The damage of fodders caused very high discomfort to farmers, as fodders supply for livestock was restricted or cut in extreme cases. Tables 1, 2 developed by Pakistan Wetlands Programme [15] give the damage assessment of crops and area in the affected areas of Nowshera and Charsadda.

Table 1. Damage assessment of crops in Nowshera and Charsadda Districts

Crops	Area damaged (000 ha)	Damage (%)	Projected Production Loss (million tons)
Rice	5.5	47	0.02
Millet	0.9	80	0.3
Maize	16.7	-	-
Sorghum	0.9	80	0.3

Table 2. Damage to sugarcane and fodder crops in affected areas

Crops	Districts	Area Damaged (000 ha)	Yield Loss (tons/ha)	Damage Factor	Projected Production Loss (million tons)
Sugarcane	Charsadda	3.7	45	0.8	0.1
	Nowshera	2	50.4	0.5	0.1
Fodder crops	Charsadda	0.46	29.1	-	-
	Nowshera	1.64	24.9	-	-

2. MATERIALS AND METHODS

The two main districts of KP named Charsadda and Nowshera were the hardest hit by the unprecedented flood.

In order to investigate the water quality and climatic implications of floods, five water samples from the selected tube wells (main source), five from intermediate points and five from consumer's ends from each district were randomly collected in December 2010. A total of 30 samples from the areas were analyzed for water quality parameters where the majority of the inhabitants were dependent on ground water from the wells.

For the analysis of physical and chemicals parameters, each sample was collected in a fresh, clean and dry plastic bottle of 1.5 liters in capacity with air tight cap. Each bottle was marked with site name, time and date. Care was taken to avoid entry of external material in bottles during sample collection. Also, bottles were not absolutely loaded, so that they could be easily shaken before examination. Before sample collection, each well was pumped for about 5 to 10 minutes approximately. It was done to ensure that the collected water sample represented a clear picture of the properties of whole aquifer. Before collecting the water sample at the consumer's end, the tap was opened fully and ran for about 3 to 5 minutes. The faucet of the tap was sterilized with burner by putting it below the mouth of tap. It was done to avoid the entrance of the foreign matters during collection of water samples.

250 ml glass bottles were used for the analysis of bacteriological contaminants. First, the bottles were cleaned with detergents and rinsed with distilled water. Then 5 ml of Sodium thiosulfate in each glass bottle was taken and covered with air tight cap and was sealed with

aluminum foil tightly. These bottles were placed in an oven for about 45 minutes at 140°C. Bottles were finally sterilized. After collecting samples at the site, bottles were kept in ice coolers to transfer it to the laboratory. Table 3 shows the techniques and instruments used for finding various water quality parameters.

Table 3. Instrumental/Chemical techniques used for analysis of different water quality parameters

Sr. No	Water Quality Parameters	Techniques/Instruments used
1	Temperature	Celsius Thermometer for Field data
2	pH	pH Meter
3	Turbidity	Nephelometric Method
4	Electrical Conductivity	Conductivity Meter
5	Total Dissolved Solids	Gravimetric Method
6	Chloride	Titration Method
7	Nitrate	Colorimetric Method
8	Sulfate	Turbidimetric Method
9	Calcium	Titration method (using EDTA standard solution)
10	Magnesium	Titration method (using EDTA standard solution)
11	Coli form	Multiple fermentation tube technique
12	Fecal Coli form	Multiple fermentation tube technique

3. RESULTS AND DISCUSSION

The consuming water products from 10 main sources (tube wells), 10 intermediate points at pipe distribution system from main source and 10 consumer's ends from main source in 30 selected sites of flood affected areas were taken. Each sample was analyzed for 12 key factors. The results are presented in table 4 and each parameter is discussed in detail.

3.1 Temperature

Measurement of temperature is the most common physical parameter in assessing the quality of drinking water. It affects the chemical and biological characteristics of water. This study was carried out in December - January 2011. Table 4 shows negligible variation between the temperature of samples collected at intermediate point and its respective consumer's end while there is considerable variation between the temperature of the samples collected from tube wells and its respective sampling points.

3.2 pH

pH is a water quality parameter which is used to evaluate the level of acidity or alkalinity of drinking water. From the point of view of effective water treatment with chlorination, pH should be controlled and preferably be below 8. Generally, high pH value causes corrosion of metallic pipes which is associated with serious health problems due to the presence of heavy metals in drinking water leached inside the pipe distribution systems through which it passes. WHO guidelines [16] recommend a desirable limit of 6.5 to 8.0 for the pH of drinking water [17]. A little variation from this range is usually not associated to health issues but the value ranging from 10 to 12.5 have been documented to cause swelling of hair fibers and skin problems, and stomach disorders. While inflammation of eyes has also been reported with pH lower than 4. As given in Table 4, pH values for Charsadda district varies from 6.5 to 8.3 at main sources, from 6.8 to 8.7 at intermediate points and 6.7 to 8.9 at consumer's ends. Whereas pH values for Nowshera varies from 5.6 to 9.2 at main sources, from 5.8 to 9.6 at intermediate points and 6.2 to 8.9 at consumer's ends. Tube wells # 1, 6, 7 and 10 should be regularly monitored for their pH increase against alkalinity and similarly tube well # 8 against acidity.

3.3 Turbidity

Turbidity is a measure of relative clarity and cloudiness of water. It is generally caused by the suspended and colloidal particles which are responsible for the obstruction of transmission of light through water. The WHO recommended value of turbidity should ideally be less than 1 NTU for effective disinfection. Table 4 shows that all the values are higher than the recommended value. For Charsadda its values varies from 11 to 16 NTU at main sources, 15 to 19 NTU at intermediate points and 18 to 22 NTU at consumer's ends. Similarly for Nowshera, its values vary from 13 to 16 NTU at main sources, 17 to 20 NTU at intermediate points and 19 to 25 NTU at consumer's ends, showing all the values are out of range and may lead to the rejection of drinking water by the inhabitants of the flood affected areas.

3.4 Electrical Conductivity and Total Dissolved solids

There exists a general relationship between Electrical Conductivity and Total dissolved solids (TDS) i.e. [TDS (ppm) = Conductivity (μScm^{-1}) x (0.5 – 0.7)] [18]. In the present study, both tests were performed separately to get the precise values of both parameters. However, the approximated value of total dissolved solids can be found out by the above mentioned formula. The analysis of data for these two parameters showed that the conversion factor for overall samples was close to a value of 0.7.

The desirable value of 600 mg/l for TDS is proposed by WHO [14] on the basis of acceptability of drinking water. TDS with value higher than 1000 mg/l provide taste to the water and cause excessive scaling in pipes, heating units and boilers.

From Table 4, TDS at the consumer's ends and intermediate points for all samples is higher than that at the main sources. All samples show higher values than desirable limit. Samples from tube well #2 and #3 shows higher values of TDS than 1000mg/l. These high values may be because of the damaged pipe systems and leaked joints running through open drains and may affect the taste of water and can be rejected by the consumers.

3.5 Chloride

Chloride content for pollutants-free and fresh water is always less than 10 mg/l [19]. According to WHO guidelines, the chloride content less than 250 mg/l is considered good on the basis of acceptability for drinking since value higher than 250 mg/l imparts considerable taste to drinking water and consumers may reject use of such water. Table 4 shows that the chloride level for all samples are less than the desirable value for the samples obtained from Nowshera district but shows higher value for tube wells # 2, 3, 4 and 5 for Charsadda district. Chloride level increases remarkably approaching the consumer's ends for all samples may be due to the contact of leaked pipes with transmitted effluents from industries, human and animal wastes and highway salts during flooding. The higher chloride concentrations also causes corrosion of pipes and taps which may further contaminate water at intermediate and consumer's ends points.

Table 4. Water Quality Parameters for Main source, Intermediate points and End points

Parameter	Temperature (°C)			pH			Turbidity (NTU)		
Tube Well No.	Main source	Intermediate point	End point	Main source	Intermediate point	End point	Main source	Intermediate point	End point
1	18.0	18.8	18.8	8.30	8.70	8.90	11	15	18
2	19.2	21.0	21.2	7.24	7.5	7.7	11	15	19
3	18.5	18.8	18.5	7.01	6.80	6.70	13	16	19
4	20.2	21.0	20.7	7.40	7.90	7.10	15	19	22
5	17.0	19.6	19.6	6.50	6.90	7.01	16	19	21
6	20.5	22.5	22.2	9.20	8.90	8.80	13	17	20
7	17.6	18.8	18.2	8.70	8.95	8.90	14	17	19
8	18.3	21.0	20.9	5.60	5.8	6.20	17	20	25
9	19.2	23.5	23.2	8.20	8.00	8.00	13	16	20
10	21.5	24.4	24.1	9.10	8.90	8.7	16	20	24
Parameter	**Electrical Conductivity (μScm^{-1})**			**Total Dissolved Solids (mg/l)**			**Chloride (mg/l)**		
Tube Well No.	**Main source**	**Intermediate point**	**End point**	**Main source**	**Intermediate point**	**End point**	**Main source**	**Intermediate point**	**End point**
1	1133	1159	1265	793	800	873	221	208	208
2	1271	1307	1507	890	902	1040	275	305	322
3	1220	1351	1396	854	932.4	963.2	245	256	270
4	1299	1471	1562	909	1015	1078	268	299	312
5	1127	1249	1306	788.9	861.7	901	259	276	291
6	928	968	1013	649.6	667.8	699	167	172	198
7	1030	1143	1246	721	789	860	145	152	171
8	1030	1217	1335	721	840	921	172	175	192
9	1173	1319	1304	821	910	900	176	189	201
10	984	1064	1157	689	734	798	159	183	189
Parameter	**Nitrate (mg/l)**			**Sulfate (mg/l)**			**Calcium (mg/l)**		
Tube Well No.	**Main source**	**Intermediate point**	**End point**	**Main source**	**Intermediate point**	**End point**	**Main source**	**Intermediate point**	**End point**
1	10	12.5	16	74	83	94	59	62	57
2	8	10	14	102	115	120	42.5	44	46.7
3	9	11.5	14	178	182	184	31.8	31	35
4	12	15	17.5	167	172	175	21	19	25
5	7	10	13.5	172	175	198	47	47.5	50.6
6	17	21	27	345	348	357	27.4	28.5	23.5
7	16	19	24	309	312	314	54	50.6	56.5
8	16	21	26	312	316	316	68	53	65
9	20	24	29	198	209	213	19.6	22	22
10	21	24	30	302	304	304	18	23	23

Table 4 Continued.....

Parameter	Magnesium (mg/l)			Total Coli form (MPN / 100ml)			Fecal Coli form (MPN / 100ml)		
Tube Well No.	Main source	Intermediate point	End point	Main source	Intermediate point	End point	Main source	Intermediate point	End point
1	48	52	53	13	15	22	7	10	16
2	32	27	28.6	16	17	20	7	10	14
3	26	27	27	14	18	22	8	9	13
4	19.8	22	22.8	26	27	29	15	17	19
5	20	17	20	17	22	28	7	14	17
6	29.8	34	32.5	22	26	32	14	19	22
7	41	47	49	18	20	25	10	13	15
8	44	49	48	24	29	32	14	17	19
9	13	15	17	26	28	31	14	16	19
10	14	15.5	16.7	20	23	27	10	13	16

3.6 Nitrate

According to WHO guidelines the nitrate content in drinking water should be less than 50 mg/l. The higher level of nitrate ions may cause an unusual illness; ingestion of such water also affects health of pregnant women [20]. From Table 4, nitrate level in all the samples falls within the desirable limit and therefore, is acceptable for drinking and other domestic purposes. The variation of nitrate level from main source to the consumer's ends may be attributed to the poor sanitation system, improper disposal of human and animal wastes and open septic tanks and their contact with the pipe distribution systems in the study area. However, the nitrate concentrations in the principle sources are higher for Nowshera district as compared to Charsadda district. Earlier studies revealed that the exceeding level of nitrate in sub surface water may be increasing due to the extensive use of nitrogen fertilizers and human and animal wastes [21].

3.7 Sulfate

No guideline value for Sulfate related to health concerns has been established by WHO. However according to WHO guidelines the desirable limit for sulfate ions in drinking water should be less than 250 mg/l. The higher concentration level of sulfate imparts bitter taste and rotten egg smell to drinking water and causes scale in water pipes [22]. Water with high content of sulfate causes severe dehydration and Diarrhea in every age of people. Young and old age people may become accustom using such water with the passage of time but care should be taken while using such water in the food preparation for infants. From Table 4, the sulfate concentration level in all samples falls within the safe limits except for tube wells # 6, 7, 8 and 10 which are greater than the permissible limit.

3.8 Calcium

While flowing through the rocks and soil, water dissolves some minerals. Water containing two important and common minerals Calcium and Magnesium in high range is known as hard water. Calcium is a key nutrient for health, structure of bones, muscular contraction, cell signaling and blood clotting [23]. In 2001, an expert consultant for the Food and Agricultural Organization (FAO) and WHO set recommended nutrient intakes (RNIs) for Calcium at 1,000 mg/day for women aged 19-50 and men aged 19-65 years, and 1,300 mg/day for women of age 51-65 years [23]. To avoid the chances of developing osteoporosis, nephrolithiasis and hypertension, it has been recommended that adequate consumption of calcium should be taken in daily dietary intake. Table 4 shows that the concentration of calcium content in samples from main sources varies from 21 to 59 mg/l for Charsadda and from 18 to 68 mg/l for Nowshera, at intermediate points it varies from 19 to 62 mg/l for Charsadda and from 22 to 53 mg/l for Nowshera and at consumer's ends it varies from 25 to 57 mg/l for Charsadda and from 22 to 65 mg/l for Nowshera.

3.9 Magnesium

Like Calcium, Magnesium is also a nutrient mineral and a vital component of water. It partially contributes to the hardness of water. It is introduced to the ground water in the wells by the rocks enriched with magnesium minerals. In 2001, an expert consultant for the Food and Agricultural Organization (FAO)/WHO set recommended nutrient intakes (RNIs) for Magnesium at 220 and 260 mg/day respectively for women and men aged 19-65 years old [23]. The concentration of Magnesium content in samples from main sources varies from 20 to 48 mg/l for Charsadda and from 13 to 44 mg/l for Nowshera, at intermediate points it varies from 17 to 52 mg/l for Charsadda and from 15 to 49 mg/l for Nowshera and at consumer's ends it varies from 20 to 53 mg/l for Charsadda and from 17 to 49 mg/l for Nowshera (Table 4). The increase in Mg level from one sampling point to another point may be attributed to the intrusion of effluent of chemical industries into the leaked and damaged pipe distribution systems. The higher values at main source may be attributed to the presence of the M g-rich rocks in the localized area.

3.10 Total Coliform bacteria and fecal Coliform

Total coliform bacteria are a group of bacteria generally present in the environment. Fecal coliform are a sub group of total coliform which are present in the human and animal feces. The hefty flooding and increased precipitation in the study area increased the occurrences of total coliform bacteria in water. Since the wells of the study area have been designed using old standards, lacking of regular maintenance of the wells and pipe distribution systems, submergence of wells and pipes in floods and rain water, open septic tanks and sewer systems and inadequate drainage systems, all these factors contributed to the presence of total coliform and fecal contamination at all sampling points and correspondently gave rise to infectious and serious water borne diseases. Any water to be used for intake or food preparation should be free of all such contamination. The flooded areas are more prone to water borne disease outbreaks. WHO guidelines recommended a value of Zero M PN per 100 ml for total coliform as well as for fecal coliform bacteria [17] [24]. Table 4 shows that all the values are more than 0 M PN per 100 ml ranging from 13 to 26 M PN/100 ml at main sources, 15 to 29 M PN/100 ml at intermediate points and 20 to 32 M PN/100 ml at consumer's ends for total coliform bacteria. In addition, the values are also higher than 0 M PN per 100 ml ranging from 07 to 15 M PN/100 ml at main sources, 09 to 19 M PN/100 ml at intermediate points and 13 to 22 M PN/100 ml at consumer's ends for fecal coliform bacteria.

4. CONCLUSION

As it is quite clear from the results, the groundwater is polluted and unfit for drinking purposes. It can be declared that floods cause a direct impact on environment and all its components. There is an intense need of environmentally sound and feasible arrangement of drinking water provision for people affected by floods in any part of the world as water quality becomes no more reliable when hit by floods. Climatic implications of floods on water quality may not only cause diseases in humans but may also destroy the aquatic life and hence disturb the complete natural cycle. The natural dynamics of floods should be investigated and a wise policy taking into consideration environmental aspects should be formulated and strictly implemented. In order to make sure that environmental concerns are judiciously and properly represented in flood management decisions, individual cases must be assessed from a basin-wide perspective.

Proactive measures should be adopted in the regular flooded-area to avoid contamination from the wastes, prior to flooding. The types of crops should be suggested that can tolerate flooded water of a specific nature and contaminants not causing rotting of organic matters, if it is required to divert flood water into farmlands during peak flood. System weaknesses and operational shortcomings should be identified based on historic floods and develop conceptual improvements, including the integration of non-structural alternatives, in anticipation of major floods instead of in response to them. This may be accomplished by approaching problems holistically. By considering the entire system, operational plans should integrate existing and proposed structural and non-structural flood control measures to compose comprehensive flood management strategies. If severe flooding remains, the major impetus for policy change, a plan in hand will allow full utilization of the post-flood window of opportunity created by heightened social awareness and disabled structural controls. The study of the quantitative environmental aspects of severe flooding should be continued. A new branch of study should include ecological and biological modeling.

ACKNOWLEDGEMENTS

We would like to thank WHO, USAID and UN-HABITAT for the kind permission to use the Flood Maps 2010. Any use of trade, product, or firm names is for descriptive purposes only.

COMPETING INTERESTS

Authors have declared that no competing interests exist.

REFERENCES

1. Hickey JT, Salas JD. Environmental effects of extreme floods. US- Italy Research Workshop on the Hydrometeorology, Impacts, and Management of Extreme Floods, Perugia Italy; 1995.
2. Erikson KT. Everything in its path: Destruction of Community in the Buffalo Creek Flood. Simon and Schuster, New York; 1976.
3. Goolsby DA, Battaglin WA, Thurman EM. Occurrence and transport of agricultural chemicals in the Mississippi River basin July through August 1993. U.S Geological Survey circular 1120-C, U.S. Government Printing Office, Washington D.C.; 1993
4. Myers MF, White GF. The Challenge of the Mississippi Flood. Environment. 1993;35(10):6-9, 25-35.
5. Gardiner J. Environmental impact of floods. G. Rossi, N. Harmancioglu, and V. Yevjevich Editions - Coping with Floods, Kluwer Academic Publishers, Dordrecht, Boston and London. 1994;529-548.
6. Gardiner J. Developing flood defence as a sustainable hazard alleviation measure. J. Gardiner, O. Starosolszky, and V. YevjevichEdiyions - Defence from floods and floodplain management, Kluwer Academic Publishers, Dordrecht, Boston and London. 1995;13-40.
7. Nachtnebel HP. Environmentally and socially sound utilization of floodplains: some Austrian experiences. J Gardiner, O Starosolszky, and V Yevjevich (eds.) Defence from floods and floodplain management, Kluwer Academic Publishers Dordrecht, Boston and London. 1995;539-554.
8. Leentvaar J, Stortelder PBM. The methods and mechanisms of establishing consensus on water management policy in the Netherlands. Gardiner J, Starosolszky O, Yevjevich V. (eds.). Defence from floods and floodplain management, Kluwer Academic Publishers, Dordrecht, Boston and London. 1995;555-564.
9. Rasul G, Chaudhry QZ, Mahmood A, Hyder KW, Dahe Q. Glaciers and Glacial Lakes under Changing Climate in Pakistan. Pakistan Journal of Meteorology. 2011;8(15):1-8.
10. Gulf News. Floods in Pakistan worse than tsunami, Haiti; 2010. Retrieved 12 August 2010.
11. MacFarquhar N. UN Warns of Supply Shortage in Pakistan. New York Times; 2010. Retrieved 18 August 2010.
12. Dawn News. Floods to hit economic growth: Finance Ministry; 2010.

13. Hanaa M, Salem, Eweida A, Farag A. Heavy metals in drinking water and their environmental impact on human health. ICEHM 2000 Cairo University Egypt. 2000;542- 556.
14. World Health Organization. Guidelines for drinking-water quality 4th edition; 2011
15. Pakistan Wetlands Programme. Rapid Assessment of Flood Impact on the Environment in Selected Affected Areas of Pakistan; 2010
16. World Health Organization. pH in Drinking water - Background document for development of WHO Guidelines for Drinking-water Quality; 1996.
17. Haydar S, Arshad M, Aziz JA. Evaluation of Drinking Water Quality in Urban Areas of Pakistan: A Case Study of Southern Lahore. Pakistan Journal of Engineering & Applied Science. 2009;(5):16-23
18. Thirumalini S, Joseph K. Correlation between Electrical Conductivity and Total Dissolved Solids in Natural Waters. Malaysian Journal of Science. 2009;28(1):55-61
19. Joseph PV, Jacob C. Physicochemical Characteristics of Pennar River, A Fresh Water Wetland in Kerala India. E-Journal of Chemistry. 2010;7(4):1266-1273
20. Self JR, Waskom RM. Nitrates in drinking water. Colorado State University; 2008. Available: http://www.ext.colostate.edu/pubs/crops/00517.html/
21. Khan M, Khwaja MA, Imdadullah. Nitrate and Nitrite Contamination of sub surface water in some areas of North West Frontier Province Pakistan. Jour. Chem. Soc. 1998; 20(2):110
22. Sharon OS, Bruce ID, Woldt W. Drinking Water: Sulfur (Sulfate and Hydrogen Sulfide). University of Nebraska-Lincoln Institute of Agriculture and Natural resources; 2010. Available: http://www.ianrpubs.unl.edu/epublic/live/g1275/build/g1275.pdf
23. Cotruvo J. Health Aspects of Calcium and Magnesium in Drinking Water (NSF/ILSI/WHO) in Baltimore, Md; 2006.
24. Unnisa SA, Rao B. Drinking water quality surveillance in Hyderabad. Journal of Chemical, Biological and Physical Sciences. 2011;1(2):402- 405.

2

Carrying Capacity and Sustainability Appraisals on Regional Water Supply Systems under Climate Change

Ching-Pin Tung[1], Tzu-Ming Liu[1*], Szu-Wei Chen[1], Kai-Yuan Ke[1] and Ming-Hsu Li[2]

[1]Department of Bioenvironmental Systems Engineering, National Taiwan University, Taiwan.
[2]Institute of Hydrological and Oceanic Sciences, National Central University, Taiwan.

Authors' contributions

This work was carried out in collaboration between all authors. Author CPT designed the study and the framework. Author TML performed the model simulation, managed the analyses of the study and completed the draft of the manuscript. Author SWC performed the model programming and wrote the first draft of the manuscript. Authors KYK and MHL managed the literature searches and performed the statistical analyses. All authors read and approved the final manuscript.

ABSTRACT

This study aims to appraise the carrying capacity and sustainability of four water supply systems which are in the same watershed under climate change. An integrated assessment tool, TaiWAP, which integrates the common procedures of impact assessment of climate change, i.e., downscaling, weather generation, hydrological model, and interface for linking system dynamics model, is used to evaluate the sustainability of regional water resources systems. The GWLF physical model is used to simulate surface water processes and Vensim (a specialized software tool) is used in a system dynamics approach to simulate Taiwan's Danshuei river watershed supply system to analyze climate impact on sustainable water resource utilization, which are both included in TaiWAP. To understand the sustainability of water supply systems, definition of a sustainable index are necessary to reveal the effects of response strategy and climate change. The results of this study could support making governmental strategies to enhance adaptive capacity, mitigate the impact of climate changes on water supplies and achieve sustainable and resilient water supply systems for the future.

**Corresponding author: Email: tedliu13@gmail.com;*

Keywords: Sustainable development; water resources; carrying capacity; system dynamics.

1. INTRODUCTION

In Taiwan, the average annual rainfall is up to 2500 mm, which is much higher than the world average of 834mm. However, most of the rainfall concentrates in the wet season (from May to October), and about 80% of the rainfall flows directly into the ocean due to the temporal and spatial non-uniform distribution [1,2]. As a result, in Taiwan, water shortages during the dry season do happen. In recent years, water shortages have seriously affected the availability of water supply. Under the future threat of climate change, efficient use of water resources becomes a vital area of study. To guard against climate change causing more extreme events with deleterious impacts on sustainable regional water resources development and increase drought risk, it is very important to consolidate visible water resource strategies and policy decisions with climate change assessment.

The impacts of climate change to water resources have been evaluated on many studies. The river flows may be changed due to climate change [3,4]. IPCC (Intergovernmental Panel on Climate Change) indicated that that flow seasonality increases, with higher flows in the peak flow season and either lower flow during the low flow season or extended dry periods [5]. It may cause more difficult on water resources management and challenge irrigation system [6].

With the threat of increasing dryness due to climate changes, finding more efficient ways to use water resources is becoming an important area of study. To estimate the sustainability of water resources under climate change, it is required to use an integrated model on assessment due to miscellaneous simulation steps and climate change scenarios. Liu et al.[7] developed an integrated assessment on water resources system to climate change, TaiWAP, which integrates the common procedures of climate change impact assessment, including scenarios of climate change released by IPCC [8], downscaling, weather generation, hydrological model , and interface for linking system dynamics model (SDM). This friendly tool merely requires users to set up a SDM for their own purpose and make a connection to the TaiWAP, the complex of assessing climate change impact is thus no longer a problem and can be download on TaiWAP Website [9].

Climate change refers to the long-term variation in climate trends that exacerbate short-term weather variability and possibly related extreme meteorological events, leading to significant impacts on both human society and the natural environment. Socioeconomic development depends on reliable water supply system, but climate change may affect the system, including changing available stream flow, groundwater, and irrigation requirement, etc. The purpose of this study is to define a sustainability indicator associated to evaluate climate change impacts on water supply systems. TaiWAP was used in this study to adopt the sustainability of regional water supply systems, and to strengthen adaptive capacity of risk management for water resources by proposing strategies for achieving sustainable development for the future.

2. STUDY AREA

The study area is the Danshuei River watershed, which crosses northern Taiwan and serves as the major water supply source for Taipei municipal areas and neighbor counties. Fig. 1 shows the Danshuei River watershed with three major tributaries, the Sindian River, the

Dahan River, and the Keelung River. The total length and area of the Danshuei River are 157.8 km and 2726 km^2, respectively. There are four main water supply districts, Taipei, Banshin, Taoyuan and Keelung. The Taipei district draws water from the Sindian River and the on-site Feitsui reservoir, while the neighboring water supply districts, Banshin and Taoyuan, utilize water from the Dahan River where the Shihmen reservoir is located. The Keelung River with only a few small reservoirs and hydraulic structures to hold water, is the source of Keelung water supply system. A conceptual diagrams of the water supply systems of the Danshuei River are given in Fig. 2 and Fig. 3.

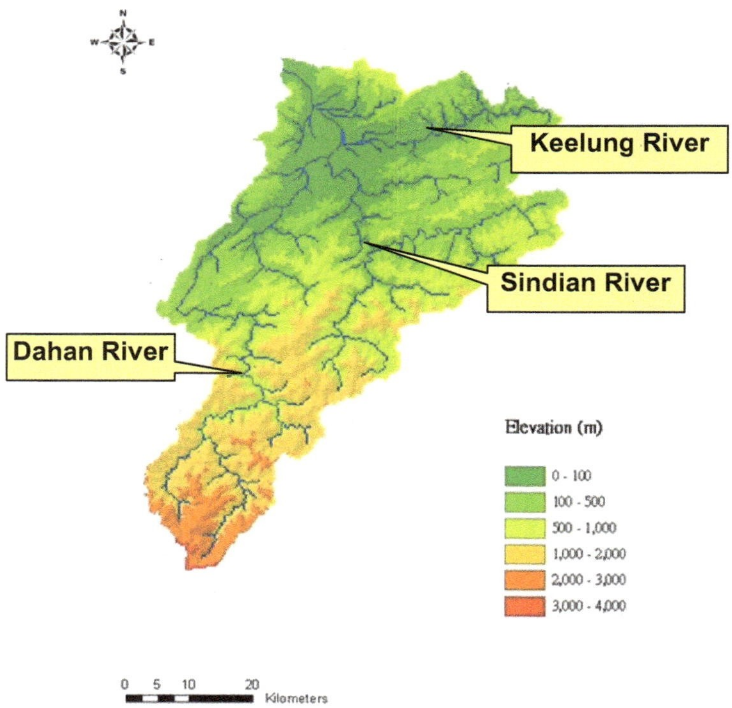

Fig. 1. The Danshuei River watershed with three tributaries

There are three water treatment plants, Chihtan, Chingtan and Gongguan, in the Taipei water supply district. Chihtan Weir delivers water to the Chihtan water treatment plant with a capacity of 2.70×106 m^3/ per day (CMD), while Chingtan Weirs store water for Changshin and Gongguan water treatment plants. Total capacity of the three major treatment plants and inlets are 3.82×10^6 and 3.80×10^6 CMD, respectively. These water treatment plants utilize water from the Nanshi River first. If this water supply is insufficient, a request will be sent to the Feitsui reservoir management bureau to release more stored water.

The major hydraulic facilities in the Dahan River include Shihmen reservoir, Houchi Weir, Yuanshan Weir, Sansia Weir, Shihmen major irrigation ditch, and Taoyuan major irrigation ditch. The Shihmen major ditch intake is located upstream of the Shihmen reservoir, while the intake of Taoyuan major ditch is installed in the Houchih Weir. The Taoyuan water supply system has four water treatment plants that withdraw water via Shihmen and Taoyuan ditches. The total capacity of water treatment plants and inlets are 1.27×10^6 and

1.35×10^6 CMD, respectively. The Banshin water treatment plant takes water from Sansia Weir first. If the water is insufficient, more water will be taken from Yuanshan Weir. The capacity of treatment and inlet for Banshin district are 1.20×10^6 and 2.10×10^6 CMD, respectively. The Banshin district also receives water supply from Taipei water supply district directly.

The Keelung water supply system includes two reservoirs, two pumping stations and seven weirs, as shown in Fig. 3. This supply system is more complicated than the other water supply systems. The Xishi and Xinshan reservoirs in the Keelung River watershed store water as much as possible and provide water without a discount water supply, which are different from reservoirs of the other water supply districts. Two pumping stations and seven weirs hold and send water to the water treatment plants to provide water to the whole Keelung district.

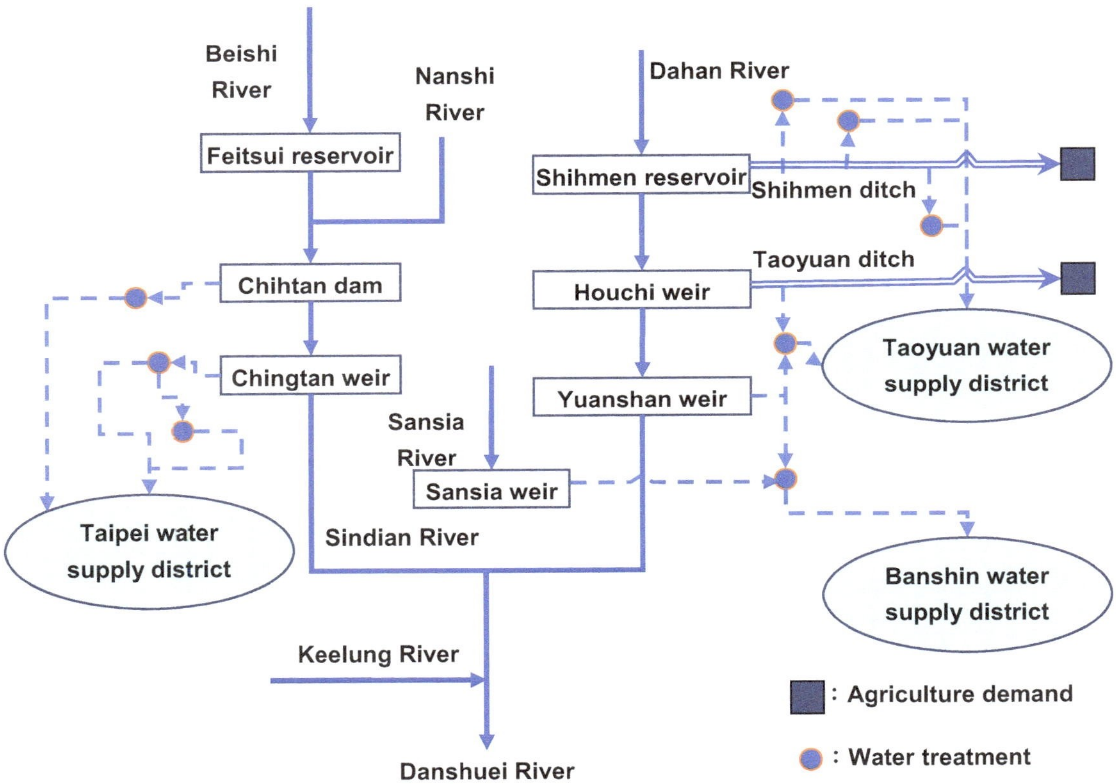

Fig. 2. Taipei, Taoyuan and Banshin water supply systems

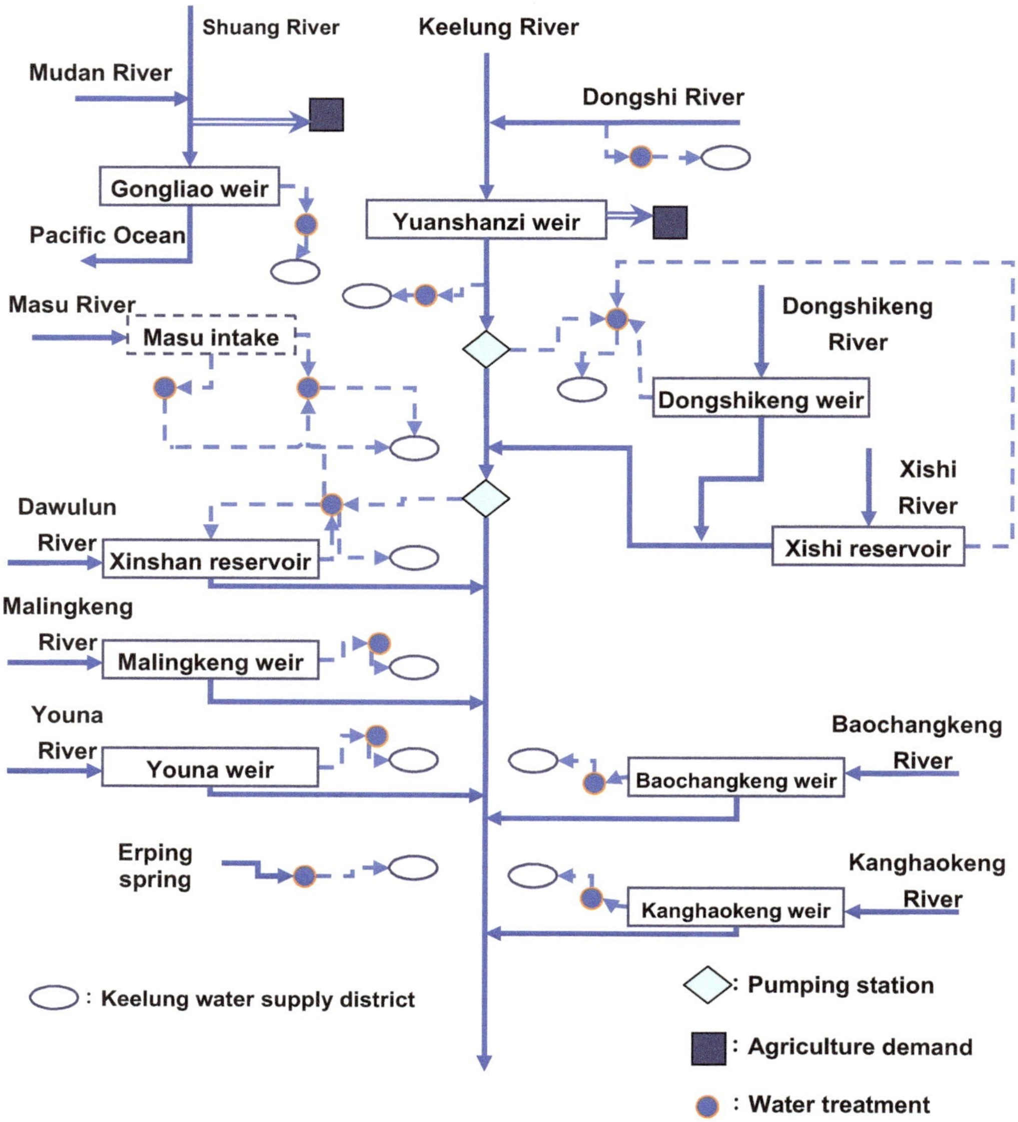

Fig. 3. Keelung water supply system

3. METHODOLOGY

The procedure of assessment was divided into five steps, as shown in Fig. 4. First, current and future weather data are prepared, which means climate scenarios are derived based on several General Circulation Models (GCMs) with downscaling skills. Then these scenarios consisting of monthly climate statistics are used to create different weather data sets by a weather generation model in the second step. The third step is to input weather data into a hydrological model to simulate the inflow discharge in the water supply system. Fourth, a water supply system dynamics model is employed for the water supply simulation and sustainability appraisal of a water supply system. In the end, determination of a sustainable index which improves understanding the status of a water supply system was applied. The whole process were included in TaiWAP which was developed to assess the vulnerability of the water resources systems in but not limited to Taiwan. It involved the climate change scenarios of precipitation and air temperature which were released by IPCC. In addition, the weather generator, hydrological model(GWLF model), and water resources system dynamics model(package built by Vensim) are within. The assessment steps as the components of TaiWAP are described below.

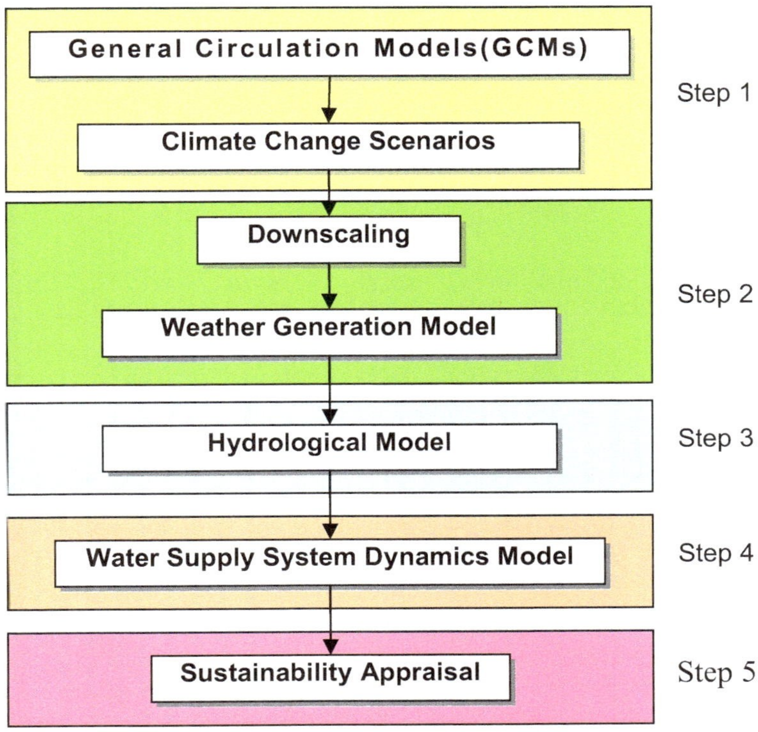

Fig. 4. The methodology diagram

3.1 Climate Change Scenarios

Research on climate change impacts are based mainly on the results of GCMs. However GCMs are global-scale simulations with uncertain validity for local-scale application. For impact assessment, the predictions of GCMs can only be used to establish climate change scenarios, and to generate input data for assessment model with the weather generator. The predictions of GCMs are large-scale results and should be transformed by downscaling to obtain local climate characteristics.

For the analysis of future climate scenarios, the A2, A1B, and B1 scenario of Special Report on Emissions Scenarios (SRES) [10] are adopted in this research. This study applies different GCMs scenarios and discusses the variation by the changes of precipitation and mean temperature in short-term (2010-2039). A downscaling method named "Simple Downscaling" was adopted herein.

Simple downscaling method directly uses the forecast data of the nearest GCMs grid and assumes the changes of temperature and precipitation will be equal in the same zone. Accordingly, the first step is getting the meteorological data, such as temperature and precipitation, of future climate-change scenarios derived from GCM predictions. To reduce the uncertainty of GCMs, it is recommended to adopt the better skill GCMs. In this study, six GCMs, MRI-CGCM2.3.2, GFDL-CM2.1, UKMO-HADCM3, MPIM-ECHAM5, INM-CM3 and NCAR-CCSM3 [4] were chose due to high correlated precipitation pattern to study area. When TaiWAP was developed, there were only ten GCMs released with three scenarios, A2, A1B, and B1. Table 1 shows the correlation coefficients of monthly mean precipitation between simulated precipitations from these ten GCMs and historical observations in study area. The six higher correlation GCMs were marked with bold font in Table 1 and the brief of 6 GCMs were shown as Table 2.

Table 1. Correlation coefficients of monthly mean precipitation between baseline of GCMs and historical observations in study area. Six higher correlation GCMs are in bold font

GCM / Station	CSIRO-MK3	GFDL-CM2.0	GFDL-CM2.1	INM-CM3	IPSL-CM4	NIES: MIROC3.2-MED	MPIM-ECHAM5	MRI-CGCM2.3.2	NCAR-CCSM3	UKMO-HADCM3
Fusan	0.451	0.150	0.648	0.565	-0.041	-0.067	0.811	0.523	0.722	0.453
Gaoyi	0.369	0.062	0.563	0.552	-0.016	-0.123	0.811	0.518	0.640	0.397
Baling	0.377	0.062	0.590	0.566	0.042	-0.086	0.815	0.483	0.586	0.376
Bihu	0.630	0.071	0.752	0.593	0.204	0.095	0.820	0.370	0.619	0.375

Table 2. The brief of six GCMs which were used in this study (IPCC, Data Distribution Center [8])

Center	Center Acronym	GCM	Atmospheric resolution (longitude x latitude)
Max-Planck-Institut for Meteorology(Germany)	MPIM	ECHAM5	Approximate 1.9 x 1.9 degrees
Geophysical Fluid Dynamics Laboratory(USA)	GFDL	CM2.1	Approximate 2.5 x 2 degrees
Institute for Numerical Mathematics(Russia)	INM	CM3	5 x 4 degrees
Meteorological Research Institute(Japan)	MRI	CGCM2.3.2	Approximate 2.8 x 2.8 degrees
National Centre for Atmospheric Research(USA)	NCAR	CCSM3	Approximate 1.4 x 1.4 degrees
UK Met. Office(UK)	UKMO	HadCM3	3.75 x 2.75 degrees

3.2 Weather Generation Model

The weather generation model [7] is required in the impact assessment of climate change to generate daily air temperature and precipitation based on different climate scenarios. To reproduce the statistic of precipitation and temperature of the observation, a quite number of generation data, at least 100years, is required [7]. Therefore, sequences of 100 years of daily precipitation and air temperature data were generated for current and future climate scenarios in this study.

In the weather generation model, the daily temperature is generated using one-order Markov Chains as equation (1).

$$T_i = \mu_{Tm} + \rho_m(T_{i-1} - \mu_{Tm}) + v_i\sigma_{Tm}\sqrt{1-\rho_m^{\ 2}} \tag{1}$$

where T_i is the temperature in day I, μ_{Tm} is the mean temperature of the target month, ρ_m is the correlation coefficient of T_i and T_{i-1} of the target month called the first order serial correlation coefficient, v_i is the normal sampling deviate which is the number between 0 and 1 generated randomly for each day and σ_{Tm} is the standard deviation of the observed temperature of target month respectively. The temperature of the first day in simulation uses the mean temperature of the first month and of the rest following days can be generated day by day.

The daily rainfall is generated in two steps, generating occurrence and generating amount. For generating occurrence, a random number, *RN*, between 0 and 1 is generated. *P(W|W)* and *P(W|D)* which are the conditional probability of a wet day following a wet day and the conditional probability of a wet day following a dry day are adopted by the historical data and used to generate the occurrence. If day *(i-1)* is a wet day and $RN \leq P(W/W)$, then the day *i* will be a wet day. Otherwise, it will be a dry day. If day *(i-1)* is a dry day and $RN \leq P(W/D)$, then the day *i* will be a wet day. Otherwise, it will be a dry day. When the occurrence is simulated as a wet day in the first step, the rainfall amount will be produced by using a formula in the next step :

$$P = \mu_P(I) \times [-\ln(1-RN)] \tag{2}$$

The distribution of precipitation is assumed as an exponential distribution. Therefore the rainfall amount can be generated with a uniform distributed random number RN which is between 0 and 1.

3.3 Hydrology Simulation Model

The hydrology simulation model GWLF [11-13] was used in this study to simulate the water budget of the watershed. In the model, the streamflow was calculated from runoff and groundwater. In water balance, rainfall brings the water into the watershed. When rainfall reaches to the ground, some of the rainfall infiltrates to the soil and the rest becomes surface runoff and flows into the river. The rainfall will supply the water content of the unsaturated zone. When the soil moisture content in the unsaturated zone is more than the field capacity, because of gravity, the water will percolate down into the shallow saturated zone and can then enter the stream as underground drainage. Streamflow contains runoff and underground drainage. The water balance in the watershed is mainly divided into three categories: surface, unsaturated zone and the shallow saturated zone. GWLF was used in this study due to its simple input requirement, i.e., only rainfall and temperature. Readers can refer to Haith et al. [14] for more details of GWLF.

3.4 The Water supply System Dynamics Model

Systems dynamics was started from simulating urban dynamics describing development of a city that finally collapses due to resource exhaustion [15]. It studies how the behavior of complex system changes through time and has been applied to many research areas. Simonovic [16] mentioned that systematic planning for water resources is complicate and requires computer programming. System dynamics modeling is a powerful measure to develop a simulation model. Forrester [17] developed a system dynamics model to describe urban dynamics, and Meadows et al. [18] used system dynamics to describe limits to growth. Many researchers have also developed their system dynamics model and applied to water resources systems (e.g., Guo et al. [19]).This study uses system dynamics in a simulation model to evaluate sustainability of water supply systems.

A system consists of many components and functions to describe the relationships between components. System dynamics modeling has three major components that include level, rate, and auxiliary and uses arrow to link components. Thus, system dynamics modeling can be applied to develop a simulation model more intuitively, especially for water resources systems. For example, a reservoir can be represented by a level component, and its inflows and outflows can be described by the rate components. Besides operational rules of a reservoir, principles of withdrawing water from a weir or other management strategies can be stated by the auxiliary component.

In this study, the system dynamics model was used to build the Danhsuei River water supply system and to evaluate the sustainability of water supply systems. This model is divided into several sub models, including three water supply systems for the Sindian, Dahan and Keelung River and water demand sub models for four water supply districts. The model was developed by using Vensim software. The water supply system dynamics model needs two kinds of input data which are upstream flow and total water demand. For the reservoir component in this model, its outflows are determined according to total water demand, inflow, and the operation rules. For the water supply to the domestic and industrial water, the water is determined by the outflow from the river or the reservoir, water demand and the capacity of the water treatment plant. The complicated water supply system is composed by many components like water treatment plants, reservoirs or dams and supply or demand components. However, the system dynamics model can simulate the realistic water supply with considering all the components and make the complex system a modular model.

3.5 Sustainable Appraisal Index

There are three steps to make a sustainable appraisal. The first is defining an environmental carrying capacity which is referred to water supply capacity in this study. According to the definition of sustainable development, “Meets the needs of the present generation without compromising the ability of future generations to meet their own needs”, it needs a critical limit to avoid degrading the environment and meet needs of future generations, which is called environmental carrying capacity. The water supply capacity is the ability of water supply and the main constraint on development, or it will lead to degraded environment. The ability of water supply is evaluated based on the goals of shortage index (SI)of 0.1, 0.5, 0.5, and 0.5 for Taipei, Banxin, Taoyuan and Keelung, respectively. Second, the cumulative impact is identified. The cumulative impact means the impact on the environment resulted from incremental effects of the project in addition to other past, present, and reasonably foreseeable future projects regardless of what person undertakes the other projects. The cumulative impact in this study is total water demands. The third step is to get a sustainable index by comparing the carrying capacity with total water demands and it helps to clarify the sustainability of water supply. A water supply system which meets sustainability needs total water demands less than the carrying capacity. On the other hand, if total water demands are beyond the carrying capacity, which means water supply cannot meet demand, continuous economic and social development will be stopped or new water resources, which normally cause impacts on ecosystems, are needed. Therefore, a good index should be able to represent above principles and can be designed as Eq.3 and shown as Fig. 5.

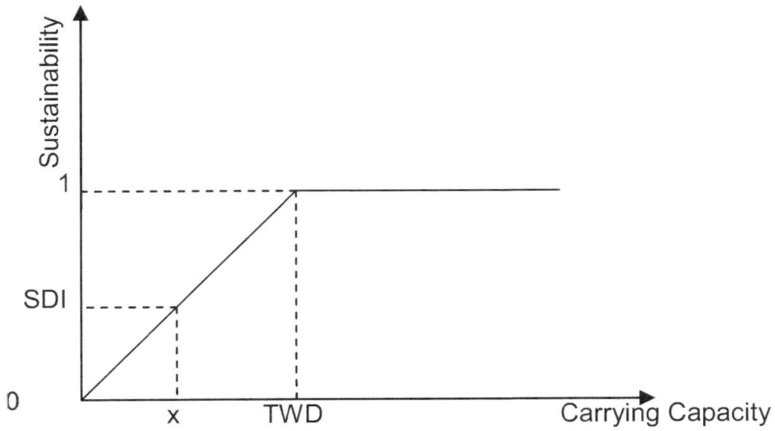

Fig. 5. The sketch of sustainable appraisal index

$$SDI = \begin{cases} 1 & \text{if } x \geq TWD \\ \dfrac{1}{TWD} x & \text{if } x < C \end{cases} \qquad (3)$$

Where SDI is a sustainable index, TWD is total water demands. When the index equals one, it means that carrying capacity is more than total water demands and the water supply system is sustainable. However, if carrying capacity is less than total water demands, the index will decrease from one to zero, a nil carrying capacity condition. A lower index value implies less sustainable water supply.

4. VALIDATION STUDY

The water supply system dynamics model was validated by assessing practical predictions for the Sindian, Dahan and Keelung River. To verify whether the proposed model can reasonably describe the management strategies and reservoir operational rules, three sets of historical stream flows recorded in the period of 1991 through 2000 were used. Upstream flows were entered into the model to simulate downstream flows which were compared with the observed flows recorded in the Hsiulung, Sanying bridge and Wudu gauge stations located downstream in the Sindian, Dahan and Keelung Rivers, respectively. The results are shown in Figs. 6 to 8, which indicate the reasonable predictions by the water supply system dynamics model. There is a significant error on 96th month (i.e., October of 1998) noted. This is likely explained by a strong typhoon with heavy rainfall that increased stream flows that month. The system dynamics model developed in this study lacks ability to describe reservoir operational rules for flood mitigation. Besides, the model assumes lateral flow proportional to its drainage area. This assumption may not be hold for a typhoon event. However, this study is mainly concerned about the ability of water supply system and its vulnerability under climate change, and hence flood mitigation is less relevant.

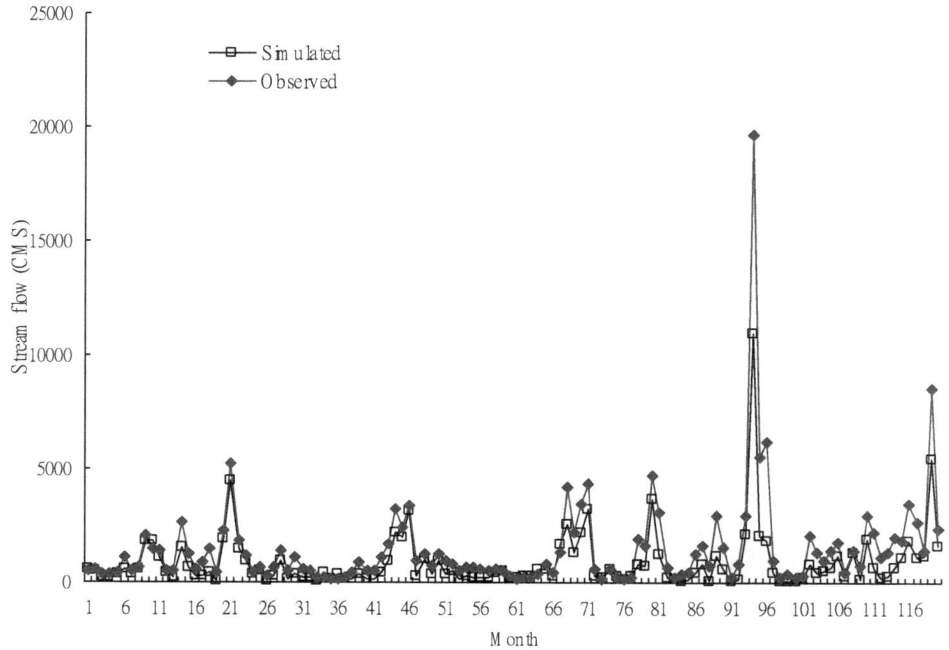

Fig. 6. Validation study – observed version simulated monthly streamflows of the Sindian River. (Correlation coefficient=0.95)

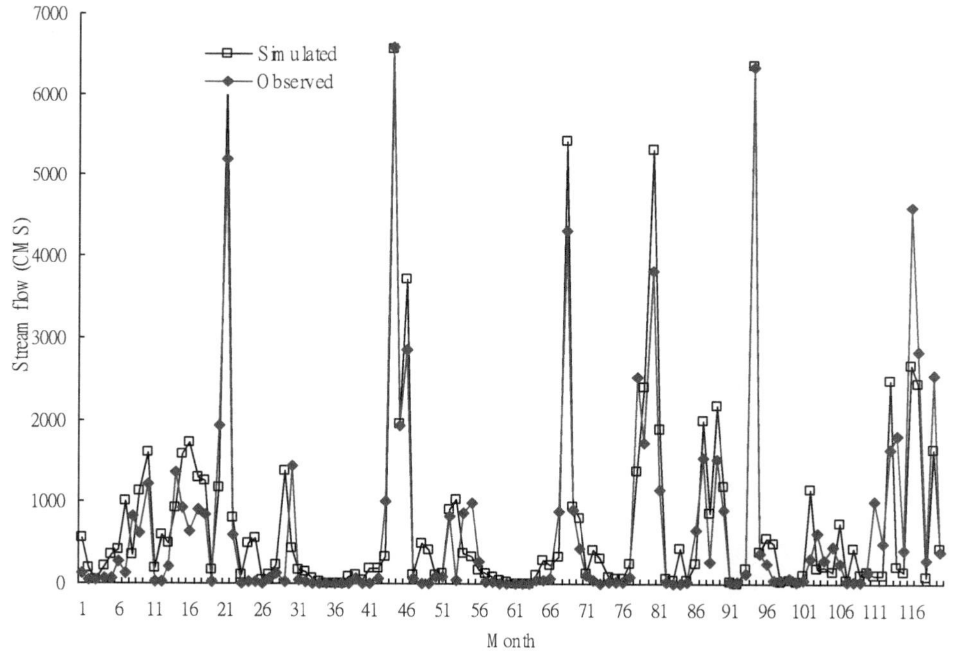

Fig. 7. Validation study – observed version simulated monthly streamflows of the Dahan river. (Correlation coefficient=0.92)

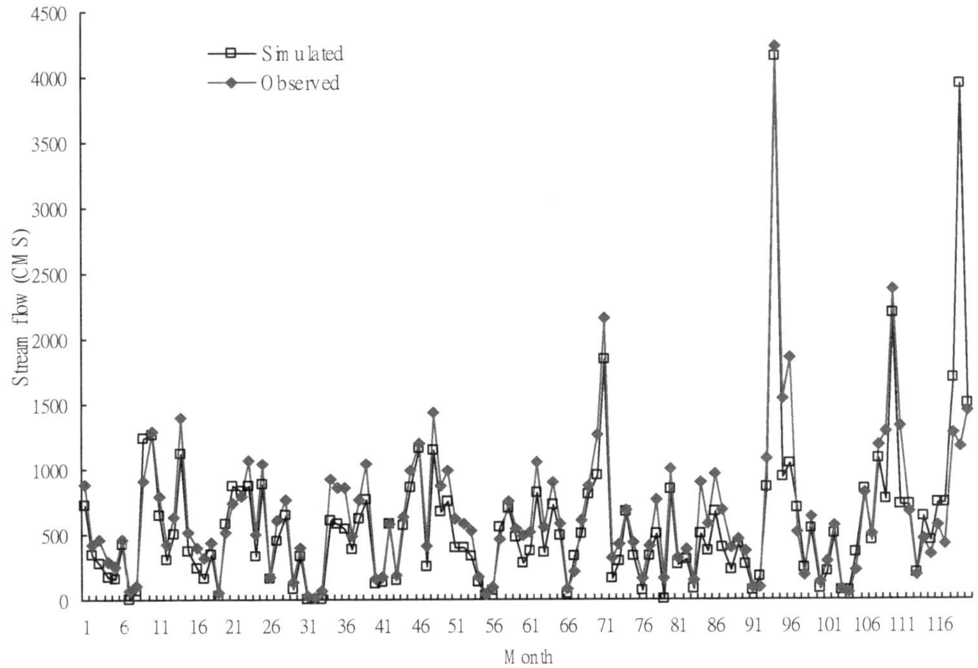

Fig. 8. Validation study – observed version simulated monthly streamflows of the Keelung River. (correlation coefficient=0.85)

5. RESULTS

There are three parts in this section to appraise the sustainability of the water supply systems, i.e. the baseline assessment, the future projection without climate change impact and the future projection with climate change impact.

5.1 Baseline Study

Baseline study is the assessment for the period of 1991 to 2001. In 2001, a water supply project which is called the First Stage of Water Supply Improvement Project for Banxin District was carried out and accomplished in 2003. The purpose of this project is to improve the water supply in Banxin district as the amount of water supported by Taipei water supply system reaches 530,000 CMD. To evaluate the effectiveness of the First Stage of Water Supply Improvement Project for Banxin District, two cases were proposed for baseline study. Case I does not consider the First Stage of Water Supply Improvement Project for Banxin District, while Case II does.

By using SDI index to indicate the sustainability of a water supply system, Table 3 shows that all water supply systems are sustainable except Banxin water supply system in the Case I, which is consistent to the real case. Because the Banxin water supply system shares the same water source from TaHan river with Taoyuan located upstream of TaHan river. Therefore, Banxin has more water shortage risk because of the lack of water intake. That's why the government undertook the water supply improvement project for Banxin District to enhance the water supply capacity of Taipei and transfer water from Taipei to Banxin. Case

II shows the benefit of the improvement project because the sustainable index in Banxin becomes 1. Also, after the accomplishment of the First Stage of Water Supply Improvement Project for Banxin District, the water supply capacity raised from 282.0 × 10^4 CMD to 348.7 × 10^4 CMD. It not just made Taipei with higher water supply capacity but also solved the water shortage problem of Banxin.

Table 3. The sustainability assessment of two cases for baseline (Unit: 10^4 CMD)

Water supply system	Case I. Disregarding the First Stage of Water Supply Improvement Project for Banxin District			Case II. Considering the First Stage of Water Supply Improvement Project for Banxin District		
	Average water demand	**Water supply capacity**	**SDI**	**Average water demand**	**Water supply capacity**	**SDI**
Taipei	220.9	282.0	1	220.9 52.6 (Supporting Banxin)	348.7	1
Banxin	70.6	52.2	0.74	52.6 (Supported by Taipei) 18.0	69.3	1
Taoyuan	66.8	108.9	1	66.8	108.9	1
Keelung	22.9	36.3	1	22.9	36.3	1

5.2 Future Projection without Climate Change Impact

In order to meet the future water demand in 2021, some water resources facilities and allocation have been planned to raise water supply capacity. To understand the future water resources sustainability, the carrying capacities of water supply systems conditioned on current and future water supply systems are appraised. The input data of baseline are based on historical upstream flow and water demand. The upstream flow in the future is assumed statistically identical as baseline. The projected future water demand data is obtained from the government's published reports. Further, the SDI index and water supply capacity of baseline and future condition were discovered for each water supply system.

In Table 4, the projected future water supply appraisal without climate change was compared with the baseline result. Based on the government's published reports, water demand in Taipei district doesn't increase too much, but water demand in Banxin, Taoyuan, and Keelung district rise significantly. Note that the Banxin supply plus a supply of 53×10^4CMD from Taipei just met total water demand in Banxin. As a result, the sustainability indices of water uses for Taoyuan and Keelung in the short-term are 0.88 and 0.84 respectively. It means that these areas will likely be unsustainable in the future. To solve the water shortfall, the government is undertaking a master plan of water resources management to enhance the stability of water supply and reduce the existing vulnerability.

5.3 Future Projection with Climate Change Impact

To estimate the climate change effects on water supply system in various future scenarios, a sequence of 100 years of daily precipitation and air temperature is generated by weather generation model for current and future climate scenarios and were input to GWLF model to simulate the flow discharge for baseline and future climate scenarios. Because of the uncertainty of GCMs [20], different GCM has different projection for the same climate change scenario. Figs. 9 to 11 shows the results of the number of GCMs for different change ratio of the water supply capacity for SRES-A2, SRES-A1B, SRES-B1, respectively. Because of the uncertainty of GCMs, it is not recommended to have quantitative assessment from GCMs. Using categories of change ratio is a better way to appraise the carrying capacity. Six categories, ">10%", "5%~10%", "0%~5%", "0%~-5%", "-5%~-10%" and "<-10%", are adopted to classify the water capacity tendency of future.

Table 4. The comparison of the sustainability of water supply system without climate change impact between the present and future (Unit: 10^4 CMD)

Water supply system	Baseline			Future		
	Average water demand during 1991 to 2001	Water supply capacity	SDI	Average water demand in 2021	Water supply capacity	SDI
Taipei	273.5	348.7	1	277.6	377.9	1
Banxin	18.0	69.3	1	39.2	66.7	1
Taoyuan	66.8	108.9	1	145.2	127.8	0.88
Keelung	22.9	36.3	1	48.4	40.4	0.84

In Taipei, three scenarios have the same change ratio among 6 GCMs which showed that the water supply capacity of Taipei will slightly increase under climate change. Banxin, Taoyuan and Keelung water supply systems don't have the consistent trend under climate change. In SRES-A2 scenario, Banxin water supply system has three GCMs on the "increase side" and three GCMs on the "decrease side". It implies no obvious trend from the result. Taoyuan and Keelung water supply systems both have four GCMs on the "decrease side" which shows that the water supply capacity might decrease under SRES-A2 scenario. In SRES-A1B scenario, Banxin, Taoyuan and Keelung water supply systems all have the trend of decreasing water supply capacity by the results of more than three GCMs on the "decrease side". In SRES-B1 scenario, Banxin has three GCMs on the "increase side" and three GCMs on the "decrease side". However, two GCMs are in ">10%" category which means it has more probability to have increasing water supply capacity due to climate change. Taoyuan water supply system might have slight increasing water supply capacity because of 4 GCMs in "0%~5%" category. Keelung has equal GCMs on both sides but slightly trend to decreasing water supply capacity due to 3 GCMs in the "0%~-5%" category.

From the results of three scenarios, it is obviously that Taipei water supply system will slightly gain more capacity from climate change. Banxin, Taoyuan and Keelung water supply systems couldn't have the consistent result from different scenarios. From the definition of each scenario, SRES-A1B is more like the path where the world is going to, which means the result of SRES-A1B will be more like our future. If that is so, then Banxin, Taoyuan and Keelung water supply systems could have the risk of decreasing water supply capacity under climate change. From the result of SDI index shown as Fig. 12, even though the carrying capacity will decrease in Banxin water supply system, the SDI index still remains 1 which means Banxin water supply system is still sustainable in the future. Otherwise, Taoyuan and Keelung water supply systems will have more water shortage risk in the future because of unsustainable water supply systems. The strategies to enhance the adaptive capacity and achieve sustainability of these two water supply systems will be needed for the future.

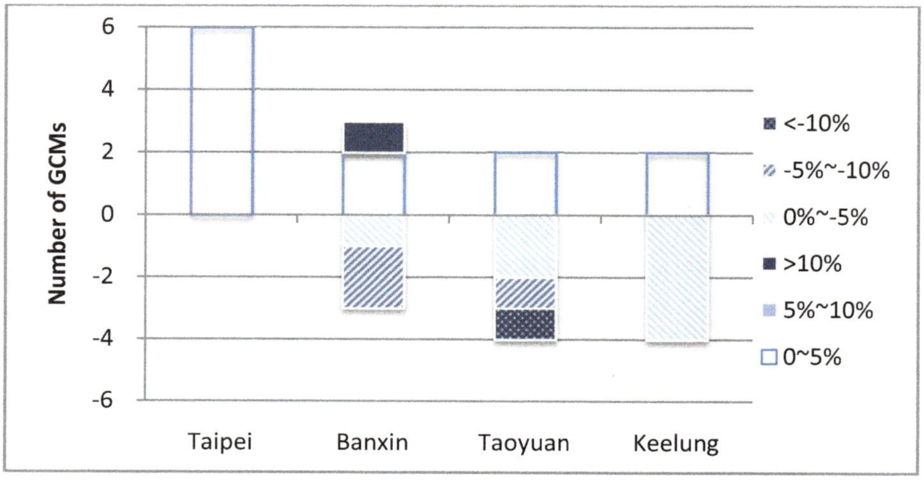

Fig. 9. Number of GCMs for different change ratio of the water supply capacity in A2 scenario

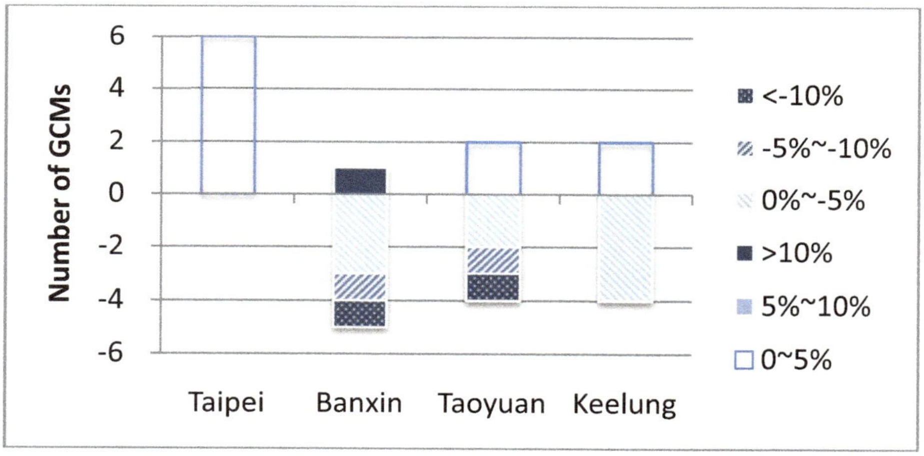

Fig. 10. Number of GCMs for different change ratio of the water supply capacity in A1B scenario

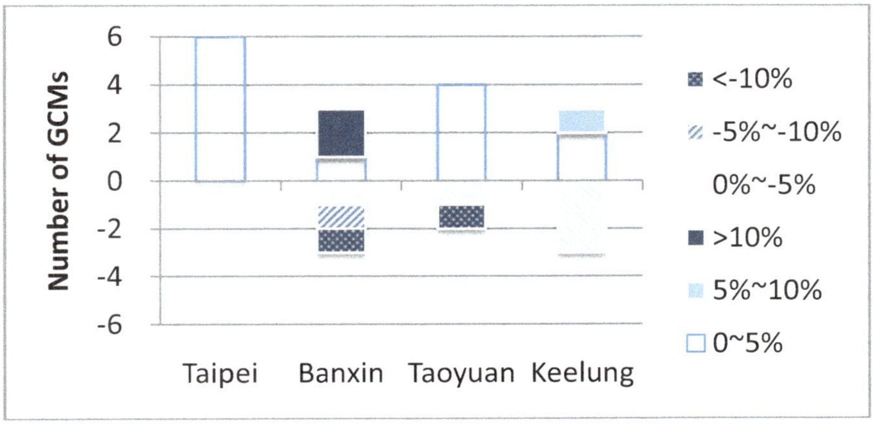

Fig. 11. Number of GCMs for different change ratio of the water supply capacity in B1 scenario

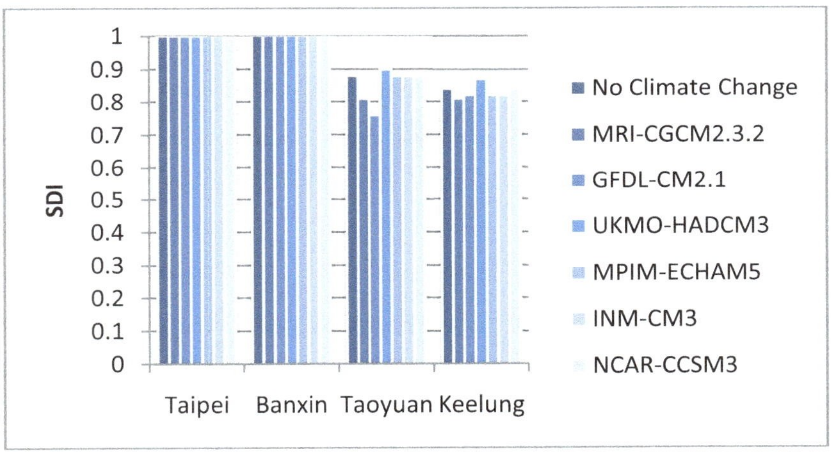

Fig. 12. Sustainable Index (SDI) of each water supply system from different result of GCM conditioned on no climate change and six GCMs' results of SRES-A1B

6. CONCLUSIONS

Under the threat of climate change, it is important to appraise the sustainability of water supply system, find whether the water supply can meet the water demand or not and make the strategies for the future. This study assesses the carrying capacity and sustainability of four water supply systems under climate change. From climate change scenario data to model programming, modeling integration, and sustainability appraisal, the whole process is very complex. Therefore, an integrated assessment tool, TaiWAP, was used to integrate the common procedures of impact assessment of climate change to evaluate the sustainability of regional water resources systems in this study. There are some major conclusions listed below:

1. For current situation, Taipei, Banxin, Taoyuan and Keelung are all sustainable water supply systems after the First Stage of Water Supply Improvement Project for

Banxin District was applied and the water shortage risk in Banxin water supply system was thus solved. In 2021, Taoyuan and Keelung water supply systems will turn into unsustainable because of raising water demand.

2. Due to the uncertainty of climate change simulation, as more as possible GCMs is recommended to be used in assessment. This study employed six GCMs for three climate change scenarios to evaluate the carrying capacity and sustainability of water supply systems. Six GCMs refer to MRI-CGCM2.3.2, GFDL-CM2.1, UKMO-HADCM3, MPIM-ECHAM5, INM-CM3 and NCAR-CCSM3. Because of the uncertainty of GCMs, it is not recommended to have quantitative assessment from GCMs. Using categories of change ratio is a better way to appraise the carrying capacity. In this study, six categories are adopted to classify the water capacity tendency of future. The more number of GCMs in the same category shows the more likely tendency of future.
3. Three climate change scenarios refer to SRES-A2, A1B and B1 scenario, which has different path for the future. Among three scenarios, SRES-A1B is more like the path where the world is heading and is recommended to choose if there is only one scenario can be used to reduce the uncertainty of climate change scenario.
4. To discover the most likely future of water supply systems, the sustainability appraisals for SRES-A1B scenario are adopted to reveal the sustainability of four water supply systems. Taipei and Banxin water supply systems will remain sustainable under climate change, otherwise Taoyuan and Keelung water supply systems will remain unsustainable. Climate change may slightly decrease their carrying capacity so that the effectiveness of strengthening water adaptive capacity might be affected by climate change. It is recommended further evaluation for future strategies under climate change.

COMPETING INTERESTS

Authors have declared that no competing interests exist.

REFERENCES

1. Wu YT, Chen MJ, Su HJ. The association of rainfall and drought between geographical distribution and infectious diseases in Taiwan, epidemiology. 2009;20(6):219-219.
2. Chen ST, Kuo CC, Yu PS. Historical trends and variability of meteorological droughts in Taiwan, Hydrological Sciences Journal. 2009;54(3):430-441.
3. Wilby RL, Harris I. A framework for assessing uncertainties in climate change impacts: Low-flow scenarios for the River Thames, UK. WATER RESOURCES RESEARCH, 2006;42:02419. dio: 10. 1029/2005WR004065
4. Wood AW, Leung LR, Sridhar V, Lettenmaier DP. Hydrologic implications of dynamical and statistical approaches to downscaling climate model outputs," Climatic Change, 2004;62(1-3):189-216.
5. Kundzewicz ZW, Mata LJ, Arnell NW, Döll P, Kabat P, Jiménez B, Miller KA, Oki T, Sen Z, Shiklomanov IA. Freshwater resources and their management. Climate Change 2007: Impacts, Adaptation and Vulnerability. Contribution of Working Group II to the Fourth Assessment Report of the Intergovernmental Panel on Climate Change, Parry ML, Canziani OF, Palutikof JP, van der Linden PJ, Hanson CE. Eds. Cambridge University Press, Cambridge, UK. 2007;173-210.

6. Chiang JL, Liu TM. Impact of climate change on paddy field irrigation in southern Taiwan, Paddy and Water Environment. 2013;11:311-320.
7. Liu TM, Tung CP, Ke KY, Chuang LH, Lin CY. Application and development of a decision-support system for assessing water shortage and allocation with climate change. Paddy and Water Environment. 2009;7(4):301-311.
8. IPCC, Data Distribution Center; Accessed on 25 July 2013. Available: http://www.ipcc-data.org/index.html
9. Liu TM, TaiWAP Website; Accessed on 25 July 2013. Available: https://groups.google.com/forum/#!forum/taiwap2
10. Carter TR, Jones RN, Lu X, Bhadwal S, Conde C, Mearns LO, O'Neill BC, Rounsevell MDA, Zurek MB. New assessment methods and the characterisation of future conditions. Climate Change 2007: Impacts, Adaptation and Vulnerability. Contribution of Working Group II to the Fourth Assessment Report of the Intergovernmental Panel on Climate Change,. Parry ML, Canziani OF, Palutikof JP, van der Linden PJ, Hanson CE. Eds. Cambridge University Press, Cambridge, UK. 2007;133-171.
11. Pickering NB, Stedinger JR and Haith DA. Weather input for nonpoint source pollution models. Journal of Irrigation and Drainage Engineering. 1988;114:674-690.
12. Tung CP and Haith DA. Global warming effects on New York streamflows. Journal of Water Resources Planning and Management. 1995;121(2):216-225.
13. Haith DA and Shoemaker LL. Generalized Watershed Loading Functions for Stream Flow Nutrients. Water Resour. Bull. 1987;23(3):471-478.
14. Haith DA, Mandel R and Wu RS. GWLF: Generalized Watershed Loading Functions User's Manual, Vers. 2.0. (Computer program manual). Cornell University, Ithaca, NY; 1992.
15. Forrester JW. Industrial dynamics, Productivity Press, PortlndOreg; 1961.
16. Simonovic SP. A shared vision for management of water resources. Water Int. 2000;25(1):76-88.
17. Forrester JW. Urban Dynamics. Pegasus Communications, Walthan, MA; 1969.
18. Meadows D, Randers J, and Behrens W. The Limits to Growth, Universe Books, New York; 1972.
19. Guo HC, Liu L, Huang GH, Fuller GA, Zou R, Yin YY. A system dynamics approach for regional environmental planning and management: a study for the Lake Erhai basin. J. Environ. Manage. 2001;61(1):93-111.
20. Kay AL, Davies HN, Bell VA, Jones RG. Comparison of uncertainty sources for climate change impacts: flood frequency in England. Climatic Change. 2009;92(1-2):41-63.

3

Are Diatom-based Indices from Europe Suitable for River Health assessment in China? A Case Study from Taizi River, Northeastern China

Xiaodong Qu[1,2], Ying Zhou[2,3], Rui Zhao[1,2], Catherine Bentsen[1,2], Xuwang Yin[3] and Yuan Zhang[1,2*]

[1]State Key Laboratory of Environmental Criteria and Risk Assessment, Chinese Research Academy of Environmental Sciences, Beijing, China.
[2]Laboratory of Riverine Ecological Conservation and Technology, Chinese Research Academy of Environmental Sciences, Beijing, China.
[3]Liaoning Provincial Key Laboratory for Hydrobiology, College of Fisheries and Life Science, Dalian Ocean University, Dalian, China.

Authors' contributions

This work was carried out in collaboration among all authors. Authors Qu, Yin and Zhang conducted the field trip including collecting sample of diatom and water quality. Author Yin conducted the identification of diatom. Authors Qu and Zhou designed the study and conducted the calculation of diatom-based indices. Author Zhao conducted main data analysis and modeling. The manuscript was prepared by author Qu and revised by CB. Author Qu oversaw the whole process as the research advisor. There were series of interactions among the authors to prepare and revise the manuscript.

ABSTRACT

Aims: Diatom-based indices are widely used for river health assessment. Many such indices were originally developed in European countries based on a specific taxa list of benthic diatoms. Thus, the transferability of these indices to other rivers and geographic locations has been questioned.
Design, Place and Duration of Study: In this study, we sampled benthic diatoms in the Taizi River, a temperate river in northeastern China during May 2009 to evaluate the applicability of eight commonly used diatom-based indices for assessing the principal water quality gradient and relationship with different water chemical parameters.

**Corresponding author: Email: zhangyuan@craes.org.cn;*

Methodology: Sensitivities of the eight indices were evaluated by applying the principle component analysis (PCA), the box-plot map and multiple comparisons of the Kruskal–Wallis nonparametric test (K-W test).

Results: The results showed that all eight of the tested indices showed significant correlations with the principal contamination gradient of both nutrient enrichment and organic pollution. The contamination gradient was extracted through principal component analysis and the first three axes explained 40.19%, 18.72%, and 9.77% of the total variation, respectively.

Conclusion: Our results confirmed that the diatom-based indices did not properly reflect the current Chinese surface water quality classes. However, these indices showed consistent trends with chemical parameters that reflected general water quality condition, such as electric conductivity (EC); organic pollution, as reflected by dissolved oxygen (DO) and chemical oxygen demand (COD) and nutrient enrichment, as reflected by total nitrogen (TN) and total phosphorus (TP). The results indicated that both BDI and SHE were the most suitable diatom-based indices among the eight tested indices, although SPI, DES and ROTT were also suitable for river health assessment in the Taizi River.

Keywords: Diatom-based index; BDI; river health; OMNIDIA; Taizi River.

1. INTRODUCTION

Benthic algae are important aquatic biota in riverine ecosystems. As primary producers, algae provide the main food resources for macro invertebrates and fish and play an important role in energy cycling [1,2]. Due to their short life span, relatively complex community composition and high sensitivity, benthic algae can integrate both short-term and long-term environmental variation and a broad range of human stressors [3]. Thus, many monitoring programs in the United States [4] and Europe [5-7] include diatoms as main elements to assess river health.

There are several diatom-based indices that have been developed in Europe using the Zelinka and Marvan formula [8]. The indices take into account the variations of relative abundance and tolerance of each taxon in the community, and have been widely accepted as important indictors in river health assessment. The Descy Index [9] was the first diatom index developed based on the pollution tolerance classification for diatoms. The standardized Biological Diatom Index (BDI) [10] was first developed and applied in France for the surveillance of water quality. The BDI index was revised (BDI-2006) [11] for better suitability to the European Water Framework Directive (WFD). Other widely used indices include the Specific Pollution sensitivity Index (SPI) [12], Generic Diatom Index (GDI) [13], Trophic Diatom Index (TDI) [14], Commission for Economical Community Index (CEE) [15], Schiefele and Schreiner's Index (SHE) [16] and Rott Saprobic Index (ROTT) [17].

Even though diatom indices are widely applied, there are two aspects that still present conflicts. The first is the suitability and applicability of different indices to different geographic regions [18]; the second is the sensitivity of diatoms to different kinds of contamination, such as ionic stresses, organic pollution, heavy metal contamination and eutrophication. Hering et al. [19] recommended that benthic diatom indices were better indicators to reflect eutrophication and land use gradients in European countries: the 25 tested diatom indices (e.g., SPI, SLA, DES, LMI, SHE, Watanabe index, TDI, BDI, DAIpo, CEE, GDI, Rott) showed similarly strong correlations with eutrophication/organic pollution and land use gradients.

However, there are also several examples of unsuccessfully applied indices. For example, Rott et al. [20] found the BDI may not be suitable for use in Austrian rivers and Gomà et al. [21] found the correlations between BDI and water quality variables were not significant in the upper Segre basin of the Oriental Pyrenees, Spain. Deng et al. [22] also compared seven indices of diatom sensitivity and applicability in the Dongjiang River, southeast China. The results showed that BDI and GDI were the most suitable indices for river health assessment in this area. Tan et al. [18] compared the usefulness of 14 indices in a subtropical river of central China. They identified strong, significant correlations of BDI and EPI to trophic status and ionic content, and strong correlation of Watanabe's index to organic pollution and conductivity gradients. Except for the Index Diatom Artois–Picardie (IDAP) [23,24], all other indices were suitable to reflect the water quality in the Han River, China during base-flow conditions.

Regarding sensitivity, Prygiel and Coste [25] confirmed six indices, including SPI, DES, CEE, GDI, Sládecek's saprobic index (SLA) [26] and Leclercq and Maquet's index (LMI) [27], as suitable indicators to detect the dominant stressor of organic pollution in the Artois–Picardie water basin in France. Among them, SPI, GDI and CEE also showed significant correlations with ionic strength and eutrophication. Rimet et al. [28] designed an experiment to evaluate the sensitivity of nine diatom indices by transferring the epilithic diatom biofilms from polluted rivers to an unpolluted stream. The results confirmed that five indices (SPI, TDI, CEE, Eutrophication Pollution Index [EPI] and ROTT) had higher sensitivities than the other four indices (GDI, BDI, LMI and SLA).

Currently, diatom-based indices are applied with greater frequency for river health assessment in China [29-31]. However, the assessments of suitability and sensitivity of these indices are rare for Chinese rivers [22]. Furthermore, there is still no research relating diatom indices to the national water quality standards, which currently represent the main aspect of the river health assessment method in China [32]. Biological river health assessment has become one of the most efficient management tools to date. Currently, both the Ministry of Environmental Protection and the Ministry of Water Resources of the People's Republic of China processed corresponding researching programs of river health assessment. And selecting suitable indices of benthic diatom, macro invertebrates and fish is one of the hot topics of the programs. It is also interested in for all the river management authorities in China. In this study, we conducted research to evaluate the applicability of eight diatom indices for water quality assessment; compare the sensitivity of each index to the Chinese national water quality and water contamination gradients; establish quantitative correlations between the stressor gradient and indices and establish suitable references values for each index for the temperate continental rivers in China.

2. MATERIALS AND METHODS

2.1 Study Region

The study was conducted in the Taizi River basin, which is the part of the Liao River basin located in northeastern China. Taizi River is located in the mid and high latitudes of China in a temperate, continental monsoon climate zone. It has a watershed area of 1.39×10^4 km^2 and stream length of 413 km. Main features of the local climate include a hot rainy summer, cold sunny winter, and a short spring and autumn. The annual temperature is around 7.9ºC and annual precipitation ranges from 650–950 mm across the basin [33]. Taizi River basin includes nine main tributaries and supplies drinking water for 5.5 million people as well as water for domestic, industrial and irrigation purposes.

Human disturbances are unevenly distributed from upstream to downstream. Forest dominates the main land cover in the eastern upper stream regions, while agriculture and urban areas dominate the main land uses in the middle and western downstream regions (Fig. 1) [34]. Across the whole watershed, the key threats to river health include clearing of riparian vegetation, construction of in-stream barriers and dams and in-stream extraction of sand and gravel [35]. In the middle and downstream areas, urban and rural non-point source pollution and industrial point source discharge have caused severe contamination of water quality in the Taizi River since the 1950s. Investments have been made for riverine restoration and protection during the last decade to improve water quality.

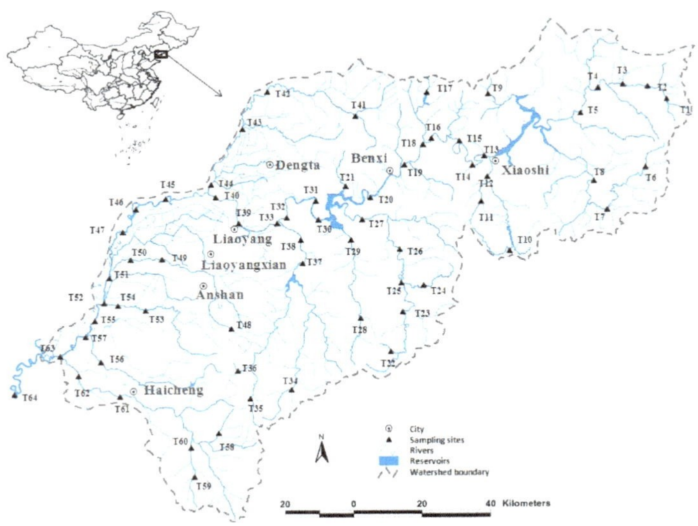

Fig. 1. Distribution of the 64 sites sampled for benthic diatoms in the Taizi River basin

2.2 Chemical Characteristics of Water Quality

Water samples were collected during the field trip before the processing of benthic diatom samples. Chemical parameters of electrical conductivity (EC) and dissolved oxygen (DO) were measured using a handheld YSI multiparameter instrument (professional plus) and pH was measured using a pen-style YSI pH tester instrument in situ. River water at each site was collected in polyethylene plastic bottles pre-rinsed three times with distilled water and kept below 4°C for laboratory analysis. In the laboratory, ammonia-nitrogen (NH_3) was measured with Nessler's reagent. Nitrate-nitrogen (NO_3) and total nitrogen (TN) were determined through the alkaline potassium persulfate oxidation–UV spectrophotometric method. Concentrations of total phosphorus (TP) and orthophosphate-phosphorus (PO_4) were measured by the ammonium molybdate spectrophotometric method. Five-day biological oxygen demand (BOD_5) was calculated by quantifying the dissolved oxygen of the samples before and after the 5-day incubation at 20°C and the permanganate index (COD_{Mn}) was measured by the potassium permanganate method. Chemical oxygen demand (COD) was measured by the standard potassium dichromate method. Suspended solids (SS) were measured by the standard operating procedure for total suspended solids analysis. The collection, transportation and analysis of water samples were performed according to the Chinese Standard Methods for Examination of Water and Wastewater [36] (Table 1).

2.3 Evaluation of Chinese Environmental Quality Standards for Surface Water

China manages its surface freshwater quality based on ambient water quality standards. Currently, national surface freshwater quality standards are used for the protection of water quality, decreasing the risk of water contamination, concerning human health of drinking water and maintenance of aquatic ecosystems. Five grades were used to evaluate from good to bad water quality conditions. The evaluation of Chinese environmental quality standards for surface water is based on the 'one-out, all-out' principle. In this way, the lowest water quality grade from any parameter represents the final water quality grade or class. According to the different water resource use aims, the water quality standards are divided into five classes in China [37]. These classes refer to headwater streams and national nature reserves (Class I); key zones for protection of surface drinking water, habitat for rare freshwater organisms and spawning regions for fish and shrimps (Class II); surface drinking water sources, wintering and migration regions for fish and shrimps (Class III); industrial and recreational water (Class IV) and agricultural and landscape water (Class V).

In this research, we selected six of the measured parameters for the water quality evaluation, excluding EC, NO_3 and TN. The EC does not have a national-level water quality standard since it is broadly affected by local geological conditions. The standard value of NO_3 is 10 mg/L, applied as a supplementary parameter for surface drinking water. The national water quality standard for TN is only applicable for lentic water in China; a TN standard for lotic waters has yet to be developed.

2.4 Benthic Algae Sampling and Identification

Benthic algae samples were collected from 64 sampling sites in the Taizi River before the monsoon season in May 2009 (Fig. 1). Nine samples were collected at each site from multiple substrates (e.g., stones, boulders, moss, vascular plants, sand or silt) and from multiple habitats (e.g., riffles, runs, shallow pools) within a 300 m sampling region [38]. All of the surface area of hard substrate within a 3.5-cm diameter corer was brushed thoroughly, and surface areas of soft substrate were collected using a core sampler with a 3.5 cm diameter, then all the samples were rinsed with a total of 500 ml distilled water and combined. The samples were preserved in a 4% formalin aqueous solution in a 500 ml plastic bottle. In the laboratory, diatom slides were kept in a glass jar to oxidize the organic material with acid disposal. Within each diatom slide, a minimum of 300 valves were counted under high magnification oil emersion. The 'soft' algae were identified directly using a 0.1-ml counting chamber. Most of the benthic algae were identified to the species level and several taxa were only identified to the genus level, according to the classification manuals of Hu and Wei [39] and Zhu and Chen [40].

2.5 Diatom Indices

Eight indices, including BDI, SHE, SPI, DES, GDI, CEE, TDI and ROTT were selected for evaluation in the river health assessment in China. The selected diatom indices mainly use weighted average values of sensitivity, indicator value and relative abundance of each taxon. The eight diatom indices (Table 2) were calculated using the OMNIDIA 7 software (Version 8.0) [24,41]. Zelinka and Marvan [8] served as a basis for the calculation of four indices, including DES, BDI, SPI and GDI, as follows:

$$ID = \frac{\sum A_i v_j i_j}{\sum A_i v_j},$$

where A_j is the relative abundance of species j; v_j is the indicator value ($1 \le v \le 3$) and i_j is sensitivity.

The DES uses 106 species of diatoms that have been classified into five categories of sensitivity. The BDI is based on the ecological profiles of 209 taxa for which presence probabilities are defined from 14 physical and chemical parameters (e.g., organic pollution, nutrients, conductivity and pH) within seven water quality classes. The SPI is calculated from the Saprobic system and comprises around 13,000 taxa, which usually includes all observed taxa [23]. The GDI also uses all observed species and five categories of sensitivity. This index was developed in order to have an easily usable index for river health assessment that utilizes genera, instead of species for cases of limited taxonomic skills [23].

The Differentiating Species System (DSS) takes into account 100 freshwater diatom species with worldwide abundance [42]. Similar to the DSS, SHE has been modified to use 386 species from seven classifications based on trophic state and pollution resistance. The CEE is based on a double entry table and 208 species are used. Low indicator taxa are horizontally ranked by increasing tolerance and high indicator taxa are vertically ranked by increasing tolerance. Correspondingly, the taxa are classified in different groups (low indicator) and subgroups (high indicator). The intersection of the groups and subgroups results in the values that correspond to the quality class. TDI uses 1603 species of diatoms, five classes of sensitivity to trophic state and three classes of reliability. This index is suitable to detect eutrophication in rivers caused by large, predominantly lowland sewage works.

Values calculated for each index were linearly adjusted on a scale from 0–20, with 0 representing the worst water quality and 20 the best water quality. In this study, we classified the index scores into five categories corresponding to different water quality statuses of excellent (16–20), good (12– <16), fair (8– <12), poor (4– <8) and critical (0–4).

2.6 Data Analysis

Spearman rank correlation analysis was used to detect the correlations among different indices. Principal component analysis (PCA) was performed to detect the major water quality gradients. Before PCA, water quality parameters with relatively high partial correlation coefficients ($r > 0.75$) and variance inflation factors ($F > 20$) were eliminated. Through the PCA, the water quality parameters were reduced to a few interpretable principal components. Simple linear regression analyses were applied to evaluate the quantitative relationship between the disturbance gradients and diatom indices. The Kruskal–Wallis nonparametric test (K-W test) was used to evaluate the variation of chemical parameters among different diatom-based health indices and the general variation of each parameter within the groups was analyzed using ANOVA. The correlation analysis, PCA analysis and K–W test were performed with STATISTICA software (Version 7.0).

3. RESULTS

A total of 109 species and subspecies of diatoms within 24 genera and 8 families were identified in the Taizi River. The three most dominant species were *Achnanthes minutissima*, *Ceratoneis arcus* and *Diatoma vulgare var. lineare* Grun with relative abundances of 33%, 8%

and 7% respectively. For the 64 sampling sites, six species—*Nizschia palea*, *Synedra ulna*, *Cymbella ventricosa*, *Synedra acus*, *Melosira varians* and *Navicula rhynchocephala* — comprised more than 60% of the occurrences.

Nine water chemical parameters were measured to evaluate the usefulness of eight diatom-based indices for river health assessment in the Taizi River. According to the surface water quality standards, eutrophication was considerable, with relatively high levels of NH_3 (average 0.92 mg/L), TN (average 5.18 mg/L) and TP (average 0.31 mg/L). Meanwhile, the organic pollution was relatively lower compared with eutrophication. Most of the sampling sites had relatively high DO (average 5.70 mg/L) and low COD_{Mn} (average 5.71 mg/L). However, the COD was high with an average 27.74 mg/L. The variation of EC was considerable, ranging from a minimum of 57μs/cm to a maximum of 1126 μs/cm (average 336 μs/cm) in the Taizi River (Table 1).

Table 1. Values of water quality parameters and diatom-based indices for the Taizi River

	Code	Mean ± SD	Min–Max	Standards*
Chemical parameters				
Suspended solids	SS	36.160±27.855	5–124	
Electric conductivity	EC	336.070±244.259 μs/cm	57–1126	–
Dissolved Oxygen	DO	5.700±2.044 mg/L	0.4–10.3	7.5–2
Chemical Oxygen Demand	COD	27.740±19.449 mg/L	0.0–100.5	15–40
Permanganate index	COD_{Mn}	5.710±5.030 mg/L	2.65–29.25	2–15
5-days Biological Oxygen Demand	BOD_5	5.440±5.207 mg/L	0.0–28.5	3–10
Ammonia-nitrogen	NH_3	0.920±1.401 mg/L	0.00–6.21	0.15–2.00
Nitrate-nitrogen	NO_3	2.880±2.620 mg/L	0.00–12.25	< 10**
Total nitrogen	TN	5.180±3.617 mg/L	0.68–17.75	0.2–2.0§
Total phosphorus	TP	0.310±0.560 mg/L	0.00–3.03	0.02–0.40
Diatom indices				
Biological Diatom Index	BDI	12.850±3.731	5.0–20.0	
Schiefele and Schreiner's Index	SHE	11.810±4.720	1.0–19.9	
Specific Pollution sensitivity Index	SPI	12.010±4.735	1.0–19.7	
Descy Index	DES	9.480±4.486	1.0–16.0	
Generic Index of Diatom assemblage	GDI	12.680±4.176	1.0–19.4	
Commission for Economical Community Index	CEE	7.930±.027	1.0–17.7	
Trophic Diatom Index	TDI	7.650±3.436	1.0–19.6	
RottSaprobic Index	ROTT	11.340±3.691	3.7–16.5	

**Chinese environmental quality standards for surface water (NEPB, 2002b)*
***Supplementary parameter of surface water for drinking water*
§TN standards are currently available only for lentic but not lotic, waters in China

Comparisons among the different diatom-based indices showed that GDI had a relatively high level of evaluation with an average value of 12.68, while TDI had relatively low level of evaluation with an average value of 7.65. Among the eight indices, BDI showed generally higher evaluation with all values higher than 4. In contrast, DES and ROTT showed generally lower evaluation with all values lower than 16 (Table 1). Significant correlations were found among the eight diatom indices using Spearman correlation analysis and all the P-values were less than 0.01 (Table 2). The correlation coefficient was highest for SPI and GDI (r = 0.951) and lowest for SHE and CEE (r = 0.666).

Table 2. Correlation coefficient matrix for the eight diatom indices (n = 64, P< 0.01)

	BDI	SHE	SPI	DES	GDI	CEE	TDI	ROTT
BDI	1							
SHE	0.829	1						
SPI	0.903	0.884	1					
DES	0.917	0.864	0.878	1				
GDI	0.857	0.872	0.951	0.857	1			
CEE	0.691	0.666	0.691	0.671	0.677	1		
TDI	0.708	0.744	0.756	0.795	0.822	0.686	1	
ROTT	0.883	0.883	0.885	0.932	0.861	0.667	0.822	1

The first three factorial axes were constructed to represent the dominant water quality gradients on the basis of the decline in eigen values in the PCA (Fig. 2). Three axes accounted for 68.68% of the total variance (Fig. 2), among which the first axis accounted for 40.19% and the second for 18.72%. The main explanatory parameters for PCA Factor 1 were NH_3, DO, EC and TP. The NH_3, EC, TP, COD, COD_{Mn} and BOD_5 were negatively correlated with PCA Factor 1, while DO was positively correlated. Main explanatory parameters for PCA Factor 2 were NO_3 and TN, both of which were negatively correlated. SS explained the largest amount of variance on PCA Factor 3 and was positively correlated.

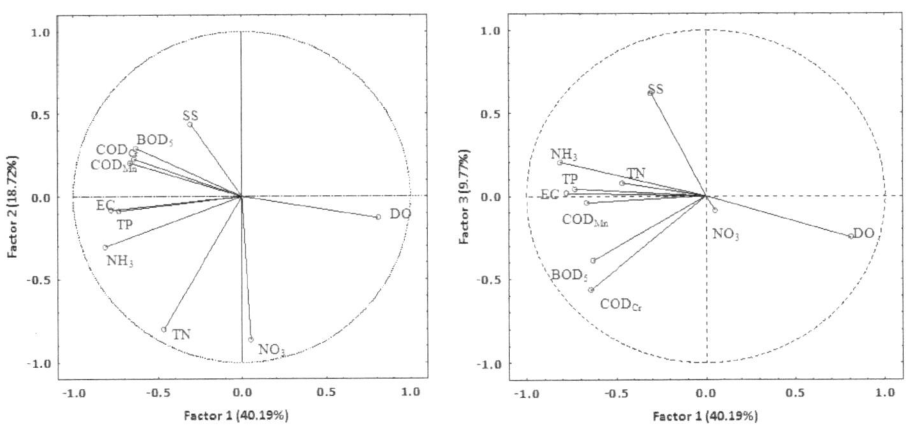

Fig. 2. Principal component analysis (PCA) ordination plots for 10 water quality parameters at the 64 sites sampled in the Taizi River basin. (a) Factor 1 × Factor 2 and (b) Factor 1 × Factor 3

The regression analyses revealed that all eight of the diatom indices could be explained significantly by PCA Factor 1. The negative direction of the PCA Factor 1 represented the main water quality stressor gradient. All indices showed positive linear relationships with PCA Factor 1. Among the eight indices, BDI, SPI and ROTT showed relatively higher levels of explanation (adjusted R^2 = 0.46–0.48). GDI also showed a significant linear model with PCA Factor 2 but at low levels of variance (R^2 = 0.06, P=0.04). GDI was positively correlated with PCA Factor 2. There were no indices that had significant correlations with PCA Factor 3 (Table 3).

Table 3. Adjusted coefficients of determination (R^2) from linear regressions used to analyze the response of eight diatom indices to the multiple stressor gradients (PCA Factors 1 and 2)

Equation	Adjusted R^2	F-value	P-value
BDI = −1.28*PCA Factor 1+12.85	0.47	56.50	< 0.001
SHE = 1.40*PCA Factor 1+11.81	0.34	33.69	< 0.001
SPI = 1.64*PCA Factor 1+12.01	0.48	58.17	< 0.001
DES = 1.33*PCA Factor 1+9.48	0.35	33.90	< 0.001
GDI = 1.29*PCA Factor 1+12.68	0.38	38.82	< 0.001
CEE = 1.22*PCA Factor 1+7.93	0.36	36.13	< 0.001
TDI = 0.98*PCA Factor 1+7.65	0.32	30.03	< 0.001
ROTT = 1.26*PCA Factor 1+11.34	0.46	54.99	< 0.001
GDI = 0.83*PCA Factor 2+12.68	0.06	4.93	0.04

In general, the variation of diatom-based indices did not correspond to water quality classes from relative clean water (Class II) to contaminated water (Class V) in the Taizi River. Among the eight indices, SHE, DES and TDI had no significant variation overall and among the different groups in relation to the decreasing water quality classes (Fig. 3). The other five indices of BDI, SPI, GDI, CEE and ROTT showed significant differences of general variation through the one-way ANOVA test (P< 0.05). However, the differences were not significant among Classes II– IV. These indices only showed significant differences between the lowest water quality grade (Class V) and the other classes using the multiple comparison K-W test (Fig. 3).

The sensitivities reflected by the correlations of chemical parameters to the river health categories of diatom-based indices varied significantly among different indices. Generally, chemical parameters all showed significant increasing water quality trends with increasing health categories of diatom-based indices except for COD_{Mn} and NO_3. Among the nine chemical parameters, EC, NH_3 and TP showed broad, significant correlations with all of the diatom-based indices (P< 0.001). DO also showed broad correlations to indices except for TDI. COD and BOD_5 were significantly correlated with five indices and TN with two indices. The multiple comparisons K-W test showed considerable variations for BDI, DES and GDI with six chemical parameters; SHE, TDI, CEE and ROTT showed variations with seven chemical parameters and SPI showed significant variations with eight chemical parameters (Table 4).

As the health categories improved, the average values of chemical parameters gradually changed in the positive direction or for DO, the negative direction. The variation was significant between the highest (i.e., very good) and lowest (i.e., critical) health class when the multiple comparison of K-W test was used. There were 31 chemical parameters in total that showed these types of significant variations, identified as Type I and Type II. These indices were suitable bio-indicators to reflect the changing concentrations of chemical parameters. Except for the gradual, significant increasing or decreasing trends, there were also 14 chemical parameters that showed abrupt variation from the best to the worst health category. These indices may not be sufficient or feasible to fulfill the assessment objectives of chemical parameter variations.

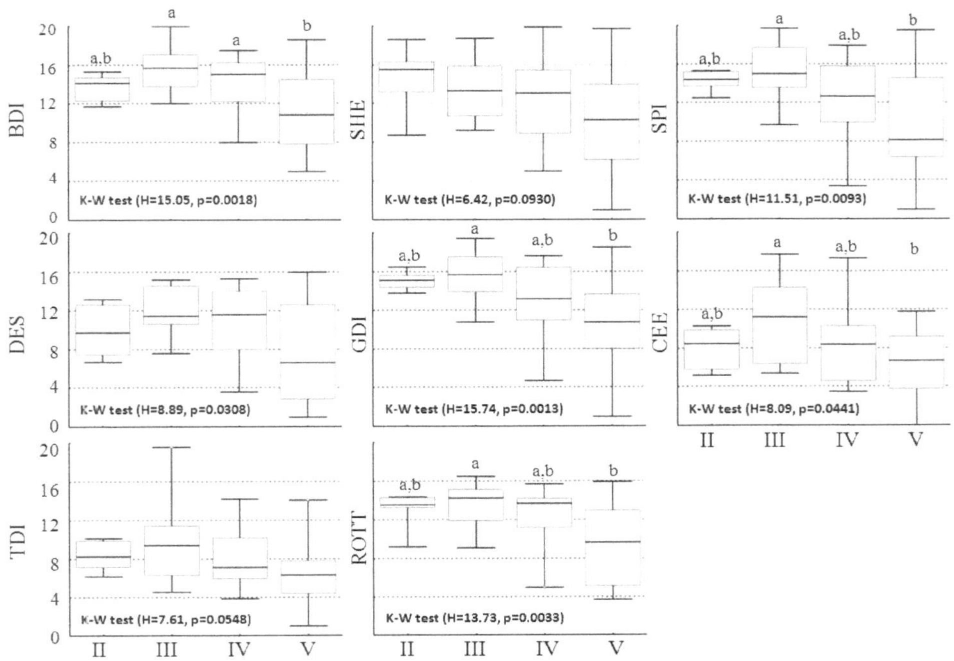

Fig. 3. Box plots for the eight diatom indices in relation to the Chinese surface water quality standards established for Classes II–V. The general variation and multiple comparisons among different water quality grades were conducted with the Kruskal–Wallis test (K-W test), with associated H- and P-values shown for each index. Different letters above the boxes indicate significant differences (P< 0.05) among the groups based on the multiple comparison K-W test. The dashed lines represented the health classes explained at the diatom-based indices in details

Eight sub-types were identified for those classifications that showed similar, significant variations between river health categories (Fig. 4). Sub-type I (Type I– I) represented when the variation was significant between the first category (very good for SHE; good for BDI, DES and ROTT) and the critical category. However, the variations were not significant among the middle three categories (or two categories for BDI, DES and ROTT) with the very good or critical categories (Fig. 4A and 4B). In total, 11 indices showed the same sub-types, including DO, COD, COD_{Mn} and BOD_5 for BDI; DO, COD_{Mn} and BOD_5 for SHE; TN for SPI and DES and EC and NH_3 for ROTT (Table 4). Sub-type II (Type I – II) showed similar variations as Type I – I, however the variation was significant between the very good category and poor and critical categories, such as TN and TP for SHE (Fig. 4C). For the four health classes of BDI and DES, the variations of EC, NH_3 and TP showed significance between the good category and poor and critical categories (Fig. 4D and Table 4). Sub-type III (Type I – III) was also similar to Type I – I but the variation was significant between the good and critical categories of DO for SPI (Fig. 4E). For the four health categories of DES and ROTT, the variations of EC and TP were significant between very good and good categories with the poor category (Fig. 4F and Table 4). For sub-type IV (Type I – IV) the variation was significant for very good and good from poor and critical, such as the EC for SPI (Fig. 4G). For the four health categories of BDI, the variation of NH_3 was significant for the very good to the poor and critical (Fig. 4H and Table 4). Sub-type V (Type I – V) showed that variation was significant between the very good category and the rest of the categories, only represented as TP for TDI (Fig. 4J). Sub-type VI (Type I – VI) had significant variation between the very good category and fair, poor and critical categories, only represented as COD for SHE (Fig. 4K). Sub-type VII (Type I – VII) showed significant variation for very good, good and fair categories with the critical category, represented as EC for GDI (Fig. 4L). Finally for sub-type VIII (Type I – VIII), the variation was significant for very good, good and fair categories with the poor category, represented as the BOD_5 for DES (Fig. 4M).

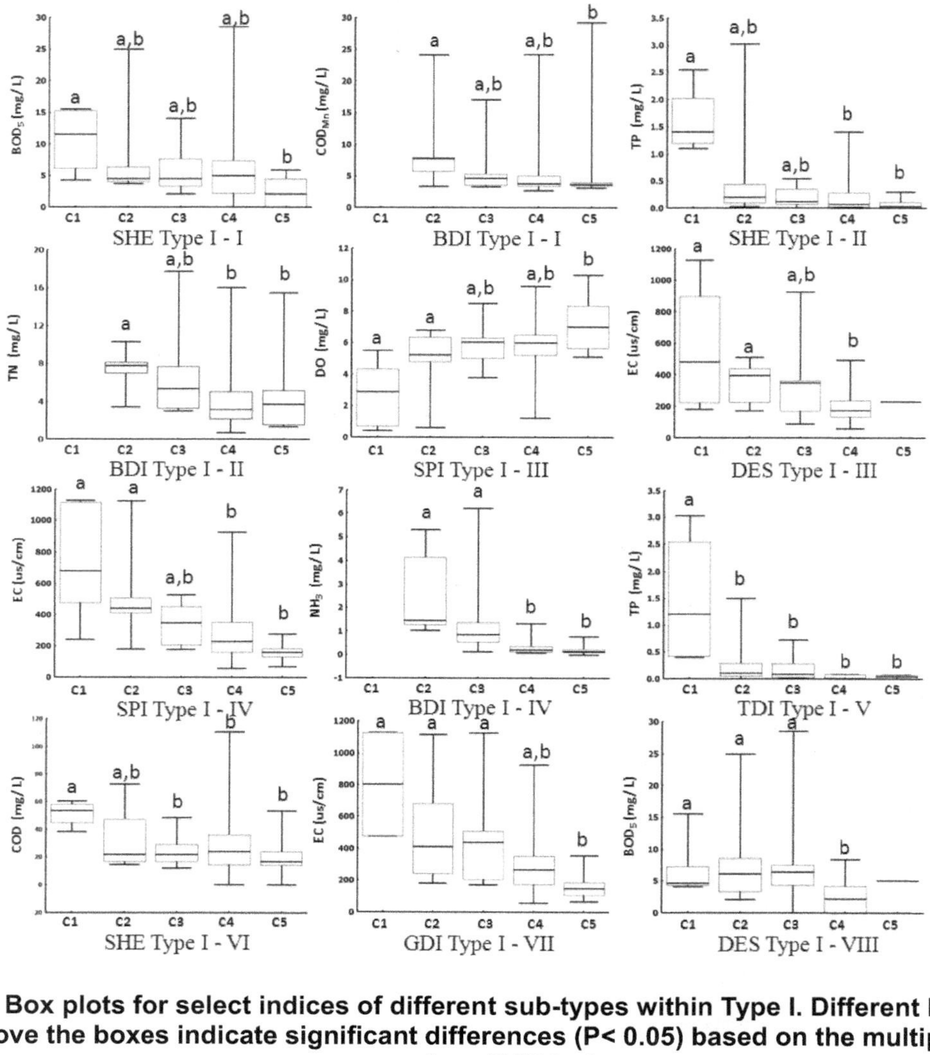

Fig. 4. Box plots for select indices of different sub-types within Type I. Different letters above the boxes indicate significant differences (P< 0.05) based on the multiple comparison K-W test

Table 4. Differences of nine chemical parameters, excluding suspended solids, among the five river health categories for each diatom-based index. Different letters indicate significant differences among different categories based on the multiple comparisons K-W test. The bold text indicates significant differences of the five categories with the K-W test, with associated F- and P-values

		C1 (very good)	C2 (good)	C3 (fair)	C4 (poor)	C5 (critical)	F-value	P-value
Sampling sites		**0**	**9**	**15**	**24**	**16**		
BDI	EC		690.90 ± 125.38^{a}	371.60 ± 51.43^{b}	255.80 ± 24.33^{b}	223.60 ± 31.33^{b}	13.42	< 0.001
	DO		3.20 ± 0.87^{a}	5.50 ± 0.40^{b}	5.90 ± 0.27^{b}	7.00 ± 0.46^{b}	9.64	< 0.001
	COD		44.30 ± 7.18^{a}	34.30 ± 7.00^{ab}	23.00 ± 2.48^{b}	19.50 ± 2.68^{b}	4.96	0.004
	COD_{Mn}		8.40 ± 2.06	5.60 ± 0.93	5.00 ± 0.87	5.40 ± 1.60	1.02	0.3910
	BOD_5		10.10 ± 2.74^{a}	6.40 ± 1.43^{ab}	4.70 ± 0.73^{b}	3.00 ± 0.72^{b}	4.60	0.006
	NH_3		2.70 ± 0.57^{a}	1.50 ± 0.48^{b}	0.40 ± 0.08^{c}	0.20 ± 0.05^{c}	12.84	< 0.001
	NO_3		2.90 ± 0.72	2.80 ± 0.72	3.10 ± 0.60	2.60 ± 0.59	0.13	0.941
	TN		7.40 ± 0.62	6.10 ± 0.96	4.30 ± 0.75	4.20 ± 0.92	2.35	0.081
	TP		1.00 ± 0.25^{a}	0.50 ± 0.20^{ab}	0.10 ± 0.03^{b}	0.10 ± 0.02^{b}	8.92	< 0.001
Sampling sites		**4**	**10**	**15**	**22**	**13**		
SHE	EC	960.50 ± 161.18^{a}	404.60 ± 51.31^{b}	297.70 ± 35.96^{b}	297.00 ± 43.72^{b}	201.50 ± 31.81^{b}	15.04	< 0.001
	DO	7.40 ± 0.53^{a}	4.90 ± 0.45^{b}	4.30 ± 0.37^{b}	6.80 ± 1.37^{b}	5.60 ± 1.97^{b}	5.99	< 0.001
	COD	51.40 ± 4.71^{a}	31.70 ± 6.30^{ab}	24.00 ± 2.52^{b}	29.20 ± 5.36^{b}	19.40 ± 3.43^{b}	2.60	0.05
	COD_{Mn}	7.40 ± 0.53	4.90 ± 0.45	4.30 ± 3.67	6.80 ± 1.37	5.60 ± 1.97	0.73	0.57
	BOD_5	10.70 ± 2.73^{a}	7.10 ± 2.05^{ab}	5.70 ± 0.80^{ab}	5.40 ± 1.29^{ab}	2.30 ± 0.67^{b}	2.73	0.04
	NH_3	3.80 ± 0.85^{a}	1.80 ± 0.58^{b}	0.60 ± 0.13^{ab}	0.60 ± 0.25^{ab}	0.20 ± 0.08^{c}	10.30	< 0.001
	NO_3	2.80 ± 1.29	2.90 ± 1.01	3.80 ± 0.71	2.90 ± 0.63	1.90 ± 0.23	0.84	0.504
	TN	8.60 ± 0.57^{a}	5.80 ± 1.45^{ab}	5.60 ± 0.90^{ab}	4.80 ± 0.82^{ab}	3.80 ± 0.66^{b}	1.59	0.188
	TP	1.60 ± 0.32^{a}	0.50 ± 0.29^{b}	0.20 ± 0.04^{b}	0.20 ± 0.07^{b}	0.10 ± 0.02^{b}	10.92	< 0.001
Sampling sites		**5**	**12**	**8**	**27**	**12**		
SPI	EC	728.00 ± 174.72^{a}	491.70 ± 72.04^{b}	339.90 ± 48.09^{bc}	269.80 ± 33.04^{bc}	163.80 ± 15.99^{c}	10.17	< 0.001
	DO	2.80 ± 0.99^{a}	4.80 ± 0.59^{b}	5.90 ± 0.50^{bc}	5.90 ± 0.31^{bc}	7.20 ± 0.50^{c}	6.69	< 0.001
	COD	34.90 ± 8.75^{ab}	42.60 ± 6.54^{a}	22.10 ± 2.70^{b}	25.70 ± 4.09^{ab}	18.30 ± 1.09^{b}	3.30	0.02
	COD_{Mn}	6.60 ± 0.86	6.40 ± 1.67	4.70 ± 1.63	6.40 ± 6.55	3.80 ± 0.12	0.73	0.57
	BOD_5	7.90 ± 1.89^{ab}	10.10 ± 2.55^{a}	3.90 ± 0.97^{bc}	4.70 ± 0.58^{bc}	2.50 ± 0.71^{c}	4.89	0.001
	NH_3	4.00 ± 0.78^{a}	1.40 ± 0.37^{b}	0.90 ± 0.16^{bc}	0.50 ± 0.21^{bc}	0.20 ± 0.03^{c}	14.64	< 0.001
	NO_3	4.60 ± 1.92^{a}	2.10 ± 0.45^{a}	4.30 ± 1.29^{a}	3.00 ± 0.51^{a}	1.70 ± 0.19^{a}	2.17	0.09
	TN	10.10 ± 2.02^{a}	5.10 ± 0.64^{ab}	6.60 ± 1.59^{b}	4.50 ± 0.66^{b}	3.80 ± 0.72^{b}	3.85	0.008

	TP	1.10 ± 0.41^{a}	0.60 ± 0.25^{ab}	0.20 ± 0.07^{b}	0.20 ± 0.06^{b}	0.10 ± 0.01^{b}	5.53	< 0.001
Sampling sites		12	13	16	22	1		
DES	EC	565.30 ± 106.65^{a}	349.20 ± 31.93^{b}	353.40 ± 56.53^{b}	195.40 ± 23.14^{b}	231.0	7.89	< 0.001
	DO	4.20 ± 0.68^{a}	5.60 ± 0.28^{ab}	5.60 ± 0.58^{ab}	6.60 ± 0.33^{b}	9.5	4.46	0.007
	COD	36.20 ± 4.52	29.70 ± 4.98	30.10 ± 7.02	20.80 ± 2.67	16.9	1.86	0.15
	COD_{Mn}	5.70 ± 0.56	4.50 ± 0.53	6.50 ± 1.45	5.90 ± 1.45	5.5	0.38	0.76
	BOD_5	6.80 ± 1.20^{a}	7.50 ± 1.74^{ab}	7.00 ± 1.58^{b}	2.40 ± 0.54^{b}	5.1	4.41	0.007
	NH_3	2.80 ± 0.55^{a}	0.70 ± 0.11^{b}	0.70 ± 0.35^{b}	0.30 ± 0.06^{b}	0.2	14.06	< 0.001
	NO_3	3.30 ± 0.93	2.90 ± 0.41	2.80 ± 0.70	2.80 ± 0.61	1.6	0.1	0.96
	TN	7.90 ± 1.06^{a}	4.70 ± 0.50^{ab}	4.90 ± 0.92^{b}	4.40 ± 0.83^{b}	1.8	2.99	0.038
	TP	1.00 ± 0.28^{a}	0.20 ± 0.06^{b}	0.20 ± 0.09^{b}	0.10 ± 0.02^{b}	0.1	10.72	< 0.001
Sampling sites		2	6	15	26	15.0		
GDI	EC	801.50 ± 324.50^{a}	507.30 ± 140.86^{ab}	428.80 ± 60.15^{bc}	307.70 ± 37.17^{bc}	162.00 ± 19.68^{c}	7.04	< 0.001
	DO	3.10 ± 2.40^{a}	3.90 ± 0.79^{b}	5.60 ± 0.46^{ab}	5.60 ± 0.35ab	7.10 ± 0.47^{b}	4.73	0.002
	COD	44.80 ± 6.30	32.80 ± 7.16	31.80 ± 4.75	27.70 ± 4.79	19.50 ± 1.60	1.35	0.26
	COD_{Mn}	6.80 ± 1.00^{a}	5.00 ± 0.95^{a}	5.00 ± 0.40^{a}	7.30 ± 1.46^{a}	3.80 ± 0.20^{a}	1.32	0.27
	BOD_5	6.20 ± 1.93^{a}	8.20 ± 2.02^{b}	7.20 ± 1.55^{ab}	5.00 ± 1.11^{b}	3.20 ± 0.70^{b}	1.64	0.18
	NH_3	2.90 ± 1.50	2.80 ± 0.93	1.20 ± 0.31	0.60 ± 0.22	0.20 ± 0.02	7.87	< 0.001
	NO_3	4.10 ± 2.95	3.70 ± 1.60	2.90 ± 0.76	3.20 ± 0.53	1.80 ± 0.21	0.95	0.44
	TN	9.20 ± 1.10^{a}	7.50 ± 2.27^{ab}	5.70 ± 0.93^{ab}	5.00 ± 0.67^{ab}	3.50 ± 0.62^{b}	2.34	0.07
	TP	1.80 ± 0.73^{a}	0.50 ± 0.19^{b}	0.50 ± 0.21^{b}	0.20 ± 0.06^{b}	0.10 ± 0.03^{b}	6.97	0.001
Sampling sites		**6**	**32**	**21**	**3**	**2**		
TDI	EC	595.30 ± 174.74^{a}	384.90 ± 37.79^{b}	244.10 ± 33.895^{b}	101.50 ± 25.02^{b}	95.00 ± 26.00^{b}	4.91	0.002
	DO	4.20 ± 1.15^{a}	5.40 ± 0.31^{b}	6.40 ± 0.41^{b}	7.10 ± 1.61^{b}	6.90 ± 1.80^{b}	2.35	0.06
	COD	44.20 ± 3.48^{a}	29.50 ± 3.80^{ab}	22.70 ± 3.88^{ab}	16.90 ± 3.74^{b}	$20.\ 50 \pm 1.03^{b}$	1.90	0.122
	COD_{Mn}	5.60 ± 0.83	5.80 ± 0.89	6.20 ± 1.35	3.30 ± 0.09	3.90 ± 0.18	0.27	0.90
	BOD_5	6.70 ± 1.78^{a}	6.30 ± 0.85^{a}	4.20 ± 1.37^{a}	4.40 ± 2.26^{a}	3.20 ± 1.00^{a}	0.70	0.60
	NH_3	2.40 ± 0.60^{a}	1.20 ± 0.29^{b}	0.30 ± 0.07^{b}	0.10 ± 0.02^{b}	0.20 ± 0.05^{b}	4.11	0.005
	NO_3	2.80 ± 1.01	3.20 ± 0.52	2.70 ± 0.54	1.40 ± 0.13	1.50 ± 0.27	0.55	0.70
	TN	7.90 ± 0.60^{ab}	5.40 ± 0.73^{bc}	4.90 ± 0.66^{bc}	1.60 ± 0.14^{a}	2.80 ± 1.27^{c}	2.02	0.10
	TP	1.50 ± 0.45^{a}	0.20 ± 0.06^{b}	0.20 ± 0.04^{b}	0.00 ± 0.02^{b}	0.00 ± 0.03^{b}	12.45	< 0.001
Sampling sites		**9**	**26**	**21**	**3**	**5**		
CEE	EC	704.90 ± 127.69^{a}	346.50 ± 29.61^{b}	237.10 ± 24.25^{b}	229.30 ± 44.88^{b}	97.70 ± 15.12^{b}	12.87	< 0.001
	DO	3.20 ± 0.85^{a}	5.60 ± 0.28^{b}	6.70 ± 0.40^{b}	7.40 ± 0.54^{b}	6.00 ± 0.69^{b}	7.07	< 0.001

	COD	45.50 ± 10.11^{a}	28.40 ± 3.54^{ab}	23.10 ± 3.09^{ab}	16.10 ± 4.86^{b}	18.80 ± 2.06^{b}	3.08	0.02
	COD_{Mn}	7.50 ± 1.32	5.10 ± 0.80	6.50 ± 1.52	3.50 ± 0.17	3.60 ± 0.18	0.86	0.49
	BOD_5	7.80 ± 1.49^{a}	6.60 ± 1.33^{a}	4.00 ± 0.61^{a}	1.00 ± 0.97^{a}	3.90 ± 1.31^{a}	1.94	0.11
	NH_3	3.60 ± 0.71^{a}	0.70 ± 0.09^{b}	0.30 ± 0.08^{b}	0.30 ± 0.19^{b}	0.10 ± 0.02^{b}	26.47	< 0.001
	NO_3	3.00 ± 1.21	3.50 ± 0.57	2.20 ± 0.23	4.40 ± 3.01	1.50 ± 0.10	1.37	0.25
	TN	8.20 ± 1.42^{ab}	5.20 ± 0.68^{abc}	4.10 ± 0.50^{bc}	8.70 ± 3.53^{a}	2.10 ± 0.49^{c}	4.48	0.003
	TP	1.30 ± 0.34^{a}	0.20 ± 0.03^{b}	0.20 ± 0.05^{b}	0.10 ± 0.02^{b}	0.00 ± 0.02^{b}	15.39	< 0.001
Sampling sites		**3**	**11**	**15**	**34**	**1**		
ROTT	EC	696.30 ± 214.85^{a}	505.50 ± 103.31^{ab}	436.50 ± 50.11^{bc}	211.50 ± 17.67^{c}	121.0	11.68	< 0.001
	DO	4.20 ± 1.77^{a}	4.40 ± 0.66^{ab}	5.00 ± 0.54^{ab}	6.60 ± 0.26^{b}	5.1	5.85	0.001
	COD	45.70 ± 3.75^{a}	37.10 ± 6.11^{ab}	33.60 ± 7.23^{ab}	20.80 ± 2.01^{b}	19.5	4.04	0.01
	COD_{Mn}	6.30 ± 0.74	5.30 ± 0.58	6.50 ± 1.56	5.50 ± 0.96	4.1	0.18	0.91
	BOD_5	5.50 ± 1.29^{a}	8.90 ± 2.05^{a}	7.30 ± 1.74^{a}	3.50 ± 0.51^{a}	4.2	4.43	0.007
	NH_3	2.40 ± 0.96^{a}	2.50 ± 0.59^{a}	1.00 ± 0.36^{b}	0.30 ± 0.04^{b}	0.2	13.56	<0.001
	NO_3	2.70 ± 2.03	3.40 ± 0.91	3.00 ± 0.72	2.70 ± 0.43	1.8	0.16	0.92
	TN	8.00 ± 1.34^{a}	7.10 ± 1.22^{a}	5.20 ± 0.92^{a}	4.30 ± 0.58^{a}	4.1	2.47	0.07
	TP	2.20 ± 0.58^{a}	0.50 ± 0.14^{b}	0.20 ± 0.09^{b}	0.10 ± 0.03^{b}	0.1	36.46	<0.001

The second main type of distribution was identified as that of increasing health categories with the average values of chemical parameters also gradually changing in a negative direction. With the multiple comparison K-W test, the variations were significant between the very good category and the poor and critical categories and between the very good and good categories with the critical category. This kind of variation was identified as Type II. Five chemical parameters showed the same variation trends, for example, NH3 for SHE (Fig. 5A) and EC for CEE (Fig. 5B).

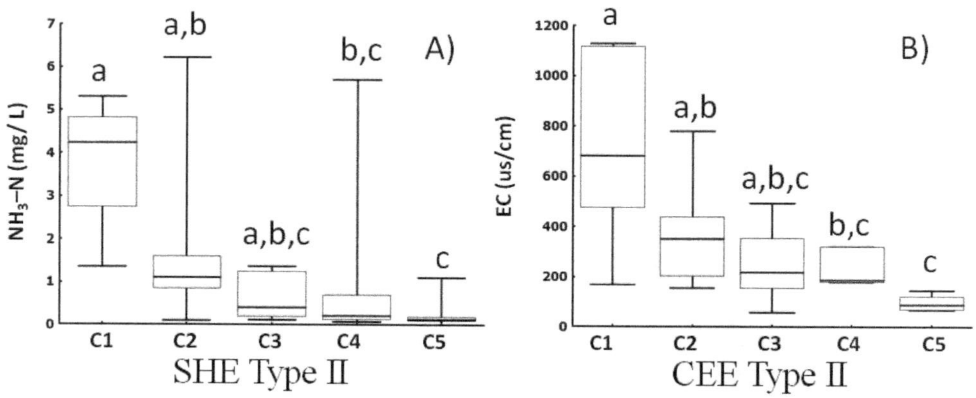

Fig. 5. Box plots for select indices within Type II. Different letters above the boxes indicate significant differences (P< 0.05) based on the multiple comparison K-W test.

4. DISCUSSION

Diatom-based indices have successfully been used as bioassessment indicators of freshwater ecosystems to reflect different kinds of human activities [11,43]. Our study confirmed that eight diatom-based indices could be used to effectively evaluate the main water quality degradation gradients (Table 3). Since these diatom-based indices were mainly developed based on the taxa found in numerous European countries, some studies have questioned the suitability and applicability of diatom-based indices in other countries, such as in China [44]. However, these indices have been successfully applied for river health assessment in several non-European countries, including Canada [45], Argentina [46], Brazil [47] and South Africa [48]. Other studies conducted in Chinese rivers, including subtropical rivers located at southern [22,29] and central China [18], in addition to our research results of temperate Taizi River in northeastern China, recommended the broad applicability of diatom-based indices in China. As the best what we know, these diatom-based indices were used for the river health assessment at the first time at temperate rivers in the Northeastern China.

This study revealed the potentially high usefulness for integrating the eight indices into the river health assessment of Taizi River and other streams in northeastern China. The linear regression analyses revealed the high explanation of the main stressor gradient (PCA Factor 1) to eight indices with relatively high R^2 (0.32–0.48, P<0.001). In our study, concentrations of the most tested chemical parameters were relatively high. For example, high concentrations of COD (maximum 110.5 mg/L), COD_{Mn} (maximum 29.25 mg/L) and BOD_5 (maximum 28.5 mg/L), as well as of TN (0.68–17.75 mg/L) and TP (0–3.03 mg/L) reflected relatively high levels of organic pollution and nutrient enrichment in the Taizi River (Table 1). Both diatom-based indices and communities have proven useful to assess organic pollution and eutrophication over the long term, either synthetically or separately [49-51]. The negative direction of those chemical parameters in the first axis of the PCA ordination biplot (Fig. 2) showed that synthetic contamination was common in the research area. Thus, the significant regression results supported the usefulness of the diatom-based indices for river health assessment. Diatom-based indices are usually developed to assess specific contaminants for example TDI was developed specifically to evaluate eutrophication. However, the results of our study showed that these indices could reflect general contamination and evaluate river health condition [25,28]. Meanwhile, the relatively high, positive inter-correlation coefficients (0.666–0.951) revealed similar variation trends to human stressors among the eight indices (Table 2).

The general patterns (Type I and Type II) were mainly identified from the box plots and multiple comparisons (Fig. 4, Fig. 5 and Table 4). The major types and sub-types identified must show consistent increasing or decreasing trends with increasing diatom-based index classes [52]. The variation trends and significance of the eight indices were not consistent across the different chemical parameters. In light of the general patterns identified (both Type I and Type II), BDI and SHE were recommended as the most suitable indices for river health assessment in the Taizi River. SPI, DES and ROTT were also identified as suitable indices for the Taizi River. Among those indices, the usefulness of BDI has been reported for other freshwater ecosystems in China [18,22,30,31]. As a result, BDI may be recommended as a national bio-indicator for river health assessment in China. SHE, SPI, DES and ROTT may be recommended as local bio-indicators for river health assessment in northeastern China.

Because the variation trends were not consistent between the water quality classes and diatom-based indices, our results recommended that integrated river health assessment use both biological and chemical parameters. Water quality standards serve as a major basis for the management and protection of freshwater ecosystems. The Chinese government manages its water quality mainly based on ambient water quality standards. Currently, water quality grades are determined based on the chemical parameter with the worst contamination level. However, since these standards are mainly derived from environmental quality standards or criteria from developed countries that consider both ecosystem and human health, the current water quality assessment may not match well with the diatom-based indices [32]. The results also supported the notion that integrated river health assessment should include both biological and chemical information for better river management [53].

From the results it was apparent that several indices were widely applicable for river health assessment in the Taizi River. However, the community composition of benthic diatoms shows high variation and succession in different seasons, which directly affect the final index scores and river health assessment [18]. Meanwhile, variations in other natural conditions (e.g., altitude, forest cover) or human activities (e.g., watershed land uses, heavy metal contamination) may also affect community compositions of benthic diatoms and subsequently the final scores of diatom-based indices in the Taizi River. Thus, continual evaluation of diatom-based indices is necessary for river health assessment on either the national or local scale.

5. CONCLUSION

Even though diatom-based indices calculated from the OMNIDIA program were developed based on taxa from European rivers, our results confirmed that the eight selected indices were suitable for river health assessment in temperate regions of China. In this study, the eight tested indices all showed significantly high correlations with the main water contamination gradient (PCA Factor 1) according to linear regression models. BDI and SHE were recommended as the most suitable indicators, and SPI, DES, and ROTT were also suitable for river health assessment in the Taizi River. We also recommended BDI as a suitable indicator for the national river health assessment in China. Considering that the health categories of diatom-based indices were not consistent with Chinese water quality classes, we emphasized the importance of a river health assessment that integrates both chemical and biological parameters.

ACKNOWLEDGMENTS

The research was supported by the National Natural Science Foundation of China (No. 51208483 & 311300101) and Chinese Major Science and Technology Program for Water Pollution Control and Management (2012ZX07501-001). The authors would like to extend sincere thanks to Dr. Liu Wei, a senior scientist at the Water Environment Monitoring Center of the Pearl River Basin, for his help in supervising the calculation of diatom indices and contributing to the results and discussion.

COMPETING INTERESTS

Authors have declared that no competing interests exist.

REFERENCES

1. Pan Y, Lowe RL. Independent and interactive effects of nutrients and grazers on benthic algal community structure. Hydrobiologia. 1994;291(3):201–209.
2. Simon KS, Benfield EF, Macko SA. Food web structure and the role of epilithic biofilms in cave streams. Ecology. 2003;84(9):2395–2406.
3. Stevenson RJ, Smol JP. Use of algae in environmental assessments. In: Wehr JD, Sheath RG (eds.) Freshwater Algae in North America: Classification and ecology. Academic Press, San Diego. 2003;775–804.
4. Hill BH, Herlihy AT, Kaufmann PR, Stevenson RJ, McCormick FH, Burch Johnson C. Use of periphyton assemblage data as an index of biotic integrity. Journal of the North American Benthological Society. 2000;19(1):50–67.
5. European Commission. Common implementation strategy for the Water Framework Directive (2000/60/EC). Overall approach to the classification of ecological status and ecological potential. Office for official publications of the European communities, Luxembourg; 2003.
6. Ector L, Rimet F. Using bioindicators to assess rivers in Europe: an overview. In: Lek S, Scardi M, Verdonschot PFM, Descy JP, Park YS, eds. Modelling Community Structure in Freshwater Ecosystems. Springer, Berlin; 2005;7-19.
7. Philibert A, Gell P, Newall P, Chessman B, Bate N. Development of diatom-based tools for assessing stream water quality in south-eastern Australia: assessment of environmental transfer functions. Hydrobiologia. 2006;572(1):103–114.

8. Zelinka M, Marvan P. Zurpräzisierung the biological classification of the purity of flowing waters. Archivfür Hydrobiologie. 1961;57:389–407.
9. Descy JP. A new approach to water quality estimation using diatoms. Nova Hedwigia. 1979;64:305–323.
10. Lenoir A, Coste M. Development of a practical diatom index of overall water quality applicable to the French National Water Board Network. In: Whitton BA, Rott E (eds.). Use of Algae for Monitoring Rivers II. Institutfür Botanik, Universität Innsbruck. 1996;29–45.
11. Coste M, Boutry S, Tison-Rosebery J, Delmas F. Improvements of the Biological Diatom Index (BDI): description and efficiency of the new version (BDI-2006). Ecological Indicators. 2009;9(4):621–650.
12. National Centre of Agricultural Machinery, Genie Rural, Eauxet of Forestry (Cemagref). Study of biological methods of quantitative assessment of water quality. Report on Water Quality Division Lyon, Agencefinanciére Basin Rhône-Méditerranée-Corse, Pierre–Bénite, French. 1982;218.
13. Rumeau A, Coste M. Introduction to the systematics of freshwater diatoms to the practical use of a indicediatomiquegénérique. Bulletin of the French Fishingand Fish Culture, French. 1988;309:1–69.
14. Kelly MG, Whitton BA. The Trophic Diatom Index: a new index for monitoring eutrophication in rivers. Journal of Applied Phycology. 1995;7(4):433–444.
15. Descy JP, Coste M. A test of methods for assessing water quality based on diatoms. Verhandlungen International Vereinigung Limnologie. 1991;24:2112–2116.
16. Schiefele S, Schreiner C. Use of diatoms for monitoring nutrient enrichment, acidification and impact of salt in rivers in Germany and Austria. In: Whitton BA, Rott E, Friedrich G (eds.). Use of Algae for Monitoring Rivers. Institutfür Botanik, Universität Innsbruck; 1991.
17. Rott E, Hofmann G, Pall K, Pfister P, Pipp E. Indikationslistenfür nursery algae in Austrian rivers. Part 1: Saprobielle indication Water Management cadastral, Bundeministerium f. Land- u. Forstwirtschaft, Wien, Germany; 1997.
18. Tan X, Sheldon F, Bunn SE, Zhang Q. Using diatom indices for water quality assessment in a subtropical river, China. Environmental Science and Pollution Research. 2013;20(6):4164–4175.
19. Hering D, Johnson RK, Kramm S, Schmutz S, Szoszkiewicz K. Verdonschot PFM. Assessment of European streams with diatoms, macrophytes, macroinvertebrates and fish: a comparative metric-based analysis of organism response to stress. Freshwater Biology. 2006;51(9):1757–1785.
20. Rott E, Pipp E, Pfister P. Diatom methods developed for river quality assessment in Austria and a cross-check against numerical trophic indication methods used in Europe. Algological Studies. 2003;110(1):91–115.
21. Gomà J, Rimet F, Cambra J, Hoffmann L, Ector L. Diatom communities and water quality assessment in mountain rivers of the upper Segre basin (La Cerdanya, Oriental Pyrenees). Hydrobiologia. 2005;551(1):209–225.
22. Deng PY, Lei YD, Liu W, Wang XT. Exploration of benthic diatom indices to evaluate water quality in rivers in the Dongjiang basin. Acta Ecologica Sinica, Chinese. 2012;32(16):5014–5024.
23. Prygiel J, Lévêque L, Iserentant R. A nouvelIndice Diatomic Practice for assessing the water quality monitoring network. Journal of Water Sciences, French. 1996;9(1):97–113.
24. Lecointe C, Coste M, Prygiel J. OMNIDIA 3.2. Diatom index software including diatom

database with taxonomic names, references and codes of 11645 diatom taxa; 2003.

25. Prygiel J, Coste M. The assessment of water quality in the Artois–Picardie water basin (France) by the use of diatom indices. Hydrobiologia. 1993;269–270(1):343–349.
26. Sládeček V. Diatoms as indicators of organic pollution. Acta hydrochimica et hydrobiologica. 1986;14(5):555–566.
27. Leclercq L, Maquet B. Two new indices chimiqueetdiatomique current qualitéd'eau. Application Samson and sesaffluents (Belgian Meuse basin). Comparison with other chemical indices biocénotiquesetdiatomique. Royal Belgian Institute of Natural Sciences of Belgium. Documents de Travail. 1987;28:1–113. French.
28. Rimet F, Cauchie HM, Hoffmann L, Ector L. Response of diatom indices to simulated water quality improvements in a river. Journal of Applied Phycology. 2005;17(2):119–128.
29. Qi YZ, Huang WJ, Luo YM, Wu LY. Evaluation of water quality of the Pearl River by using diatom assemblage index (DAIpo) and river pollution index (RPId). Journal of Tropical and Subtropical Botany, Chinese. 1998;6(4):329–335.
30. Xin XY, Yang WJ, Zhang H, Wang CH, Xu JH. Diatom evaluation of water pollution in the Daihai Lake, Inner Mongolia. Journal of Shanxi University (Natural Science Edition), Chinese. 2000;23(3):263–266.
31. Zhao XG, Cai DS, Liu W, Wang XT. Diatom biomonitoring method study on water quality in Lijiang River. Journal of Guangxi Normal University (Natural Science Edition), Chinese. 2009;27(2):142–147.
32. Wu FC, Meng W, Zhao XL, Li HX, Zhang RQ, Cao YJ, Liao HQ. China embarking on development of its own national water quality criteria system. Environmental Science and Technology. 2010;44(21):7992–7993.
33. You Z (ed). Annals of Liao River (1st version). Jilin Renmin Press, Changchun; 2004.
34. Kong WJ, Meng W, Zhang Y, Gippel C, Qu XD. A freshwater ecoregion delineation approach based on freshwater macroinvertebrate community features and spatial environmental data in Taizi River Basin, northeastern China. Ecological Research. 2013;28(4):581–592.
35. Meng W, Zhang N, Zhang Y, Zheng BH. Integrated assessment of river health based on water quality, aquatic life and physical habitat. Journal of Environmental Sciences. 2009;21(8):1017–1027.
36. National Environmental Protection Bureau, China. Standard methods for the examination of water and wastewater (version 4). China Environmental Science Publish Press, Beijing, Chinese; 2002.
37. National Environmental Protection Bureau, China. Chinese environmental quality standard for surface water, GB3838-2002, Chinese; 2002.
38. Barbour MT, Gerritsen J, Snyder BD, Stribling JB. Rapid bioassessment protocols for use in streams and wadeable rivers: periphyton, benthic macroinvertebrates and fish. 2nd ed. EPA 841-B-99-002. U. S. Environmental Protection Agency, Office of Water, Washington, D.C; 1999.
39. Hu HJ, Wei YX. The freshwater algae of China: systematics, taxonomy and ecology. Science Press, Beijing, Chinese; 2006.
40. Zhu HZ, Chen JY. Bacillariophyta of the Xizang Plateau. Science Press, Beijing, Chinese; 2000.
41. Lecointe C, Coste M, Prygiel J. 'OMNIDIA': A software for taxonomy, calculation of diatom indices and inventories management. Hydrobiologia. 1993;269(1):509–513.
42. Lange-Bertalot H. Pollution tolerance as a criterion for water quality estimation. Nova Hedwigia. 1979;64:285–304.

43. Pan Y, Stevenson RJ, Hill BH, Herlihy AT, Collins GB. Using diatoms as indicators of ecological conditions in lotic systems: a regional assessment. Journal of the North American Benthological Society. 1996;15(4):481–495.

44. Deng DY, Lu W, Tao M. Recent advances and indicative role of diatoms in water environment monitoring. Environmental Science and Technology. 2009;32(12D):222–225. Chinese.

45. Lavoie I, Hamilton PB, Wang YK, Dillon PJ, Campeau S. A comparison of stream bioassessment in Québec (Canada) using six European and North American diatom-based indices. Nova Hedwigia. 2009;135:37–56.

46. Gómez N, Licursi M. The Pampean Diatom Index (IDP) for assessment of rivers and streams in Argentina. Aquatic Ecology. 2001;35(2):173–181.

47. Bere T, Tundisi JG. Applicability of borrowed diatom-based water quality assessment indices in streams around São Carlos-SP, Brazil. Hydrobiologia. 2011;673(1):179–192.

48. Taylor JC, Prygiel J, Vosloo A, de la Rey PA, van Rensburg L. Can diatom-based pollution indices be used for biomonitoring in South Africa? A case study of the Crocodile West and Marico water management area. Hydrobiologia. 2007;592(1):455–464.

49. Rott E, Duthie HC, Pipp E. Monitoring organic pollution and eutrophication in the Grand River, Ontario, by means of diatoms. Canadian Journal of Fisheries and Aquatic Sciences. 1998;55(6):1443–1453.

50. Stevenson RJ. Epilithic and epipelic diatoms in the Sandusky River, with emphasis on species diversity and water pollution. Hydrobiologia. 1984;114(3):161–175.

51. Hill BH, Herlihy AT, Kaufmann PR, DeCelles SJ, Vander Borgh MA. Assessment of streams of the eastern United States using a periphyton index of biotic integrity. Ecological Indicators. 2003;2(4):325–338.

52. Feio MJ, Alves T, Boavida M, Medeiros A, Graça MAS. Functional indicators of stream health: a river-basin approach. Freshwater Biology. 2010;55(5):1050–1065.

53. Bunn SE, Abal EG, Smith MJ, Choy SC, Fellows CS, Harch BD, Kennard MJ, Sheldon F. Integration of science and monitoring of river ecosystem health to guide investments in catchment protection and rehabilitation. Freshwater Biology. 2010;55(suppl.1):223–240.

4

Impact Assessment of Household Demand Saving Technologies on System Water and Energy Use

Aisha Bello-Dambatta[1*], Zoran Kapelan[1] and David Butler[1]

[1]Centre for Water Systems, College of Engineering, Mathematics and Physical Sciences; University of Exeter; North Park Road; Exeter EX4 4QF, United Kingdom.

Authors' contributions

This work was carried out in collaboration between all authors. All authors read and approved the final manuscript.

ABSTRACT

Climate change, population growth, migration, urbanisation, and ageing infrastructure will all impose significant strains on the urban water services in Europe, and cities across Europe will experience increasingly frequent shortfalls in supply/demand balance. It is widely accepted that the mitigation of these and other emerging challenges should be sensitive to increasing energy prices, the environment, and the desire for low carbon intensity solutions. This paper presents the development of a new methodology for assessing the impact of household water savings from different water demand management interventions based on their water-related energy use and cost, as well as their impact on the supply/demand balance. The methodology has been applied to the water distribution system of a European city to demonstrate its application using different water demand management interventions for different types of water savings. Sensitivity analysis for different population growth rates that are representative of the different growth rates across the EU was carried out. The results show different degrees of water, energy, and cost savings can be achieved depending on the type (s) and proportion of household micro-component appliances and fittings considered. In all the intervention strategies considered, there are important trade-offs to be made between the different performance indicators as not all interventions will result in water savings and/or reductions in water-related energy use and costs or have a positive impact on supply/demand balance.

**Corresponding author: Email: a.bello-dambatta@exeter.ac.uk;*

Keywords: Cost; impact assessment; micro-components; supply/demand balance; water demand management; water-related energy use.

1. INTRODUCTION

The quality and the quantity of fresh water resources are increasingly facing challenges in many parts of the world as a result of climate change, population growth, migration, urbanisation, and ageing infrastructure. In the EU, environmental regulations and national legislations are also forcing Water Service Providers (WSP) to comply with increasingly stringent limits on chemical contaminants in drinking water, which further reduces the potential sources of supply [1,2]. Seasonal or inter-annual variations in the availability of water also lead to water stress [3].

However, despite these challenges the underlying trend is that the demand for water is increasing [4]. If the current population and demand trends persist, there will be an increase in global water demand of about 64 billion cubic metres per year [5]. It is estimated this demand will surpass availability by 56% by 2025, and it is likely that water resource development will not keep up with population growth [6]. In order to sustainably meet future water demand, various types of interventions need to be taken into consideration.

Historically, efforts to satisfy demand of water have often been expended principally on increasing the supply of resources, which were available abundantly and at relatively low cost [7,8]. However, supply augmentation such as the development of new reservoirs, dams, treatment plants, desalination plants and large scale water-transfer infrastructures are costly, and the over reliance on the development of new supply systems to respond to increasing demand often encounters public opposition as they are viewed as a potential cause of environmental degradation. Supply augmentation also tend to be unresponsive to economic, environmental, social, and political constraints and the important contributions to supply that can be obtained from comprehensive demand interventions [1].

However, in the face of growing demand, current uncertainties, and change; reducing the specific demand for water is the best source of 'new' water [9]. Water Demand Management (WDM) aims to sustainably reduce water use to conserve the resource, save water, and reduce negative environmental impacts whilst still satisfying the needs of consumers. WDM interventions can also provide energy and cost savings have a positive impact on supply/demand balance, extend the life of existing Water Distribution Systems (WDS), and reduce the total volume of wastewater arriving at treatment works.

The water industry is energy intensive and on average between 2-3% of the world's energy use is used to treat water to potable quality, deliver it to consumers, and to process and dispose of wastewater [10]. In the UK for example, energy use in the water industry rose 10% between 2001 and 2011 to 9, 016 GWh [11]. However, this represents only 11% of the total water-related energy use, with 89% attributed to domestic water use, particularly hot water use which constitutes 95% of household water-related energy use [12]. Water supply and wastewater management operations alone is therefore a poor indicator of the energy use associated with the urban water sector [12] and an assessment of household water-related energy use is needed to better determine energy use in the urban water sector.

Most of the urban water use in the EU is for household water use [13] to satisfy basic needs for drinking water and sanitation and other needs such as house cleaning, dishwashing, clothes washing, and landscaping [14]. The amount of water use in households depends on a wide range of factors, including household occupancy, type, income, and water prices [14]. This amount is increasing as a result of increasing use of water using appliances due to increasing standards of living and personal hygiene [13]. One way of reducing household demand is to make existing homes more water efficient [15], which could be accomplished by using efficient household micro-component appliances and fittings, such as lower flush WCs, lower flow showers and taps, and efficient white goods [16].

As household water and energy use are inextricably linked, reducing household water use and using efficient household appliances and fittings, especially hot water using appliances can also significantly reduce overall energy demand and household energy cost [12,17]. On average, around a third of household energy use is for heating and hot water use. Also, because the level of water-related energy use has a direct relationship with GHG emissions, WDM interventions can be viewed not only as a potential means of aiding the security of future water supplies but also as a means of reducing emissions [12].

However, not all WDM interventions will result in reduction in water-related energy use and some interventions could increase energy use. For example, reducing mains water use with rainwater may result in increasing water-related energy use as a result of energy required to pump rainwater. In general, if water use is to be reduced then some other factor must change to accommodate this, and where a reduction of water use or water-related energy use result in an increase of the other, it is not necessarily clear which is the most sustainable outcome [18]. An understanding of these water-energy-cost savings trade-offs is therefore essential in order to plan for sustainable WDM interventions.

Several methodologies have been developed for assessment of WDM interventions, including AQUACYCLE [19], UWOT [20], CWB [21] and UVQ [22]. These are all based on water balance modelling which assess the performance of integrated urban water supply, demand, and wastewater systems. Other methods that assess the performance of household water and energy use such as the Code of Sustainable Homes [23] and the Sustainable Building Alliance [24] do so in isolation of each other and provide no means of assessing the inevitable trade-offs that could result between different performance indicators. These methods also do not consider water-related energy use as a component of the performance of either energy or cost savings and cannot be used to assess the extent to which household water savings can fully affect energy use.

This paper presents a new methodology for the impact assessment of household water savings from different WDM intervention strategies based on their water-related energy use, cost and their impact on the supply/demand balance of a WDS. The methodology uses 2006 average EU per capita water use data and household micro-component water use breakdown, as well as assumptions from previous work. The methodology differs from previous methodologies in terms of understanding the impact of household water savings on water-related energy use and cost and the resulting water-energy-cost savings trade-offs that can arise from different types of household WDM intervention strategies.

2. METHODOLOGY

The methodology uses the 2006 average EU water use of 150 Litres/Capita. Day (LCD) [25,26] and household micro-component water use breakdown as a baseline for the impact assessment (Table 1). Although average figures mask variations of water use across countries, regions and even within communities and over time [3,1], they provide a basis for comparing the baseline impacts of WDM interventions on different performance indicators [7] where case study data is not fully or readily available. Moreover, the per capita figures used in the methodology are based on 2006 figures, the last accurate EU average figures we could find. This may have changed over time, especially given the recent near-drought conditions experienced in some parts of Europe.

Table 1. Average 2006 EU household micro-component water use, % [26]

Household micro-components	Water use, %
WC	35
Bath	15
Shower	5
Washbasin tap	8
Kitchen sink tap	25
Outdoor tap	6
Washing machine	12
Dishwater	4
TOTAL	150

2.1 Water Demand Management Options

Different types and combinations of household micro-component appliances and fittings can be used to reduce water use, as well as provide savings in household water-related energy use, costs, and have an impact on supply/demand balance of WDS. The WDM options considered in the methodology include different types of household micro-component appliances and fittings, domestic Grey Water Recycling (GWR) and Rainwater Harvesting (RWH) systems (Table 2). In the baseline scenario in the methodology, it is assumed that 100% of household water use is from mains water supply – i.e. no GWR and RWH systems are in use. In the other scenarios, grey water from baths and showers can be used by GWR systems to flush WC's, and rainwater can be used by RWH systems for WC flushing, clothes washing, and outdoor use as household WDM interventions. Social impacts have not been considered due to lack of data on the social impact of WDM interventions.

Efficient household micro-component appliances and fittings can offer the potential for significant water savings at point of use and can potentially reduce water use from 150-80 LCD [27]. Retrofitting households and efficiently fitting new developments is considered the least sensitive WDM intervention to issues of human interface, as no lifestyle changes are required, and are therefore the easiest intervention to implement [26]. Some of these technologies can also have short payback periods which can further enhance their uptake possibilities [27]. Three types of efficient household appliances and fittings have been considered in the methodology (Table 2):

i) Conventional household appliances – these are the current household appliances, and have been assumed to represent 100% baseline household asset ownership.
ii) Water efficient household appliances – these are the current BATNEEC (Best Available Technologies not Entailing Excessive Costs) appliances and fittings – such as new WC's, taps, baths, showers and white goods.
iii) Retrofit devices and fittings for household appliances – these are a comparably low-cost alternative to replacing households with efficient appliances which provide a low-cost means of water, energy and costs savings. These include cistern displacement and interruptible flush devices for WCs and flow restrictors and aerators for taps.

Table 2. Types of household micro-component appliances and fittings

Household appliance and fitting	Type
WC - 10 litres/flush	Conventional
WC - 7.5 litres/flush	Efficient
WC - 6 litres/flush	Efficient
WC - 6/4 litres/flush	Efficient
WC - 6/3 litres/flush	Efficient
Cistern displacement device - HIPPO	Retrofit
Cistern displacement device - SAVE-A-FLUSH	Retrofit
Retrofit flush device	Retrofit
Bath	Conventional
Medium bath	Other conventional
Large bath	Other conventional
Gravity mixer shower	Conventional
Electric shower	Efficient
Power shower	Efficient
Retrofit shower head	Retrofit
Retrofit shower flow regulator	Retrofit
Wash basin tap	Conventional
Washbasin tap insert – aerator	Retrofit
Washbasin tap insert – flow regulator	Retrofit
Kitchen sink tap	Conventional
Kitchen sink tap – aerator	Retrofit
Kitchen sink tap – flow regulator	Retrofit
Outdoor tap	Conventional
Washing machine	Conventional
Efficient washing machine	Efficient
BATNEEC washing machine	Efficient
Dishwasher	Conventional
Efficient dishwasher	Efficient
Grey Water Recycling system	Efficient
Rainwater Harvesting system	Efficient

2.1.1 Household appliances and fittings

The baseline water use for each household micro-component appliance is based on the volume of water use and frequency of appliance use per person per day (Table 3). For showers, washbasin taps, kitchen sinks taps and outdoor taps, the volume of water use is based on event duration and appliance flow rate. The factors affecting the volume of water used for bathing are the type of bath, its capacity, and its usage pattern [28]. The mean life spans used are in line with current industry standards for household micro-component appliances and fittings [28]. Although ideally energy appliances should be modelled with increased efficiency over the long-term, all energy appliances considered have been assumed to have constant efficiency for the duration of the planning horizon as it is difficult to anticipate relevant technological developments and estimate how much more efficient appliances will become in the future and when.

Table 3. Baseline assumptions for household micro-component appliances water use [26,28]

Household micro-components	Volume of water/use, litres	Frequency of use/ day	Event duration, minutes	Flow rate, litres/ minutes	Water use, LCD	Lifespan, year	Asset ownership, %
WC – 10 litres/flush	10	5.25	n/a	n/a	52.5	15	100
Bath	80	0.28	n/a	n/a	22.4	15	100
Shower	15	0.5	n/a	n/a	7.5	12	100
Washbasin tap	6	2	0.7	9.2	12.0	15	100
Kitchen sink tap	10	2.25	0.7	15.4	22.5	25	100
Outdoor tap	9	1	n/a	n/a	9	n/a	100
Washing machine	100	0.18	n/a	n/a	18	13	100
Dishwasher	29	0.21	n/a	n/a	6.1	13	100
TOTAL					150		

2.1.2 Alternative water systems

Alternative systems such as GWR and RWH can limit the amount of mains water use for non-potable uses and water abstraction needs, as well as have a positive effect on wastewater by delaying/reducing peak inflow into wastewater system. Current domestic systems can be high value, low cost, and low energy use systems that can also be utilised to reduce water use by limiting the amount of mains water use for non-potable uses – such as WC flushing, clothes washing, and outdoor uses, thereby reducing household water use. If used for WC flushing for example, a well-designed and fully functional system could potentially save about a third of the mains water used in households. Further reductions can be made if used for non-potable uses such as clothes washing and outdoor use [29].

The domestic WDM systems considered are domestic GWR and RWH systems. The GWR system has been assumed to collect grey water from baths and showers, which then is treated and reused for WC flushing. The quantity of household water from washbasins, showers and baths is similar to that used for household WC flushing. This means supply of water for household WC flushing will roughly equate to demand, as each person will generate their own water. This will in turn result in minimal to no reliance of mains water top-up and smaller tank size. The RWH system collects rainwater which can be used for WC flushing, outdoor water use (garden watering, car washing), and washing machines, and can therefore potentially provide further water savings compared to GWR systems. RWH systems are heavily reliant on rainfall, and therefore when it does not rain for some time and water cannot be collected, the system will revert to mains water supply to top-up.

2.2 Performance Indicators

Three performance indicators have been used to assess the impact of water savings from different WDM intervention scenarios: (i) energy use; (ii) cost; and (iii) supply/demand balance.

2.2.1 Water-related energy use

All household appliances that use hot water have associated water-related energy use. This energy use depends on both the volume of water used, and the temperature difference between hot and cold water use [18,12]. The assumptions for household hot water use are provided in Table 4. This includes the temperature at which mains water is delivered to households, mean temperature at point of use, and the percentage split between hot and cold water mix at point of use for baths, showers, washbasin taps, and kitchen sink tap. Boiler efficiency refers to the amount of energy that is lost from the boiler [18]. For example, 80% efficiency means that 20% of the energy is lost. It has been assumed that all washing machines and dishwashers considered internally heat cold water.

Table 4. Household hot water assumptions [18]

Variable	Assumptions
Average temperature of mains water	9°C
Shower temperature	41°C
Bath water temperature	42°C
Washbasin tap temperature	42°C
Kitchen sink tap temperature	55°C
Washbasin tap hot : cold water contribution to total volume	50:50 split
Kitchen sink hot : cold water contribution to total volume	50:50 split
Bath hot : cold water contribution to total volume	2 parts hot : 1 part cold
Shower hot : cold water contribution to total volume	2 parts hot : 1 part cold
Heating method	Electricity
Boiler efficiency	80%

Annualised household water-related energy use is calculated using the sum of energy used to produce hot water use, Q (KWh) (Equation 1) and the energy used for GWR and RWH, E (Equation 2).

$$Q = \frac{C_P \times m \times \Delta T}{c \times \eta_b} \qquad \text{...} \qquad (1)$$

where C_p is the specific heat capacity of water (J/kg/^{0}C), m is the mass of water (kg), ΔT is the change in temperature between mains water and water use temperature (^{0}C), the constant c is the conversion factor from Joules to KWh, and η_b is the boiler efficiency (MTP 2011a, Fidar et al 2010). The energy use for GWR and RWH pumping E is calculated using:

$$E = p \times \left(\frac{m}{60}\right) \times h \times d \times \varphi \qquad \text{.......................................} \qquad (2)$$

where p is the number of times the system cycles in an hour, m is the duration of the cycle (minutes), h is the number of hours in a day, d is the number of days a month and φ is the running wattage of the GWR or RWH system (KWh).

2.2.2 Cost

The cost of an intervention, C (EUR/year), is based on annualised total cost over an intervention interval calculated using the sum of annualised CAPEX of an intervention over the lifetime of the household micro-component appliances and fittings over their life time (Equation 3) and the annual operational cost of the appliances and fittings (i.e. energy use for GWR and RWH pumping and treating systems and hot water production).

$$C = K_0 \times \frac{r(1+r)^{n}}{(1+r)^{n-1}} \times W \qquad (3)$$

where K_0 is the CAPEX at the beginning of the intervention interval (EUR), r is the discount rate (%), n is the life span of the appliances and fittings (yearsand W is the proportion of households in the WDS with the micro-component appliances and fittings (%). It is assumed that the total cost of interventions is borne at the beginning of an intervention. However, CAPEX has been annualised because of the different life spans of the appliances and fittings.

2.2.3 Supply/demand balance

The supply/demand balance is calculated indirectly in terms of headroom, H (ML/year), available between supply (S) and demand (D) (Equation 4). Negative headroom indicates supply/demand deficit.

$$H = S - D \qquad (4)$$

3. CASE STUDY

The methodology has been applied to a case study of the WDS of a European city by running different WDM interventions. The aim was to assess the impact of water savings from the different WDM intervention strategies (i.e. different types and combinations of WDM options) based on their energy use, cost and impact on supply/demand balance of the WDS.

3.1 Case Study Description

The case study is of a water supplier which is operating on a 25 year concession contract that began in 2000. The WDS consists of 3 treatment plants, 8 pumping stations and 2, 400 km of water pipes and sewerage pipes each. Since privatisation, the water supplier has invested 20 million EUR and has seen significant improvement at all system levels: Non Revenue Water (NRW) has decreased from 300-90 million m^3 per year and water use has decreased from 400-150 LCD, perhaps as a result of universal metering coverage and the higher water rates. A summary of existing household conditions in the WDS is given in Table 5. Assumptions in Table 6 have been used where case study data is not available.

Table 5. Existing household conditions in the WDS

	Data
Permanent population	1, 900, 000
Maximum supply capacity, m^3/day	1, 600, 000
Total system supply capacity, m^3/year	520, 000, 000
Total domestic water demand, m^3/year	225, 000, 000
Total household demand m^3/year	102, 000, 000
Water rate, EUR/m^3	0.8
Total embodied energy (electricity), kWh/year	38, 000, 000
Domestic embodied energy, kWh/year	23, 000, 000
No of domestic metered connections	108, 000
Domestic metering coverage, %	100

Table 6. Household assumptions in the WDS

Assumption	Data
Average household size, people	2.5
No of households	760, 000
Population growth rate, %	-0.2
Unit cost of energy, EUR/kWh	0.12
Discount rate, %	0.02
Interest rate , %	0.04

The water supplier has 100, 000 contractual customers, serving a population of 1.9 million. With no major industrial use, water use in the city is mainly domestic. The current total water demand is around 225 million m^3 per year, about 45% (102 million m^3 per year) of which is for domestic use and about 40% (90 million m^3 per year) of which is NRW. The remaining 15% is for commercial and municipal water use. The NRW of 40% represents losses only to point of metering, as the water supplier's responsibilities are limited to point of customer metering. In terms of supply/demand balance, the water supplier has the capacity to produce around 520 million m^3 per year, of which only about 225 million m^3 per year is sold. Therefore the water supplier is currently producing is twice as much water as it sells.

Despite the huge water savings made since privatisation, water use can be further reduced by using WDM interventions, which can also lead to reduction in energy use and cost, as well as have an impact on supply/demand balance which could improve the security of future water supply without corresponding investment in supply infrastructure. However, despite universal metering of domestic use in the WDS, around 1,000 of the water supplier's connections are to housing blocks serving up 1,000 inhabitants each. These connections represent around 80% of domestic water use and access to them is only through household associations. There is therefore currently no data available for per capita or household micro-component breakdown and as such, assumptions based on average EU water use and household micro-component breakdown, as well as other assumptions from previous work have been used.

3.2 Case Study Assumptions

Assumptions have been made about average household data, such as the household occupancy, household size, and the number of households (Table 6). These are assumed to remain constant over the planning horizon (2010–2050). The figures for population growth rate (PGR) and the unit cost of energy have been obtained from World Bank and Europe's Energy Portal respectively [30,31]. It is assumed that the negative population growth in Table 6 will remain constant over the planning horizon. Assumption for interest rate is based on the European Central Bank's long-term interest rate for EU Member States [32].

3.3 Intervention Scenarios

Different degrees of water savings can be achieved depending on the WDM intervention scenario considered, and impacts of water saving on water-related energy use, cost and on supply/demand balance can vary significantly. Three different scenarios have been considered on the case study WDS to assess the impacts of different water savings in terms of their energy use, cost and supply/demand balance. All the scenarios considered in the case study assume the current negative PGR remains the same over the planning horizon:

i) Business as usual (BAU) scenario – where the current EU average per capita water of 150 LCD remains the same, with household micro-component appliances and fittings in line with average product lifespan.
ii) Alternative systems (ALT) scenario – where household micro-component appliances and fittings are replaced in line with average product lifespan, and GWR and RWH systems are introduced at 5% of households at each intervention interval, which will result in overall reduction of volumetric water use.
iii) Aggressive (AGG) scenario – where water use is reduced by 30 LCD, which will result in water use of 120 LCD by 2050. For example, a reduction of per capita water use to 120 LCD at 2050 by reducing water use by 10 LCD between 2012–2020, 2020–2030 and 2030–2040 by using different combinations of household appliances and fittings and GWR and RWH systems.

3.4 Results and Discussion

In the BAU scenario, there is a decrease in per capita water use by 9 LCD to 141 LCD at the end of the planning horizon (2050), which decreases overall water demand from 104 ML/yr to 83 ML/yr (Fig. 1). Headroom increases as a result of both the reduced per capita water use that will arise from the negative PGR and replacement of appliances and fittings; indicating no water resource stress issues even without increase in supply infrastructure or dedicated WDM strategy due to headroom as shown in Fig. 1. The result also shows a decrease in total household water-related energy use and costs, even with the CAPEX for the micro-component appliance and fitting replacement in line with average product life spans. The proportion of household micro-component water use also changes in the BAU intervention given the types and proportion of the appliances and fittings used over time. Less water is used for WC flushing, washing machines and dishwashers as a result of the type (s) and proportions of micro-component appliances and fittings considered (Table 7).

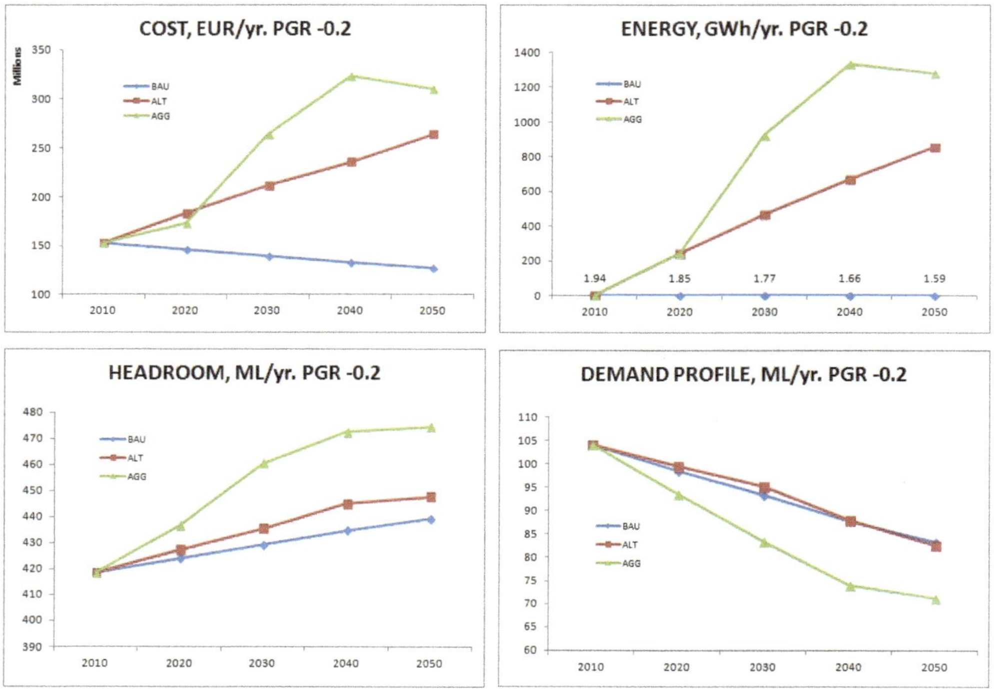

Fig. 1. Results for the three WDM intervention strategies under current negative PGR

Table 7. Household micro-component water use at the end of the planning horizon (2050)

Household micro-components	Baseline, LCD	BAU, LCD	ALT, LCD	AGG, LCD
WC	35	28	29	25
Bath	15	25	25	15
Shower	5	6	6	8
Washbasin tap	8	8	8	9
Kitchen sink tap	15	15	15	18
Outdoor tap	6	6	6	8
Washing machine	12	9	8	10
Dishwasher	4	3	3	3

In the ALT scenario, the result shows an overall reduction in volumetric mains water use and an increase in costs and headroom, with a substantial increase in energy use given the introduction of GWR and RWH systems at 5% of households at each interval (Fig. 1). Energy use significantly increases in this scenario given the energy use of GWR and RWH systems even with improved energy use of current GWR and RWH systems. The proportion of water use by source changes over time given the introduction of GWR and RWH systems, as this reduces the per capita mains water use by 10 LCD, a litre more than in the BAU scenario. However, this scenario will cost significantly more than BAU given both the CAPEX required for GWR and RWH systems and the cost of additional energy use for pumping grey water and rainwater. The proportion of micro-components water use also changes over time given the different appliances and fittings that will be used over time (Table 7). As with the

BAU scenario, less water is used for WC flushing and white goods. Water use for taps remains the same whilst that for baths and showers increases.

In the AGG scenario, there is a reduction in per capita by 30 LCD to 120 LCD at the end of the planning horizon (Fig. 1). The cost of this scenario is significantly higher than that of the ALT scenario because of the types and proportion of household micro-component appliances and fittings considered. The proportion of water use by source also changes over time given the introduction of GWR and RWH systems, as this reduces the amount of volumetric mains water use. The proportion of household micro-components water use also changes over time given the proportion and types of household micro-component appliances and fittings that will be used over time (Table 7).

Overall, the BAU scenario performs the best in terms of energy use and cost, and performs the worst in terms of headroom. This could become an issue if the current situation changes, for example if demand increases from an increase in population or industrial water use, or if supply is affected by climate change without a corresponding increase in supply resources. However, a reduction in demand will result in less revenue for WSPs and the unit cost of water may have to increase to reflect this reduction. This could make water saving and uptake of efficient household micro-component appliances and fittings to help with water savings more appealing for households, as the increase in water cost will shorten the payback period of appliances and fittings. Moreover, because water-related energy use and cost are directly related to water savings, any increase in energy costs will increase the cost of an intervention further.

Overall, the AGG scenario performs the best in terms of headroom, but is significantly more energy intensive and costly as a result of the CAPEX required. The ALT scenario performs almost the same as the BAU (1 LCD saving) in terms of demand and performs only slightly better in terms of headroom, at a considerably higher cost and energy use. If the current negative population trend continues, the BAU scenario will be the best option. However, if the population increases then the ALT or AGG scenario may become the best interventions to implement depending on the WSPs objective(s).

Sensitivity analysis was carried out to assess the impact of increased demand on the three scenarios with respect to PGR of 0.2, 0.5, 1 and 2% (in line with typical PGRs in Europe) to determine which scenario will be more appropriate under different PGRs (Figs. 2, 3, 4 and 5). The results show that water demand only decreases at the -0.2 and 0.2% PGR in all three scenarios. However, despite the 0.2% and consequent decreasing demand, energy use and cost increase in all ALT and AGG scenarios.

At PGR 0.2% demand marginally increases in BAU and ALT scenarios as a result of which headroom decreases. Demand increases at almost the same rate for BAU and ALT scenarios as there is only 1 LCD difference in water savings between the two. However, headroom decreases more in BAU than in ALT scenarios due to the water savings achieved from GWR and RWH in ALT scenario. For the AGG scenario at 0.2%, demand decreases until the last interval (2040–2050), as a result of which headroom decreases in the last interval.

Energy use increases in the ALT and AGG scenarios even with the negative PGR and improved energy efficiency of current GWR and RWH systems. Energy use and cost increase in all three scenarios at PGR 0.2%. Although demand is actually decreasing in the AGG scenarios, cost increases due to both population growth and the CAPEX of the scenario. Both cost and energy use increase substantially more in ALT and AGG scenarios than in BAU scenario for most of the PGRs (Figs. 2-5.), with the exception of both cost and energy use decreasing in PGR -0.2% and energy use decreasing at PGR1%. The difference in cost and energy use increase between BAU – ALT and between ALT – AGG is significant because of the difference in CAPEX and water saving between the scenarios. Only 1 LCD is saved between BAU and ALT, whereas 30 LCD is saved in between ALT and AGG scenarios.

At PGR 0.5, 1 and 2%, demand increases in all three scenarios, the same as at PGR 0.2% but only at a higher rate. The highest increase in demand is in the BAU scenario at PGR 2% where demand doubles at 2030 and increases 4 fold to almost the full supply capacity of the WDS at the end of the planning horizon. This indicates there will be significant implication for the security of future water supply in BAU and ALT scenario if no further action is taken. In the AGG scenario, headroom decreases only by around half at the end of the scenario at PGR 2% as a result of the water saving (30 LCD) that is made from 418 ML/year to 38 ML/year in BAU and 87 ML/year in ALT at PGR 2%. Although this indicates an increase in demand, it is not an immediate water security issue as the WDS will still have the capacity to produce a third more water that it is currently producing. At all positive PGRs, water saving is directly proportional to an increase in energy use and costs in the three scenarios because water saving is mostly achieved by GWR and RWH systems which cost more and use more energy.

In all three scenarios the BAU scenario performs the least in terms of water saving and impact on headroom, but performs best in terms of energy use and cost at all PGRs considered. In the BAU scenario, demand only decreases at PGR -0.2% and increases at all the positive PGRs. Demand increases the most at PGR 2 to almost the current full supply capacity of the WDS. However, the BAU scenario performs better in terms of all the performance indicators for the current negative PGR and 150 LCD water use (assuming current conditions remain the same for the duration of the planning horizon). Therefore the BAU scenario will be the best option if current negative population trend continues even without a reduction in per capita water use.

The AGG scenario offers the most water saving at all PGRs considered, but also has the most energy use and cost. However, because of the energy use and cost involved this scenario will only be appropriate at PGR of more than 1%. The ALT scenario can be a compromise between the BAU and AGG scenario in terms of all the performance indicators, but at a relatively higher energy use and cost than in BAU with respect to water savings. However because of the reduction in headroom in the BAU scenario from PGR 0.2%, the ALT scenario will be the most appropriate to consider for PGRs 0.2% - 1%.

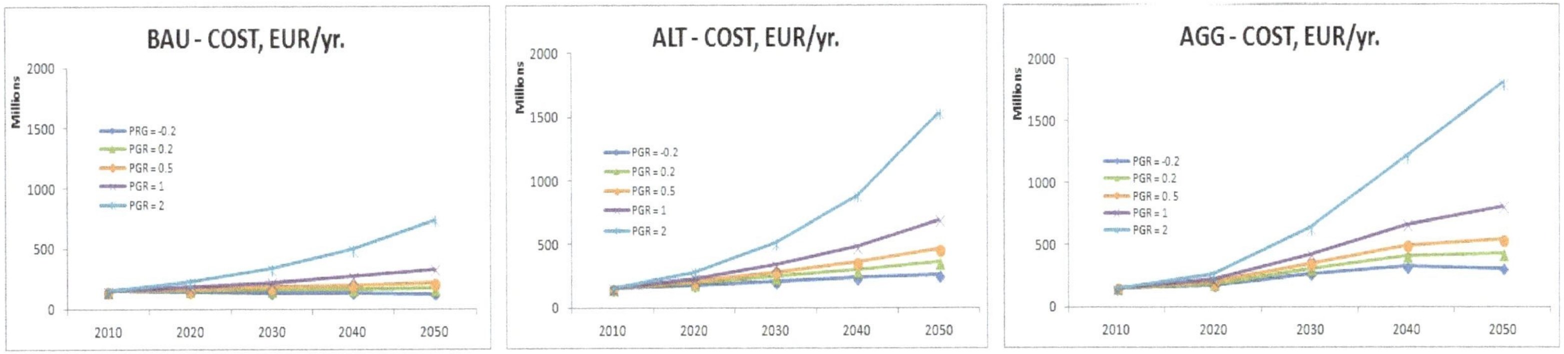

Fig. 2. Sensitivity analysis for the three intervention strategies with respect to cost (EUR/yr)

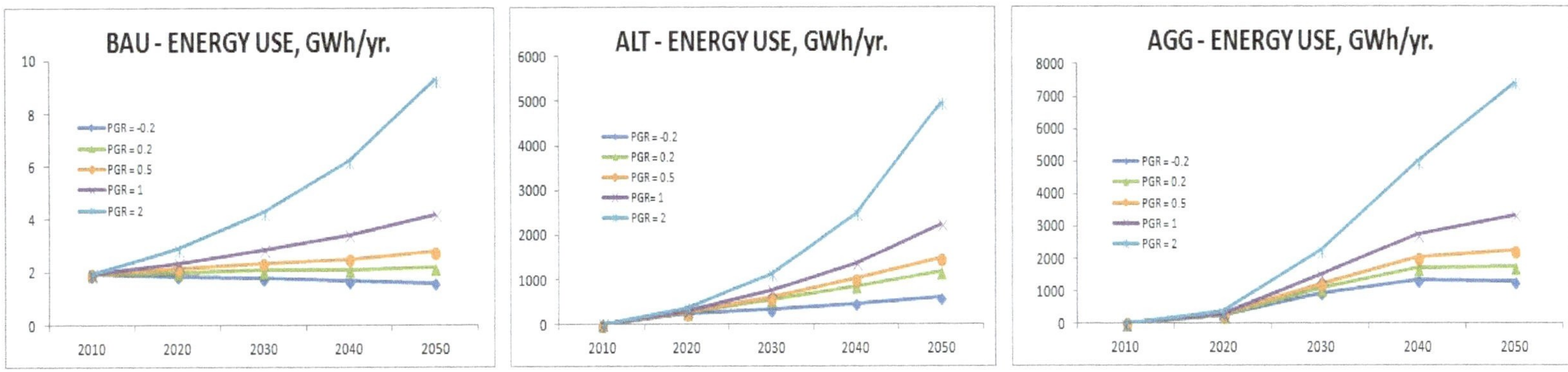

Fig. 3. Sensitivity analysis for the three intervention strategies with respect to energy use (GWh/yr)

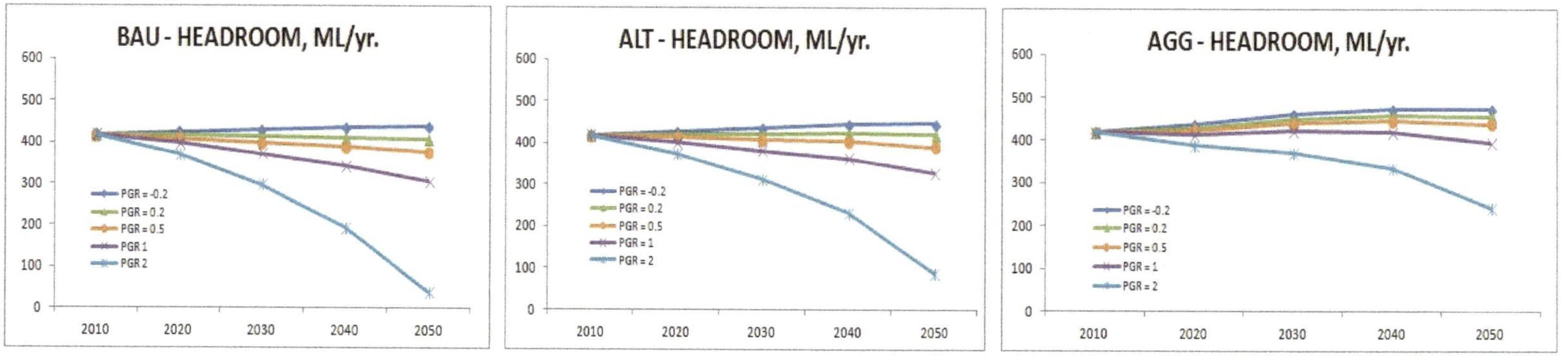

Fig. 4. Sensitivity analysis for the three intervention strategies with respect to headroom (ML/yr)

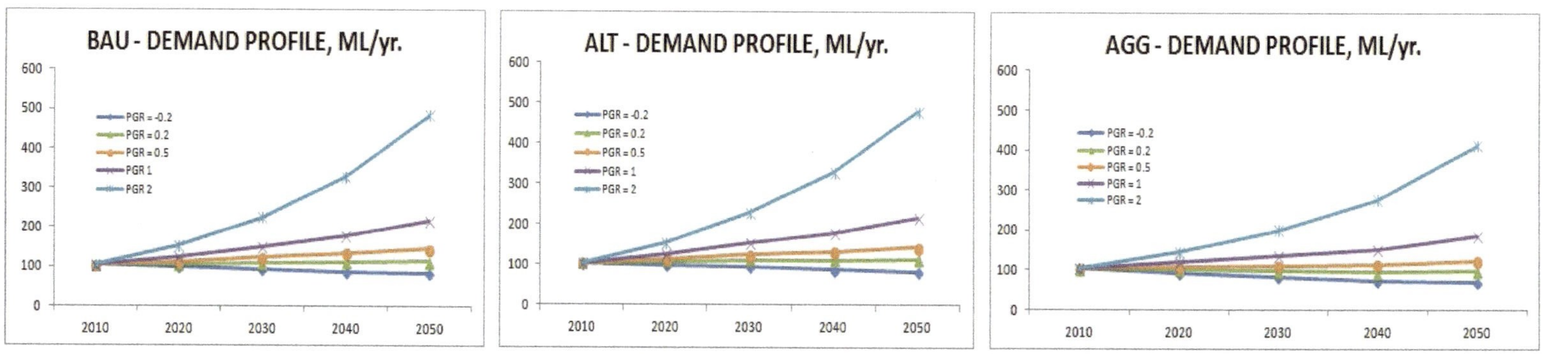

Fig. 5. Sensitivity analysis for the three intervention strategies with respect to demand (ML/yr)

4. SUMMARY AND CONCLUSIONS

WDM interventions are increasingly being used to reduce water use and water-related energy use, reduce cost and negative environmental impacts and have a positive impact on supply/demand balance, whilst still satisfying the needs of consumers. Given that most of the urban water use in Europe is in households, reducing household water use can reduce overall water demand, which can also lead to a reduction in household energy bills, overall water-related energy use, as well as associated GHG emissions and have a positive impact on supply/demand balance. However, not all WDM interventions will result in reduction of water-related energy use and some could increase both energy use and costs.

This paper presented the development and application of a new methodology for the impact assessment of household water savings from different WDM interventions based on their energy use, cost and impact on supply/demand balance on WDS. The methodology has been applied to the case-study of the WDS of a European city using three WDM intervention strategies. The result of the case study shows different degrees of water savings can be achieved depending on the type(s) and proportion of household appliances and fittings considered. Demand will decrease in all three scenarios if the negative population trend continues for the duration of the planning horizon. This indicates adequate security of future supply without any corresponding investment in supply infrastructure. However, a lot of uncertainties could arise over the planning horizon – increase in unit cost of energy and/or water, climate change could impact the availability of water resources and there could be an increase in industrial (i.e. non-household water use) over time or an increase in population and these could severely impact on supply/demand balance and potentially lead to periods of water shortages if no WDM intervention is implemented and/or new supply infrastructure is developed to mitigate this.

Sensitivity analysis was carried out with respect to increasing demand using different PGRs that are representative of PGRs across Europe. The result of the sensitivity analysis indicates although the BAU scenario performs the least in terms of water saving and impact on headroom, it performs best in terms of energy use and cost at all PGRs considered and will be the best option if current negative population trend continues even without a reduction in per capita water use. The AGG scenario offers the most water saving at all PGRs considered but also has the most energy use and cost. However, because of the energy use and cost involved in the AGG scenario, it will only be suitable at PGR over 1%. The ALT scenario is a compromise between the BAU and AGG scenarios in terms of all the performance indicators but at a relatively higher energy use and cost than in BAU with respect to water savings. However because of the reduction in headroom in the BAU scenario from PGR 0.2%, the ALT scenario will be the most appropriate to consider for PGRs more than 0.2% - 1%.

ACKNOWLEDGEMENTS

The methodology presented in this work has been developed under the EU FP7 Transition to Urban water Services of Tomorrow (TRUST) project. The authors acknowledge the contribution of TRUST partners from the University Of Applied Sciences Northwest Switzerland (FHNW). The authors thank the reviewers for their valuable and helpful comments which improved the quality of the paper.

COMPETING INTERESTS

Authors have declared that no competing interests exist.

REFERENCES

1. Baumann DD, Boland J, Hanemann W. Urban water demand management and planning, McGraw-Hill Professional. 1998;350.
2. Water UK. Water facts: Water Resources; 2011.
3. European Environment Agency. Sustainable water use in Europe: Part 1 Sectoral use of water, Environmental assessment report no. 1, EEA, Copenhagen; 1999.
4. United Nations. Water in a changing world, The United Nations World Water Development Report 3, World Water Assessment Program; 2009.
5. The Population Institute. Population and Water Fact Sheet; 2010.
6. European Environment Agency. Sustainable water use in Europe: Part 3 – Extreme hydrological events: floods and droughts, Environmental issue report no. 21, EEA, Copenhagen; 2001.
7. European Environment Agency. The European environment – state and outlook 2010: synthesis, State of the Environment Report no 1/2010 EEA, Copenhagen; 2010.
8. Brandes OM, Brooks DB, Gurman S. Making the Most of the Water We Have: The Soft Approach to Water Management, London: Earthscan Books; 2009.
9. Alliance to Save Energy and USAID (2002): Taking Advantage of Untapped Energy and Water Efficiency Opportunities in Municipal Water Systems. 2002;144.
10. Water UK. Sustainable Indicators 2010/2011; 2011.
11. Fidar A, Memon FA, Butler D. Environmental implications of water efficient microcomponents in residential buildings, *Sci Total Environ*. 2010;408(23):5828-5835.
12. Environment Agency. Conserving water in buildings, Crown Copyright; 2007.
13. European Environment Agency. European outlook on water use, EEA, Coppenhagen; 2004.
14. Environment Agency. Water efficiency in the South East of England – Retrofitting existing homes, Crown Copyright; 2007.
15. Environment Agency. International comparisons of domestic per capita water consumption, Prepared for the Environment Agency by Aquaterra, Crown Copyright, 2008.
16. Beal CD, Bertone E, Stewart RA. Evaluating the energy and carbon reductions resulting from resource-efficient household stock, Energy and Buildings. *2012;*55:422-432.
17. Defra Market Transformation Programme. BNWAT07 Water and Energy Use Supplementary briefing note. 2011;1.0.
18. Mitchell VG, Mein RG, McMahon TA. Modelling the Urban Water Cycle. Journal of Environmental Modelling & Software. 2001;16(7):615-629.
19. Makropoulos CK, Natsis K, Liu S, Mittas K, Butler D. Decision support for sustainbale option selection in integrated urban water management. Environmental modelling & software. 2008;23:1448-1460.
20. Mackay R, Last E. SWITCH city water balance: a scoping model for integrated urban water management. Rev Environ Sci Biotechnol. 2010;9:291–296.
21. Mitchel G, Diaper C. UVQ User Manual; 2010
22. Department for Communities and Local Government. Code of Sustainable Homes, A step-change in sustainable home building practice, Crown Copyright; 2006.

23. Crowhurst D, Cunha A, Hans J, Huovila P, Schmincke E, Visier J. A Framework for Common Metrics of Buildings 2010, Sustainable Building Appliance; 2010.
24. Memon F, Butler D. Water consumption trends and demand forecasting techniques, In Water Demand Management, Butler D, Memon FA. (*Eds.*), IWA Publishing, London. 2006;1-25.
25. Grant N. Water conservation products, In Water Demand Management, Butler D, Memon FA. (Eds.), IWA Publishing, London. 2006;82–106.
26. European Union. EU Water saving potential (Part 1 –Report), Ecologic - Institute for International and European Environmental Policy; 2007.
27. Defra Market Transformation Programme. BNWAT08 Modelling projections of water using products. 2011;1(0):
28. European Environment Agency. Greywater for domestic users – an information guide, EEA, Copenhagen; 2011.
29. World Bank. Population growth (annual %); 2012. Available: http://tinyurl.com/22nrlz3
30. Europe's Energy Portal Historical Energy Prices; 2012. Available: http://tinyurl.com/nz4lyc9
31. European Central Bank Long-term interest rate statistics for EU Member States; 2012. Available: http://tinyurl.com/nk4vdbz

5

Climate Change Resilience and Public Education in Response to Hydrologic Extremes in Singapore

Chew-Hung Chang[1*] and Kim N. Irvine[1]

[1]National Institute of Education, Nanyang Technological University, 1 Nanyang Walk, Singapore.

Authors' contributions

This work was carried out in collaboration between all authors. Both authors came up with the conceptual approach to the issue and provided data from aspects of their work for the paper. Chew-Hung Chang provided the meteorological aspects of the data while Kim Irvine provided the hydrological data. Both authors read and approved the final manuscript.

ABSTRACT

Aims: In February and March 2014, more than 300,000 households were affected by water rationing in Kuala Lumpur, Malaysia and the surrounding State of Selangor. Further south, reservoir levels in Singapore were dropping, prompting the government to raise the water conservation rhetoric, but falling short of implementing water rationing schemes. The region experienced a dry spell that was unprecedented in the last 30 years. Preparedness for storms has been the "talk of the town" since the 2001, 2006 and 2007 extreme high precipitation events in the southern parts of the peninsula and in Singapore resulted in costly flood damage. While resilience has been a concept used frequently in climate change adaptation, it is derived from ecology, where it refers to the capacity of the system to respond to a disturbance and resist the impact or recover from the damage of the disturbance. This paper examines the case of Singapore as an urban area in responding to a similar extreme hydrologic phenomenon by examining the climate change resilience of the small city-state, with a view to recommending some

Corresponding author: Email: chewhung.chang@nie.edu.sg;

considerations in designing climate change adaptation strategies.
Place of Study: Singapore and peninsular Malaysia.
Methodology: The paper reviews the rainfall extremes statistics covering the last 30 years for Singapore and then takes a hydrologic event-based case study approach to more closely examine the impact of record storms and the drought of March 2014 to discuss aspects of resilience that can serve as lessons for tropical cities in future adaptation to a climate-changing world.
Results: Extreme rainfall events have become more frequent in Singapore over the past 30 years, while February, 2014 was the driest February since 1869. February, 2014 also had the lowest recorded daily relative humidity at 74.5%. Tropical cyclones are not expected to hit Singapore because of its location near the equator, yet Typhoon Vamei made history by delivering 210 mm of rain on 27 December, 2001. Between 19 and 20 December, 2007 Singapore received 366 mm of rain and within the same week another storm deposited 140 mm of rain in a 24 hour period. While there were some environmental and health impacts related to the February 2014 drought, including low dissolved oxygen levels in water and a localized fish kill, as well as reports of greater human respiratory problems, Singapore was able to weather the drought by requesting voluntary conservation measures, prudent reservoir management, and increasing the output of NEWater and desalinized water. Recent extreme rainfall events have produced localized flooding, but Singapore has progressively pursued a program of improved drainage, stream naturalization, and implementation of Low Impact Development (LID) technology to reduce flood-prone areas from 3,200 ha in the 1970's to 36 ha today.
Conclusion: We do not suggest that all countries need to have NEWater or desalinated water to solve drought problems. We do suggest that in managing rainfall related hazards, droughts and extremes have been treated rather independently. Based on the case study of extremes presented for Singapore we propose the importance of establishing a three-step preparedness program for extremes that includes Preparation (vulnerability and risk identification, adaptive capacity building, and monitoring), Response (information dissemination and relief action), and Recovery.

Keywords: Singapore; climate change resiliency; drought; rainfall extreme; urban flooding; water resource planning.

1. BACKGROUND

There is spatial and temporal variation in how climate has changed in Southeast Asia, which is a region that spans more than 4,000,000 km² and is home to more than 600 million people. In particular, the variation in the changes of Southeast Asian precipitation event frequency, intensity and duration has been established in previous studies [1,2,3,4]. Any increase in atmospheric or ocean temperatures will impact the hydrological cycle as water moves through the earth-atmosphere system. The latest IPCC reports have affirmed yet again the increasing average global temperatures. Droughts and floods will be more severe and will occur more frequently and last longer, over many places. "There will be more rain at high latitudes, less rain in the dry subtropics, and uncertain but probably substantial changes in tropical areas" [5]. The IPCC Fourth Assessment Report (AR4) suggested differences in water availability between regions [6] that have caused storms and droughts to become more pronounced. Indeed, there has been an increased ratio of rainfall in the wet to the dry season between 1955 and 2005 for archipelago Southeast Asia [7]. In other words, there are more droughts during the dry season and more storms in the wet season in the recent decades.

Stern [5] has warned that any changes in precipitation patterns across Monsoon Asia would "severely affect" millions of lives. While countries may receive more water, much of this will arrive during the wet season bringing potential problems of flooding [8]. Further, this extra volume of water is useful for agriculture only if there is effective storage for use during the dry seasons [5]. Standard design practice for urban infrastructure assumes stationarity of the rainfall time series for analysis of event magnitude and frequency, but there has been a great deal of discussion recently about whether such assumptions are appropriate. Some [9] suggest that at this juncture, due to high uncertainty, there is no reason to abandon concepts of non-stationarity, while others [10] suggest extensions or correction factors to existing flood probability approaches that potentially would provide more reliable forecasts. A similar correction factor approach was developed by Dormuth [11] to update intensity-duration-frequency (IDF) curves for design work, rather than entirely updating the IDF curves. Hirsch [12] recommended that research related to climate change and water resources should follow two paths simultaneously, with one path using climate change models to drive hydrologic models (i.e. a linked model approach) and the other path being more extensive analysis of long term hydrologic records (i.e. an empirical approach). In the developed world, some long term hydrologic records exist, but such is not the case for most urban areas in developing countries. Most urban runoff studies that have examined climate change issues, to date, take some form of the linked model approach [e.g. 13,14,15,16] and with a few exceptions [e.g. 17,18,19] most work has focused on cities in temperate, developed countries. Tait et al. [20] looked at how a variety of drivers, including land use change, environmental legislation, energy and resource stress, and climate change might influence sewer system design and operation in the U.K. between 2020 and 2080. Using an example sewershed and increasing currently used synthetic design storms by a factor of between 1.2 and 1.4, modelled CSO volumes and surcharging were expected to increase by 2020. Hirschman et al. [21] did not undertake modeling exercises but provided practical examples of how existing "hard" stormwater management designs (e.g. bioretention cells, rainwater harvesting practices) and "soft" adaptive management strategies can be implemented to accommodate the uncertainties of climate change. Most of the studies reviewed above were carried out in areas expecting increasing rainfall volumes and intensities. It is important to recall, however, that there are areas throughout the world that are expected to become drier by 2080 (Food and Agriculture Organization (FAO), 2008) [22].

It has been suggested [e.g. 23,24,25,26] that droughts historically have received less attention with respect to prediction and planning because they are "fuzzy", "creeping" phenomena, difficult to define until one is experiencing them, often difficult to define the exact end, and in fact conceptually difficult to identify (i.e. is it a rainfall drought, a runoff drought, a crop drought?). A drought policy and National Drought Policy Commission have been in place in the U.S. since 1998 [24], but recent reviews have concluded, for example, that while most western states have drought plans, the scope and depth of the programs is uneven and few states conduct post-drought assessments [27]. At the local level Fu and Tang [28] found coordination and implementation strategies to address drought under most community comprehensive plans needed improvement. The European Union has a standard drought plan and Spain has implemented a drought indicator system [26]. Isler et al. [29] noted that Australia fcouses more on supply than demand side management to address drought, with the main innovation here being the introduction of large scale desalination, which they argue "...distracts from the need to build adaptive capacity and resilience into urban water systems through an integrated, whole-of-cycle approach."

While frequently climate change studies have focused on heavy precipitation events, with a view to understand how the impacts can be managed better, there clearly is a need to consider the extremes of wet and dry conditions, since drought also can have large economic, social, and environmental impacts. In the perceived water-rich equatorial region of Southeast Asia, Peninsular Malaysia and Singapore experienced a record dry spell in February-March 2014 and this raised some interesting questions for climate change researchers and policy makers. How different are precipitation patterns for the region as a result of climate change and what are the impacts on society? Are the societies in the region sufficiently prepared to manage these impacts? Are they resilient? This paper therefore sets out to begin addressing these questions by focusing on two case studies of severe storms and one case study of a dry spell which are used to illustrate the impacts of the precipitation changes. A conceptual critique of storm preparedness and drought resilience to answer the question of how are people in the region affected by extreme events is provided as a starting point to help direct attention and dialogue to the importance of learning to adapt to climate change in urban areas of Southeast Asia.

"Resiliency" refers to the ability of a system to respond and recover from disasters and includes attributes that allow the system to absorb impacts [30,31]. Many have noted that resiliency in fact may be a key ingredient of sustainability as well as vulnerability reduction, such that resilient systems (including biological and socio-economic systems) are able to survive, adapt and grow in the face of uncertainty and unforeseen disruptions [e.g. 30,31,32]. Fiksel [31] argued that a systems approach is necessary to better understand the linked behavior of complex social, economic, and biophysical systems so that we can improve our decision-making with respect to resilience and sustainability and provided a number of examples of complex, ongoing, modeling efforts, including thermodynamic life cycle analysis. While Sahely et al. [33] also noted the possible use of life cycle assessment as a method to evaluate sustainability, specifically for urban infrastructure systems, ultimately they used an indicator approach in their case study for the city of Toronto and there seems to be a groundswell of support for using such indicator approaches [e.g. 32, 34,35,36]. Our study takes a simpler, qualitative, case study approach to discuss resiliency as within the last 8 years we have been afforded clear examples of the extreme swing in events and it seems timely to use a case study approach given the record drought that was experienced in Singapore and Malaysia in 2014.

2. METHODS

2.1 Study Area

This paper focuses on Singapore, with some comparisons to its northern neighbor, Malaysia (Fig. 1). Singapore is a small island nation-state located at the southern end of the Malaysian peninsula. It has a total area of 715.8 km^2, with an east–west length of only 49 km and a north–south length of 25 km. At a current total population of 5.3 million (up from 1.6 million in 1960), Singapore's population density of 7, 257 people/km^2 is ranked the third most dense in the world, after Macau and Monaco. Singapore's per capita GDP in 2011 (adjusted for purchasing power parity and based on current international dollars) was higher, at $60,688 than either the United States ($48,112) or Canada ($40,370) according World Bank statistics [http://data.worldbank.org/indicator/NY.GDP.PCAP.PP.CD]. Land use characteristics in Singapore are well described by Ng [37], but an interesting aspect of Singapore's fabric is that traditional housing essentially has vanished and more than 80% of the population lives in high-rise apartments as part of the Housing Development Board's estate plans.

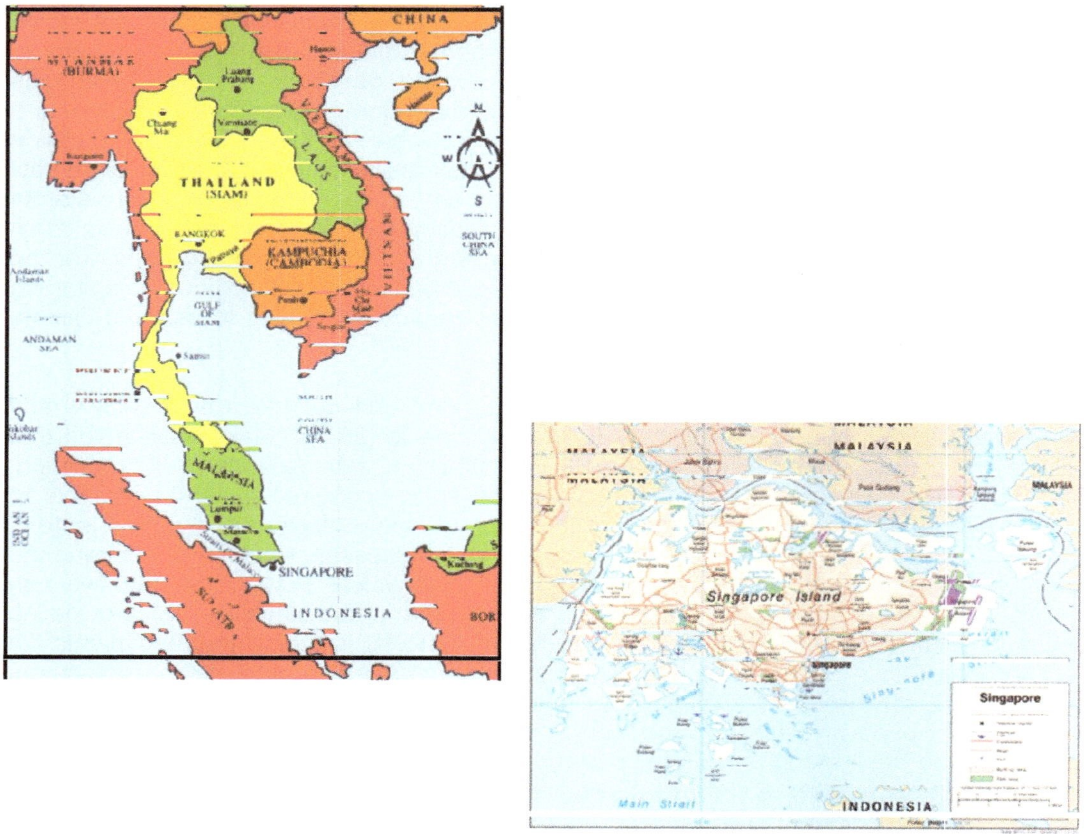

Fig. 1. Peninsular Malaysia and Singapore

The climate of Singapore is classified under the Koppen system as a tropical rainforest (Af) with no true dry season. Annual mean rainfall is approximately 2,343 mm and annual mean temperature is 27°C. The annual maximum 60 min rainfall intensity, 1980–2010, ranged between 70 mm/h and 130 mm/h. In keeping with its tropical rainforest classification, the mean number of days/month with rain ranges between 11 and 19, and the minimum number of rain days per month is in the range 1 to 3 [http://app2.nea.gov.sg/weather_faq.aspx;http://app2.nea.gov.sg/weather_statistics.aspx].

2.2 Data Collection

This study used both primary and secondary data sources to explore the hydrologic extremes issue. First, a literature survey of various works on storms in Asia was done to determine the changes to precipitation patterns in the region and a trend series analysis of data collected for Singapore is presented to provide some preliminary empirics on the issue. The average monthly rainfall data from the Changi Meteorological Station, Changi International Airport (Fig. 1), 1984-2014, were used in the trend series analysis.

As part of an ongoing water quality study being conducted by the second author, YSI 6920 data sondes (YSI Inc., Yellow Springs, OH, http://www.ysi.com/index.php)have been deployed in a wet storm water detention pond that takes runoff from a park in northern Singapore and also serves as an ornamental water feature for a restaurant. The data sondes record dissolved oxygen, temperature, conductivity, pH, and turbidity at 15 minute time intervals. Data are downloaded on a weekly basis, and routine maintenance of the sondes is done at the same time. Most pertinent for this study (since we examine the relationship between drought and dissolved oxygen for ecosystem health), the dissolved oxygen sensors are calibrated weekly, using the 100% air saturation method recommended by the manufacturer. This calibration approach also has been used routinely in other field efforts and environments throughout the world [e.g. 38,39,40].

3. RESULTS AND DISCUSSION

3.1 Rainfall Pattern Changes Over the Last 30 Years

Lau and Wu [2] reported that the extreme high (top 10%) and low (bottom 5%) tropical rainfall events are becoming more frequent between 1979 and 2003. This is corroborated by research showing increasing variability in extreme dry and wet conditions of both the East Asian Summer monsoon and the Western North Pacific Summer monsoon [3,4,41,42]. A summary of the different types of changes to precipitation for the Southeast Asian region is included in Table 1. While there may not be consensus on the absolute intensity of precipitation events, there is general agreement on the increase in frequency of both extreme rainfall and drought periods.

Table 1. Changes in tropical precipitation in the recent decades

Works cited	Period of study	Study Area/Region	Frequency of low precipitation events	Frequency of extreme high precipitation events	Intensity of extreme events
Aldrain&Djamil, 2008 [7]	1955 - 2005	Java, Indonesia	Increase	Increase	Droughts and storms are more pronounced
Chen, Yen &Weng, 2000 [41]	1979 - 1993	East & Southeast Asia	Increase	Increase	Heavy rainfall events in Eastern China
Kharin et al. 2007 [1]	2046-2065 2081 - 2100	Tropical Pacific	-	Increase	-
Lau & Wu, 2007 [2]	1979 - 2003	Tropics	Increase	Increase	Total precipitation amount increased two folds from1980s to 1990s
Yim et al. 2008 [43]	1979 - 2005	East Asia and Southeast Asia	Increase	Increase	-
Zhou & Chan, 2005 [44]	1979 - 2001	South China Sea	Increase	Increase	-

(Source: Adapted from Chang [45])

Chang [45] reported an increase in extreme wet events in Southeast Asia based on the rainfall records from 10 stations from 1978 to 2008. The monthly precipitation from January 1978 to December 2007 was normalized for the 30-year average monthly precipitation and its standard deviation was used to analyze the extreme rainfall events. Normally 30-year rainfall for any area is collected for a calendar year but the period chosen in this study was from 1 May 1984 to 30 April 2014 so that the period covers the dry spell in Singapore in 2014, one of the case studies in this paper. Using an index developed by Chang [45], the values of $Z_{m,y}$ were calculated to represent the deviation from the distribution of monthly rainfall normalized by the mean rainfall of each month and its standard deviation. Higher $Z_{m,y}$ values indicate more extreme precipitation while values lower than -1 indicate extreme dry months.

$$z_{m,y} = \frac{p_{m,y} - \bar{p}_m}{S_m}$$

where,

$z_{m,y}$ = Normalized monthly precipitation for month $\boldsymbol{m}$ in year $\boldsymbol{y}$

$p_{m,y}$ = Precipitation for month $\boldsymbol{m}$ in year $\boldsymbol{y}$

$\bar{p}_m$ = Mean monthly precipitation for month $\boldsymbol{m}$ over a 30 year period

S_m = Standard deviation of monthly precipitation for month $\boldsymbol{m}$ over a 30 year period

Using updated rainfall records from 1984 to 2014, the index $Z_{m,y}$ was used to compare against the trend reported in Chang [45].

Based on the updated records for Singapore, there were more extreme wet events than extreme dry events in the last 30 years. Fig. 2 shows the normalized monthly rainfall for Singapore in the last 30 years. There were more months with rainfall higher than the 95th percentile of all monthly rainfall than the number of months below the 5th percentile of all monthly rainfall. In analyzing higher intensity events that could cause flooding damage, the PUB shows the annual frequency of occurrence of hourly rainfall total exceeding 70 mm to be about 8 days per year, which has been increasing at an average rate of 1.8 days per decade since 1980 (Fig. 3). These observations affirm findings from various studies in the literature about more frequent high precipitation events (Table 1). However, the tail end of Fig. 2 shows a noticeable dry event, which caught Singapore by surprise. It is this unexpected turn in precipitation trend and the adaptive ability in this small island state that has raised the issue of climate system resilience in this paper. In addition to a range of impacts that high precipitation events can bring, dry spells can also affect the economy and its people adversely. The following sections will outline the range of impacts that extreme precipitation events bring, with specific focus on three case studies. The authors are not implying that every single storm or dry spell is due to climate change, but rather use examples of existing extreme weather events to discuss how society can be more adaptable and resilient to changes in weather patterns that can be potentially brought about by climate change.

3.2 Two High-Precipitation Events

In any meteorology textbook, *Coriolis force* will be listed as a key ingredient for tropical cyclone formation. It has been considered theoretically improbable for cyclones to form near the equator. However Typhoon Vamei made history and challenged the common understanding of tropical cyclone formation as it made landfall in Singapore on 27 December 2001, hardly one and a half degrees north of the equator [47,48]. A persistent north-south pressure gradient across the equator created by a storm surge of extended duration over the South China Sea has been cited as the reason for this anomalous formation [49].

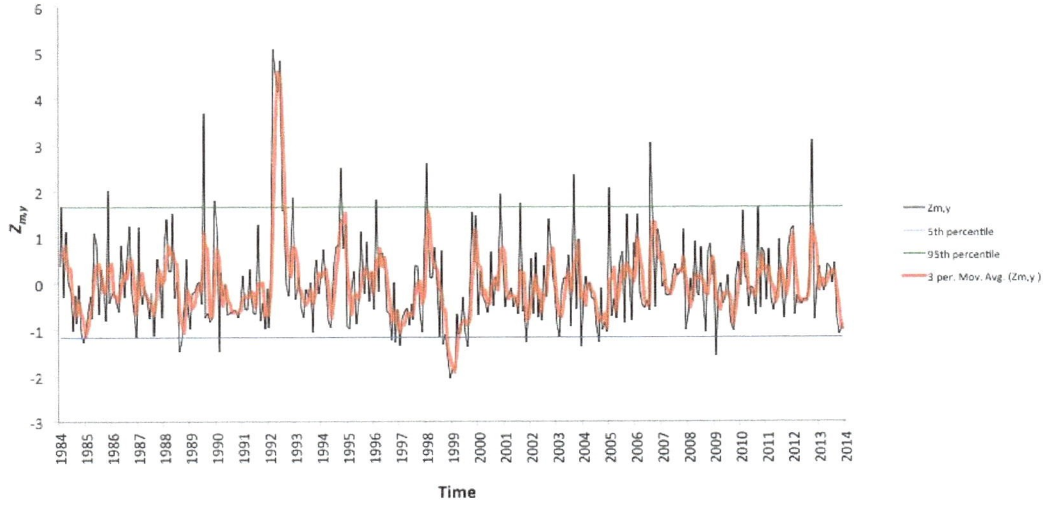

Fig. 2. Normalised average monthly rainfall recorded at Changi Meteorological Station, Singapore from 1984 to 2014.
Compiled from National Oceanic and Atmospheric Administration, n.d. [46]

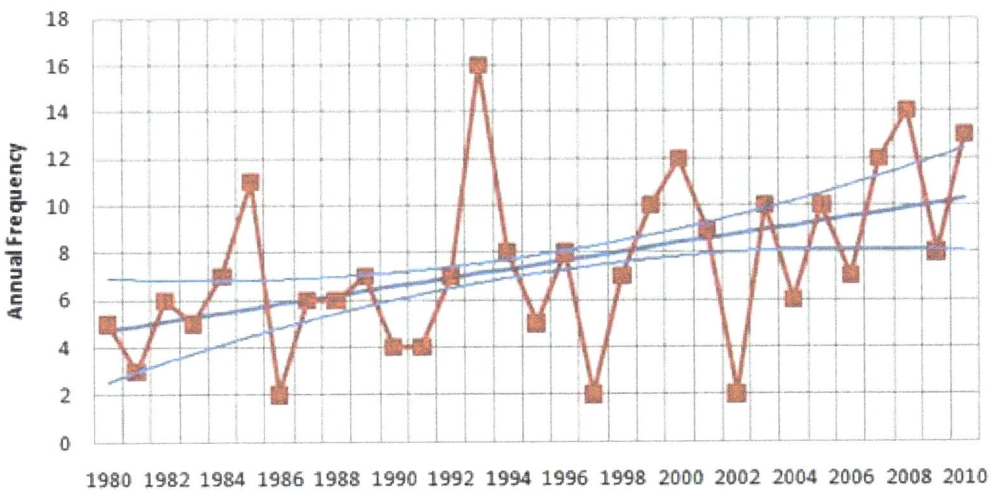

Fig. 3. Annual frequency of occurrence of hourly rainfall total exceeding 70 mm
[from http://www.pub.gov.sg/managingflashfloods/FMS/Pages/FloodResiliencePlan.aspx].

In the weeks leading up to the event, a low-pressure vortex was hovering over the South China Sea in an area with Borneo to the east and the Malay Peninsula and Sumatra to the west [50]. This vortex blocked the northeasterly monsoon winds and deflected the winds around the vortex, causing the winds to blow from a northwesterly direction. Coupled with the prevailing Northeast Monsoon winds and the northwesterly wind, these two prevailing bodies of air "wrapped around the vortex and the net result was a spinning up of a rapid counter-clockwise circulation that is similar to the spinning of a top played by a child; this led to the development of Typhoon Vamei" [49].

In Singapore, 240 mm of rain fell over a 24-hour period on 27 December 2001, which is equivalent to 10% of the entire 2001 rainfall amount. At Senai meteorological station in Johor, Malaysia, the recorded rainfall exceeded 210 mm on the same day. Further, heavy rainfall was received in the states of Johor, Kelantan, Terengganu and Pahang [50]. Fig. 4 illustrates this sudden increase in rainfall over a short period of time. As the average climatic rainfall for the month of December in the region is 225.5 mm, the rainfall within a single day when Vamei made landfall was equivalent to all the rain that is expected for the whole month. The anomalous amount of rainfall received resulted in flooding and mudslides in the Johor and Pahang States. More than 17,000 people were evacuated and there were 5 fatalities [51]. A landslide occurred at Gunung Pulai, destroyed four houses and killed five people. This event cost the Malaysian government an estimated 13.7 million Malaysian Ringgit ($4.3 million USD) damage to crops, education, transportation, and health-care facilities due to river flooding [52].

This key impact of the storm was not due to the typhoon itself but rather the lack of preparedness for such an unexpected event. One may argue that the storm was a theoretical impossibility until it happened, and people will not know what they do not know. However, the region experiences high rainfall during this time of the year annually and one could argue that flooding due to high rainfall must be something that the local government and people should be prepared for. Indeed, the lesson learnt from the storm event in 2001 should have been a warning sign, but ironically in December 2006, just a few years after Typhoon Vamei, floods once again hit Johor and this time the high rainfall event displaced no fewer than 200,000 people [53,54].

Between 17 to 20 and 24 to 28 December 2006 and from 11 to 14 January 2007 there were three occurrences of extreme high rainfall over South Johor and Singapore (see Fig. 5). These three events were due to the moisture brought in by strong northeasterly winds over the South China Sea [54]. The highest recorded rainfall of 366 mm occurred between 8 pm on 19 December 2006 and 8 pm on 20 December 2006. Within the same week, another storm brought 140 mm of rain in a 24 hour period on 26 December. In December 2006, a total of 765.9 mm of rain fell over Singapore, making it the highest ever recorded for December since 1869. Compounded with the 19-20 December storm, the events resulted in landslides, disrupted traffic, caused flooding, damaged buildings, uprooted trees and collapsed quarry walls, among other impacts [55]. The total economic loss was estimated at $500 million USD [54].

Floods are usually found in the northeastern parts of Peninsular Malaysia but not in the south during the Northeast Monsoon season. The government and people were unable to cope with the abnormally high levels of river discharge in this one in a hundred year flood. Interestingly, another flood of similar intensity and magnitude struck Johor less than three weeks later, but the rhetoric of another one hundred year flood "returning in a space of three weeks" quickly disappeared [53]. The key point about climate change resilience in these two case studies is that the lesson from Typhoon Vamei was not well learnt by the time the floods of December 2006 happened. Whether global warming is or is not a direct cause of the increase in frequency of extreme rainfall events, the fact that a theoretical impossibility was refuted had not helped the local and national governments, as well as the people cope better with subsequent storms. Given a more certain outlook in the literature on more erratic precipitation events, there is an urgent need to address the capacity of a society to adapt to changing weather patterns brought about by climate change.

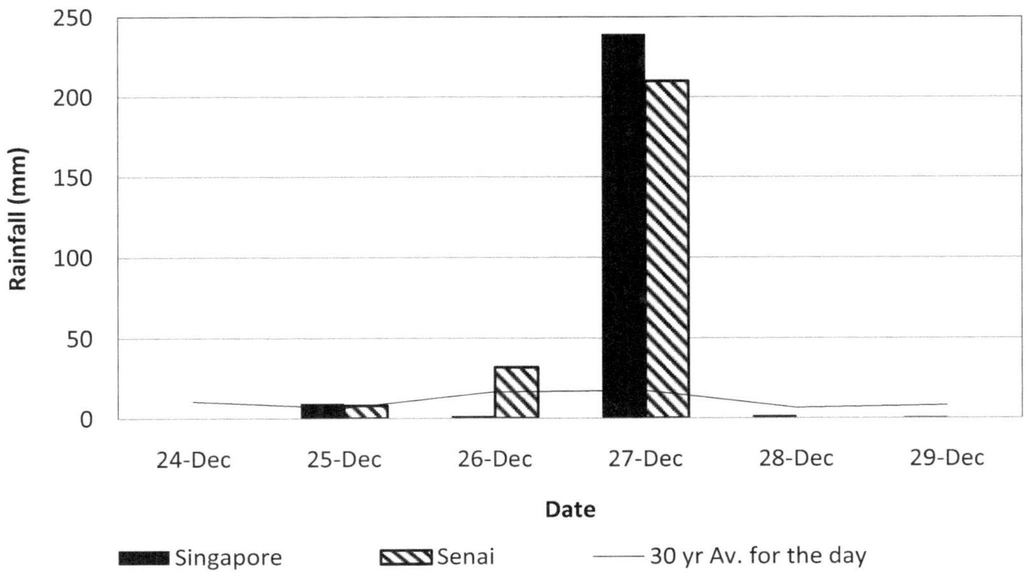

Fig. 4. Rainfall for South Johor and Singapore during Typhoon Vamei in 2001
Source: Adapted from Chang [45]

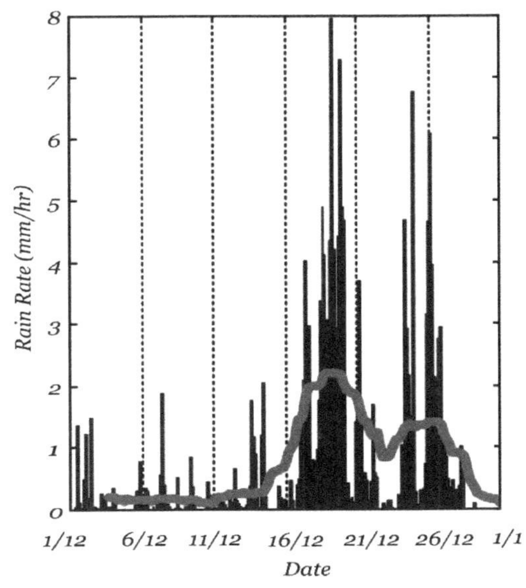

Fig. 5. Rainfall rate per hour over South Johor and Singapore
Note: Area-averaged (1°N–2.5°N, 102.5°E–105°E) precipitation rate (mm/hr) for December 2006. The thick grey line indicates the five-day running means (Adapted from Tangang et al. [54])

3.3 A Dry Spell

In the 61 days between 14 January to 16 March 2014, Singapore experienced an unusually dry period where the overall rainfall was 0.2 mm at the Changi Meteorological Station [46]. The National Environmental Agency (NEA) called it the driest February since 1869[56]. In February 2014, there were a very few scattered showers between February 7 and 19 over localized areas in the western part of the island. The total precipitation across all the rainfall stations monitored by the NEA ranged from 45 to 100 percent below the long-term average February rainfall of 161.0 mm [http://app2.nea.gov.sg/corporate-functions/newsroom/advisories/dry-spell-advisory-(4-mar) Fig. 6]. In fact, close to half the stations recorded less than 10 mm of rain for the entire month with the highest rainfall recorded being19.2 mm on 17 February at a station in the northwestern part of the island. The persistent dry conditions also resulted in the lowest recorded average daily relative humidity of 74.5 per cent for the month [56].

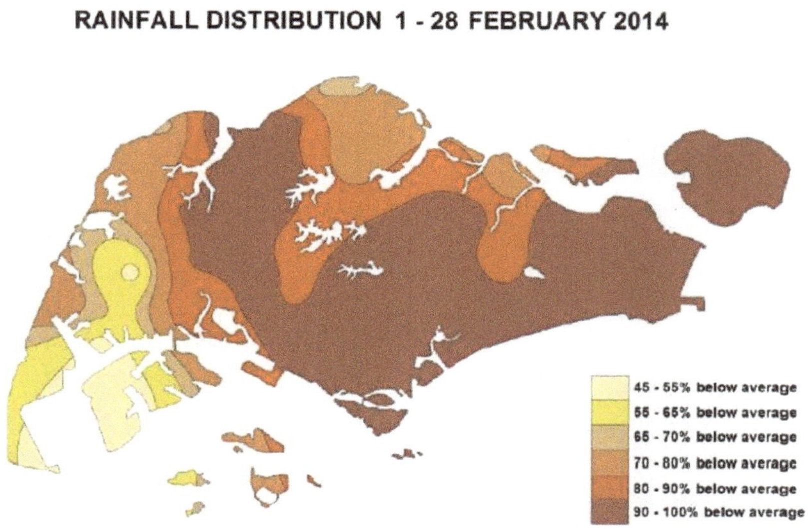

Fig. 6. Deviation from rainfall average for Singapore in February 2014
Source: National Environment Agency [http://app2.nea.gov.sg/corporate-functions/newsroom/advisories/dry-spell-advisory-(4-mar)]

While a dry period during the weakening phase of the Northeast Monsoon season is normal, what was abnormal was the extended duration of this dry spell [57,58]. Initial speculations by the authorities and local scholars point to possibly cooler than usual sea surface temperatures over the South China Sea which would have reduced evaporation and deprived the region of moisture that is needed to form rain [57]. Besides, the winds were still predominantly blowing from the northeast with average daily wind speeds of 13.3 km/h in February 2014, also coincidentally the windiest month in the last 30 years. With the combination of steady northeasterly winds and the cooler sea surface temperatures, a break in the monsoon winds by the middle of March or what is called the inter-monsoon period was expected to end the dry spell with the return of local convection rain [56]. Indeed, the heavy rains returned to Singapore on 16 March, attributed by the NEA to a tropical easterly disturbance, with the highest rainfall recorded in an hour being 33.2mm at Sentosa Island in the southern part of Singapore [59].

Of course, the key impact of the dry spell for Singapore was on the supply of potable water. However, other environmental and social impacts also were observed during this time. Apart from an evidently browning landscape, a single episode of mass death of fish at an offshore fish farm was recorded [58]. Fish kill episodes such as this periodically are observed in Singapore and have been attributed to warmer temperatures and low dissolved oxygen, as was the case in the Kallang River in June 2014 [http://www.straitstimes.com/breaking-news/singapore/story/hundreds-dead-fish-found-bishan-ang-mo-kio-park-river-20140205]. Water quality monitoring in a storm water detention pond during this drought period illustrates this potential impact clearly (Fig. 7). General practitioners saw about 20 to 30 percent more patients with colds, sore throats and cough symptoms during the dry spell [http://www.straitstimes.com/st/print/2111849].

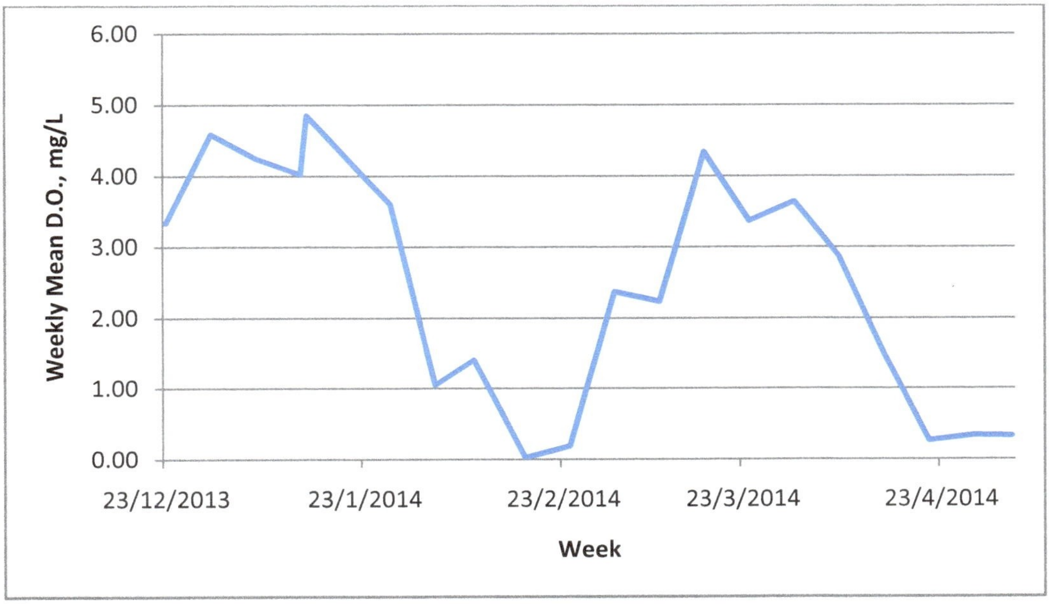

Fig. 7. Weekly mean dissolved oxygen (d.o.) levels near the outlet of a stormwater detention pond in Singapore

Weekly mean values were based on measurements recorded every 15 minutes with an YSI datasonde 6920. Low d.o. levels in April also occurred during a dry period, but it was not as severe as the February drought. Rainfall recorded at the tipping bucket gauge adjacent to the pond was: 16 March – 28.8 mm; 17 March – 34.4 mm; 18 March – 24 mm; and 19 March – 29.6 mm.

The dry spell also affected Malaysia where water levels in reservoirs dropped below 50% in Selangor, the country's most populous state, resulting in the implementation of water rationing. In peninsular Malaysia, 15 areas did not experience rainfall for more than 20 days [58]. The state of Negeri Sembilan, just south of Kuala Lumpur, even had to declare a "state of crisis" on 26 January 2014 [60]. Consequently the Malaysian government imposed water rationing on 300,000 households in the Federal Capital of Kuala Lumpur and the adjacent state of Selangor, as well as another 50,000 households in Southern Johor [61]. Apart from water rationing, the dry spell was blamed for forest fires, dengue fever outbreak as well as lower agricultural productivity [58,61].

The irony is that part of Singapore's water supply comes from buying treated water sold by Malaysia. Despite this dependence on imported water from her neighbour, Singapore did not have to resort to water rationing. Indeed, the environment ministry assured the public during the dry spell that 55 per cent of the country's demand could be met by the water produced at the country's own desalination and used-water (NEWater) treatment plants "regardless of the amount of rainfall", which affords Singapore "a safety margin" [http://www.straitstimes.com/breaking-news/singapore/story/singapore-will-be-alrightdespite-dry-spell-conserve-water-vivian-bala]. Despite this safety margin, the National Environment Agency issued a dry spell advisory which encouraged the public to adopt water conservation measures. The national water agency, called the Public Utilities Board (PUB) sent circulars to 25,000 non-domestic customers advising on water conservation measures by 8 March 2014 [http://app2.nea.gov.sg/corporate-functions/newsroom/advisories/dry-spell-advisory-(4-mar)]. Nevertheless, the city-state was able to cope with the dry spell coming through relatively unscathed and demonstrates what the authors would describe as resilience to climate change. By discussing how Singapore has managed the three historical wet and dry events, the authors will critically examine resilience within the large concept of climate change adaptation.

3.4 Preparedness and Resilience in Storm and Drought Events

The last time Singapore was as dry as February 2014 was in 1869, a good century and a half ago. Typhoon Vamei was probably a one in four hundred year storm [49] and the high rainfall event in December 2006 was a one in a hundred year storm. Although the chances that such extreme events will occur again in the immediate future are low, the cases described in this paper point to a need to be prepared and resilient for these unexpected occurrences. Many different analytical approaches and assessment frameworks have been proposed to help communities plan and prepare for climate change and associated greater extreme events, including use of indicators, input-output modeling, life cycle assessment modeling, and stochastic, risk-based simulations [e.g. 32,33,62,63]. To the extreme, Susskind [64] describes an interesting planning approach that included joint fact finding, scenario planning, and role play simulation. A pilot role play simulation, alone, took a study team 8 months to prepare and subsequently involved"...150communityleaders (including local and county elected and appointed officials, farmers, real estate developers, state legislators, environmental group heads, religious leaders, chamber of commerce members, fisherman, boating interests, and others)..." in Annapolis, Maryland. The effort also required guidance from 20 professional mediators during the 90 minute role playing exercise. While definitely interesting and inclusive, the planning exercise described by Susskind [64] points out the gap between developed and developing countries. Although we believe it unlikely that the developing countries of Southeast Asia would have the capacity to undertake this level of planning, Francisco [65] did note an example of four communes and eight villages in Thua Thien Hue Province, Vietnam, that used a scenario-building and participatory planning approach to implement climate change adaptation measures including construction of an inter-commune road and a multipurpose school to be used as an emergency shelter.

Rather than take one of the more advanced analytical and planning approaches here, instead we propose a simple framework for an adaptation program that fits and explains the Singapore context. As we will see, there is not a one size fits all solution, with respect to resiliency and water resource planning, and in fact the approaches will vary spatially and by political scale (i.e. country to country; federal to regional to local government). We propose the following key phases and steps to developing a useful preparedness program as an adaptation strategy:

1. Preparation
   a. Identify vulnerabilities and risks
   b. Adaptive capacity building (e.g. improve drainage systems)
   c. Monitoring
   d. Education and public outreach
2. Response
   a. Information dissemination (e.g. effective communication of warning and advisories)
   b. Relief action (e.g. evacuation)
3. Recovery
   a. Post-event recovery

These phases and steps in the Singaporean context are discussed in detail in the following paragraphs.

3.5 Preparation

3.5.1 Identify vulnerabilities and risks

Singapore has been lauded as a "model city for water management" and an "emerging global hydrohub" [66,67]. Luan [68] noted that the Singaporean government established legislation that goes hand-in-hand with a well-coordinated institutional framework to provide an integrated "whole-of-government" approach to land use planning, water management, a sound built environment and pollution control. Sanchez-Rodriguez [69] identified key ingredients for large U.S. cities that have been progressive in developing climate change adaptation measures include political leadership and central coordination of adaptation strategies. These characteristics certainly are true for Singapore and are discussed in more detail by Irvine et al. [70].

From the early years of independence, Singapore has had to deal with problems of droughts, flooding, serious water pollution and more importantly a lack of physical space to collect water. Singapore is barely 716.1 km^2 in size [71] with hardly any big rivers. There is also hardly any groundwater due to its geology [67]. Looking for other sources of water has become a matter of non-traditional security and has driven the country to identify the vulnerabilities, risks, and come up with alternative management and technology options. Singapore and Malaysia have a long history of water import agreements that stretches back to 1927. In 1961 and 1962, Singapore signed two water agreements with the state of Johor in Malaysia that allowed up to 86 million gallons per day (mgd)(390,963 cubic meters per day) and 250 mgd (1,136,520 cubic meters per day) of water to be purchased at the price of at 0.03 Malaysian Ringgit ($0.009 USD) for every 1,000 gallons up till the year 2011 and 2060, respectively [http://www.time.com/time/magazine/article/0,9171,1921608,00.html]. A new round of water negotiations began in 1998, which was linked to economic packages focusing on recovery from the Asian financial crisis. By 2003, Singapore had begun to look seriously for alternative water sources, including importing from Indonesia, and ultimately the negotiations with Malaysia ended without agreement [72]. With the expiry date of 2060 for the second agreement approaching, this has become the *de facto* planning horizon for Singapore to advance and diversify its technologies as it moves towards the goal of water self-sufficiency. Singapore began to look for alternatives to increase its water supply in the 1990s, largely out of security concerns, resulting in the creation of a "robust, effective and cost-efficient" way to meet Singapore's water needs - better known as the 'Four National Taps'. These four national taps include imported water, water from local catchments, intensely purified reclaimed water through used-water treatment known as 'NEWater', and desalinated water [67].

Apart from the natural catchment areas found in the center and western parts of the island, Singapore collects urban runoff from a network of "drains, canals, rivers, water collection ponds and reservoirs before it is treated for drinking" [73]. Singapore has progressively increased the area for water catchment from half to two-thirds of its land surface with the completion of the fifteenth to seventeenth fresh water reservoirs, namely the Marina, Punggol and Serangoon Reservoirs. A remarkable man-made reservoir, Marina Reservoir is the brainchild of the then Prime Minister of Singapore, Mr Lee Kuan Yew more than 20 years ago. With the construction of the Marina Barrage, a fresh water reservoir was formed out of the Marina Bay, which was previously open to the sea. The barrage collects runoff from a 10,000 ha area having a population of around 1,000,000 people [74] and also acts as a tidal barrier against possible inundation during high spring tides. In addition, the constant level of water maintained by the barrage has allowed water-based activities to be available right in the heart of the city [73]. The PUB intends to capture water from remaining streams near the shoreline, which will increase Singapore's water catchment area to 90% by 2060 [http://www.pub.gov.sg/water/Pages/LocalCatchment.aspx; 74].

NEWater, as its name implies provides a new life to "used" water. An acclaimed pillar of Singapore's water sustainability success [67,73,75,76], NEWater refers to the high quality reclaimed water produced from treated used water that is further purified through microfiltration, reverse osmosis, and ultraviolet disinfection. Since 2010, NEWater has been able to meet 30% of Singapore's water needs. There are plans to expand this capacity so that NEWater can meet up to 55% of future water demand. Another innovation in seeking alternatives to Singapore's water supply is desalinated water. Once considered a costly last option for water scarce countries, desalination technology has advanced dramatically over the past decade [77,78,79].Singapore's first seawater reverse-osmosis plant produces 136,000 cubic meters of water a day. A second and larger desalination plant with a capacity of 318,500 cubic meters of desalinated water per day, was opened in September 2013. Together, desalinated water can meet up to 25% of Singapore's current demand [73]. The plan is to ensure that the desalinated water sources are able to sustain 25-30% of the needs in the longer term [67,73].

The Singapore Ministry of Environment and Water Resources established an Expert Panel on Drainage Design and Flood Protection (EPDDFP) measures in 2011. This panel, consisting of local and international academics and consulting engineers, concluded that the PUB had done considerable good work on flooding issues over the past 30-40 years, but also made recommendations with respect to risk assessment, adaptive capacity, and education. Several of the recommendations will be noted in the following sections, but to begin, EPDDFP [80] recommended that "Singapore now needs to move towards a more integrated risk-based approach based on dynamic modelling and comprehensive monitoring" of its water resources. The PUB has begun to more fully develop its dynamic modelling capabilities both for runoff and reservoir management. The risk-based approach would consider the probability of flooding as well as the consequences of flooding (both tangible and intangible impacts). The EPDDFP [80] also recommended that PUB begin to "plan for the potential consequences of global megatrends and ensure that long term drainage solutions do not compromise Singapore's capacity to respond to their impacts."

3.5.2 Adaptive capacity building – the case of low impact development and drainage improvement

The use of stormwater runoff as a water source creates a balancing act for Singapore. The country wants to capture the maximum runoff practical, but also needs to balance the capture with concerns about localized flooding and quality of the runoff. One approach that increasingly has been implemented worldwide for stormwater quantity and quality management [cf. 81,82,83,84,85,86,87] and has been aggressively pursued in Singapore [70] is implementation of Low Impact Development (LID) technology.LID technology includes green roofs, rain gardens, rain planters, grassed swales, constructed wetlands, rain barrels, and pervious pavement. The intent is to manage rainfall at the source using uniformly distributed decentralized micro-scale controls. The LID goal is to mimic a site's predevelopment hydrology using design techniques that infiltrate, filter, store, evaporate, and detain runoff close to its source [http://www.lid-stormwater.net/background.htm]. LID addresses stormwater through small, cost-effective landscape features located at the lot level and therefore makes the system more resilient in that it does not rely on large, centralized control structures. Modeling efforts that assess the effects of these technologies increasingly are available, to the point that they are now explicitly included as an option in the Stormwater Management Model [88].

However, some important gaps in our understanding of LID operations remain, as, for example, there seems to be very little empirical data at a large sewershed scale, across multiple storms and seasons, to confirm performance of the technologies and models [89]. Modeling efforts to date seem to suggest that LID technologies are less helpful in managing larger events [90,91], although an unpublished report by Drexel University for a test area in Cambria Heights, New York City, showed that LID technology performed exceedingly well in controlling runoff from Super Storm Sandy and Hurricane Irene. Hopefully, the data gaps are starting to change with projects described, for example, by [92], although more work in tropical environments is needed.

Adapting to qualitative changes in climate and climatic and environmental transitions refers to transforming existing systems to ensure that development is viable and sustainable under future climatic uncertainty and the possibility of new climate hazards. This requires transformational change that occurs at a system level, such as by relocating activities away from flood prone zones [93].

Finally, other preparation activities undertaken by the PUB for extreme rainfall events include efforts to reduce localized flooding through a program of ongoing drainage improvement from about 3,200 hectares in the 1970s to 36 hectares today (Fig. 8).

3.5.3 Monitoring

The National Environment Agency's Meteorological Services Singapore (NEA MSS) currently monitors rainfall at 93 sites using real time automatic weather stations. Prior to the 1970s network coverage in Singapore was quite limited, and between 2012 and 2013, 31 of the 93 stations were added [80]. Rain gauge density in Singapore, as of 2013, is one gauge per 7.7 km^2, which well exceeds the density recommended by WMO [94] for urban areas at one gauge per 10-20km^2. The PUB maintains a network of 166 water level sensors in drainage canals and rivers around Singapore that update every 10 minutes and more than 40 of these sites also have CCTVs to visualize the site [http://www.pub.gov.sg/managingflashfloods/WLS/Pages/WaterLevelSensors.aspx]. The storm water level sensor data can be used to monitor flood potential in real time as well as to help optimize reservoir management for flood control and to determine how the production of NEWater and desalinated water need to be increased to meet the needs during a dry period. The monitoring system that now is in place is consistent with the recommendations made by the EPDDFP [80].

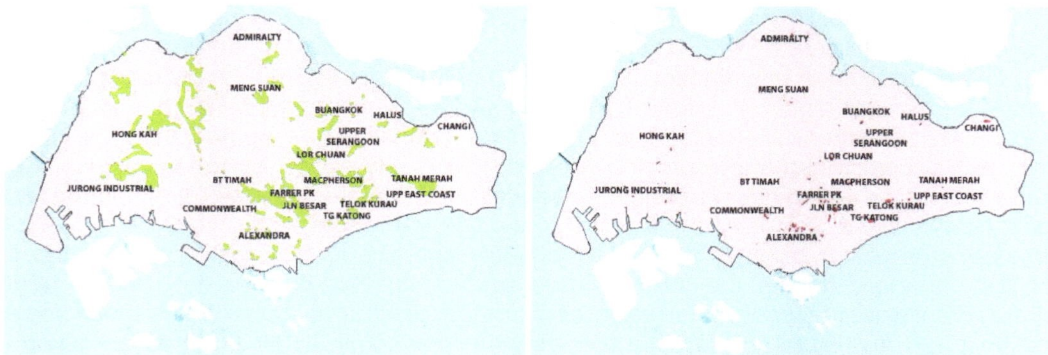

Fig.8. Flood-prone areas of Singapore in the 1970's (left) and now (right)
[fromhttp://www.pub.gov.sg/managingflashfloods/Pages/PastPresent.aspx].

3.5.4 Education and public outreach

Typhoon Yolanda (Haiyan) made landfall on 8 November 2013 as one of the strongest tropical cyclones ever recorded [95], killing at least 6,268 people in the Philippines [96]. The strong winds and heavy rains were accompanied by storm surges that measured at least 5 meters in height, destroying buildings including the Tacloban Airport terminal[97]. While countries in the western tropical Pacific are not unfamiliar with the heavy rains and winds of typhoons, the concept of a storm surge was largely unknown to many people in the Philippines as there is no equivalent word that would convey the dangers of a storm surge in the Tagalog language [98]. This vocabulary issue was used as an explanation for why people did not evacuate from the coastal areas even though the authorities had warned of 5 meter storm surges [99]. With this issue raised, it is hoped that public education can bring about greater awareness of the phenomenon and mitigate future life loss when a similar phenomenon occurs. This is perhaps only one facet that could shed light on how communities can be better prepared for storms. There are countless other examples of people and places impacted badly by storms, even when they occur on a yearly basis. Much can be learned from these lessons. Using the example from Hurricane Katrina, which also had a high death toll and displaced countless individuals, a good quarter of the people ignored evacuation orders even though ample warning was given [100]. In the case of Typhoon Vamei and the storms of December 2006, both events came as a surprise, which was largely due to a lack of preparedness. Perhaps this absence of preparedness can be explained by the prevailing political and scientific discourses on climate change in 2001. The IPCC Third Assessment Report (TAR) was unable to communicate its prediction of the impact of global warming as certainly as the Fourth Assessment Report (AR4) did. By 2007, the AR4 was "virtually certain" of the changes to precipitation patterns. There is little reason now for countries not to be prepared for heavy precipitation events such as Vamei-like storms, since there is general consensus in the literature that high precipitation events will occur more frequently.

Singapore which places high emphasis on climate change adaption and efficient drainage nonetheless was caught off guard in the 2010 Orchard Road floods. As the main shopping and tourist belt in the central business district, Orchard Road was inundated when 100 mm of rain fell within 2 hours, coupled with the fact that one of the canals serving the area was choked by rubbish and leaf litter [101]. In fact, the extensive greenery in Singapore promoted under the Garden City philosophy [102] poses some considerable risk to localized flooding due to clogged road inlets. These examples only go to show that there is still much more room for improvement in preparing for a more stormy and climate changed world. Following the 2010 (and 2011) flood events, the EPDDFP [80] noted that the PUB adopted a more proactive approach to sharing and communicating flood information, including the use of flood updates on radio, Facebook, twitter, and the PUB website. Yet, the EPDDFP [80] also noted that public preparedness might be enhanced through greater public engagement (e.g. through input to comprehensive flood response and recovery plans, or crowd-sourced reporting of flood locations).

Tortajada and Joshi [103] observed that the Singaporean school curriculum has played a central role in educating future generations with respect to water and the environment. For example, in 1987 a water conservation course was introduced at the secondary level to explain Singapore's water challenges and the importance of conservation. The PUB's Active, Beautiful, Clean, (ABC) Programme was initiated in 2006 to reconnect Singapore's population to its water. One of the signature components of the ABC Programme is the Waters Learning Trails which seek to provide unique, field-based learning experiences both for the general public and schools. To enhance the educational experience for schools, the PUB has keyed Learning Trails topics and information to the curriculum, with a multidisciplinary focus on the lower secondary level [http://www.abcwaterslearningtrails.sg/web/faq.php]. There are 11 Learning Trails located throughout Singapore, each with its own individual characteristics and opportunities and the PUB very much encourages teachers to get their students into the field and explore Singapore's water environment. The PUB also seeks to educate the public and schools about water conservation through their website [http://www.pub.gov.sg/conserve/WACProgramme/Pages/default.aspx].

3.6 Response

The common adage of saving for a rainy day needs to be reconsidered in the case of how Singapore responded to the dry spell of 2014 by saving water for a dry day. In fact, Biswas [104] argued that recent localized flooding events in Singapore have produced an unbalanced public dialog about water. He suggests that Singapore should be more concerned about droughts rather than floods and calls for the preparation of a comprehensive plan that would allow the island to withstand a prolonged drought.

Interestingly, while Malaysia responded to the 2014 dry spell by initiating water rationing to parts of Johor, Negiri Sembilan, Selangor and the Federal territory of Kuala Lumpur, Singapore adapted rather differently. There was hardly any restriction on the supply of water except for the few government advisories posted on the National Environment Agency website [http://app2.nea.gov.sg/corporate-functions/newsroom/advisories/dry-spell-advisory-(4-mar)].The most widespread impact of the dry spell was perhaps manifested in the trivial posting by netizens on the need to drink more water due to the lower relative humidity. There was a reported 20 percent rise in sales of cold drinks during the month of February [58].

Singapore has evidently developed sufficient resilience in responding to the onset of a dry spell, one hydrologic extreme event in the range of possible changes to the weather patterns due to climate change. Whether it is due to a reflexive response to the constraints that the tiny city-state faced due to the initial water supply conditions, or if it was part of a larger sustainability blue-print which the city has tacitly been implementing, the key features in the storm resilience concept that have been discussed above can be seen in this dry spell event as well.

In preparing for the inevitable shortage of water once the water agreements expire in 2011 and 2060, Singapore started planning its expansion of natural and urban catchment areas. The development of 17 fresh water reservoirs, the introduction of NEWater and desalinated water all were key ingredients to ensuring a sustainable supply of water for a future date. Coincidentally, the dry spell occurred after the completion of the second desalination plant in September 2013. This increased the capacity of desalinated water from 10 to 25% [73]. Had the dry spell occurred earlier, the outcome might have been different.

The response strategies observed during the 2014 drought episode were largely in terms of dissemination of information and providing an advisory to voluntarily mange the water demand. In addition to the advisories, the PUB dedicates a section of its website to "Water Reduction Measures During Dry Spells" [http://www.pub.gov.sg/conserve/WaterReductionMeasures/Pages/default.aspx#sthash.pFqlQA9H.dpuf]that provides information on conservation measures primarily for the domestic, commercial, and landscape maintenance sectors. Furthermore, in 2010, PUB introduced the Water Efficiency Management Plan (WEMP) program as a voluntary initiative to help non-domestic users better manage and conserve water. The WEMP involves development of a water balance chart to help identify areas for improved water conservation and establishment of an action plan which identifies measures in water savings, priorities and implementation timelines. Starting in 2015, it will be mandatory for all large water users (water consumption ≥ 5,000 m^3/month) to install water meters at the major water use areas in the facility and submit their WEMPs to the PUB on an annual basis [http://www.pub.gov.sg/conserve/WEMP/Pages/default.aspx].

3.7 Recovery

Once the rains returned to Singapore in March, 2014, it was business as usual [59] and as there was not much impact on the water supply, there was hardly any recovery work that needed to be done.

In examining the steps taken by Singapore in responding to the dry spell of 2014, it is evident that the key phases of preparing for and responding to the hydrologic hazard were in place to ensure that the city-state remained resilient to the dry spell. Given the similarity in the concepts of adapting to high precipitation events, there is scope to explore a common framework in planning the steps to take in ensuring climate change resilience to changing precipitation patterns.

Mitchell and Tanner [105] suggested that individuals, groups and natural systems should prepare for and respond to changes in climate or their environment, as a holistic approach to climate change adaptation by distinguishing risk factors between "hazard" (in terms of geographical location, intensity and probability) and "vulnerability" (in terms of susceptibilities and capacities) [105]. Further, the storm-related risk any society is subjected to is dependent on the social-cultural conditions of the people and the biophysical risks in the geographical region. Brooks, Adger and Kelly [106] proposed that an assessment of storm risk, the building of adaptive capacity and the actual adaptation strategies must be included when designing a storm preparedness program. Moreover, 3 main types of climate change adaption can be considered [93]. Policy makers and planners can endeavor to address the adaptation deficit by increasing the capacity of communities to cope with and recover from the impact of existing climate variability. Secondly, societies can adapt to incremental changes in existing climate related risks by increasing the capacity of communities to cope with extremes and variability in order to accommodate increased variability and more frequent and severe extremes, such as by climate proofing the drainage network in anticipation of more intense rainfall. Brooks et al. [93] propose that the success of any adaptation should be evaluated for its feasibility of implementation, efficiency, efficacy, and legitimacy with stakeholders, equity and sustainability of the program. However, these frameworks often provide extended discourse on the issue without outlining clear steps to operationalize the framework.

While being prepared requires an intimate understanding of the risks and vulnerabilities of a community within its geographical context [32], the adaptation strategy requires the three phases of preparing, responding and recovering. Only when a community has been able to adapt to the hazard can we begin to assess how resilient the community has been. Consequently, resilience requires that any community be able to prepare for, respond to and recover from a hydrologic hazard such as a high precipitation event.

As a case in point, prior to the 2006 storm events in Southern Johor, the Malaysian Meteorological Department (MMD) had already documented anomalous storms in several states in 2005 in addition to the data collected for the 2001 Vamei event. MMD noted that there had been "increasing number of days of extreme rainfall event (exceeding 90th percentile of total rainfall) for several stations over the Peninsular Malaysia" since the 1980s [50]. Although MMD had in place weather prediction systems as part of its monitoring system, the adaptation strategy was hardly effective as there was not yet a plan to identify vulnerabilities and risks. After the 2006 storm event, the Malaysian Ministry of Natural Resources and Environment (MNRE) started to develop "risk maps for vulnerable areas" [108]. Since then, a national level effort was undertaken to increase the adaptive capacity of the country through improving risk assessment, in order to formulate effective adaptation strategies.

4. CONCLUSION

In managing rainfall related hazards, droughts and extreme storms have been treated rather independently, perhaps due to the polarity in the amount of rainfall that each event brings. While the rainfall record has only shown more incidences of extreme storms with fewer examples of extremely dry events, the global rhetoric on climate change converges on an agreement on the certainty of changes to precipitation patterns across the world. There clearly will be areas that will be prone to extreme storms and others drought. Indeed, these two phenomena can occur in the same place at different times, with varying frequency. The examples outlined in this paper go to show that resilience is a concept that can be unraveled to include the components of preparing for, responding to and recovering from the extreme hydrologic events. In ecology, resilience refers to the capacity of the system to respond to a disturbance and resist the impact or recover from the damage of the disturbance. Indeed, with proper planning and preparation in building a system's capacity, there is opportunity to resist the impact of climate change. Failing which, the planned responses and recovery will ensure that the impact is minimized.

The authors are not suggesting that all countries need to have NEWater or desalinated water to solve drought problems but rather explicate the concepts in the steps that will make an adaptation strategy effective. For any community, the assessment of the risk, the building of adaptive capacity and the actual adaptation strategies must be included when designing an adaptation program for a particular hazard. It is also important to consider the feasibility of implementation, efficiency, efficacy, and legitimacy with stakeholders, equity and sustainability of the program [93]. There is not a one size fits all solution to sustainable water resource management.

The often quoted saying "that which does not kill us makes us stronger" by German philosopher Friedrich Nietzsche provides some insight on the concept of climate change resilience. As geographers, the authors subscribe to the view that the human-environment relationship is dynamic in nature. This relationship is not maintained at a static equilibrium in which things do not change. Indeed, the concept of adaptation for human-induced climate change requires an understanding of how climate change will impact humans. In building resilience, the fundamental premise is to either resist the impact, adapt to the impact, or to recover from the impact. The fundamental motivation for resilience is survival and it seems counter-intuitive that communities would subscribe to inaction as a way to respond to the changing weather patterns brought about by climate change. If humanity is to prevail despite adjustments to the dynamic equilibrium in our earth-atmosphere system, then we must remain resilient and take action for it.

COMPETING INTERESTS

Authors have declared that no competing interests exist.

REFERENCES

1. Kharin VV, Zwiers FW, Zhang X, Hegerl GC. Changes in temperature and precipitation extremes in the IPCC ensemble of global coupled model simulations. Journal of Climate. 2007;20(18):1419-1444. DOI: 10.1175/JCLI4066.1
2. Lau KM, Wu HT. Detecting trends in tropical rainfall characteristics, 1979 – 2003. International Journal of Climatology. 2007;27:979–988. DOI: 10.1002/joc.1454.
3. Yim SY, Yeh SW, Wu R, Jhun JG. The Influence of ENSO on decadal variations in the relationship between the East Asian and Western North Pacific summer monsoons. Journal of Climate. 2008;21(13):3165-3179.
4. Zhou W, Chan JCL. Intraseasonal oscillations and the South China Sea summer monsoon onset. International Journal of Climatology. 2005;25:1585-1609.
5. Stern N. The economics of climate change: The Stern Review. Cambridge: Cambridge University Press; 2007.
6. Lu J, Vecchi GA, Reichler T. Expansion of the Hadley cell under global warming. Geophysical Research Letters; 2007. DOI:10.1029/2006GL028443.
7. Aldrain E, Djamil YS. Spatio-temporal climatic change of rainfall in East Java Indonesia. International Journal of Climatology. 2008;28:435-448. DOI: 10.1002/joc.1543.
8. Milly PCD, Dunne RT, Wetherald KA, Delworth TL. Increasing risk of great floods in a changing climate.Nature. 2002;415:514–517.
9. Lins HF, Cohn TA. Stationarity: Wanted dead or alive? Journal of the American Water Resources Association. 2011;47(3):475-480

10. Stedinger JR, Griffis VW. Getting from here to where? Flood frequency analysis and climate.Journal of the American Water Resources Association. 2011;47(3):506-513.
11. Dormuth DW. A method to assess the impacts of climate change on the reliability of stormwater infrastructure components. In: W James K Irvine, J Li, E McBean, R Pitt, S Wright eds. Dynamic Modeling of Urban Water Systems. Monograph 18; 2010. CHI, Ch. 25.
12. Hirsch RMA perspective on nonstationarity and water management.Journal of the American Water Resources Association. 2011;47(3):436-446.
13. Waters D, Watt WE, Marsalek J, Anderson BC. Adaptation of a storm drainage system to accommodate increased rainfall resulting from climate change. Journal of Environmental Planning and Management. 2003;46(5):755-770.
14. Berggren K, Olofsson M, Viklander M, Svensson G. Tools for measuring climate change impacts on urban drainage systems. NOVATECH 2007. 2007;239-246.
15. Semadeni-Davies A, Hernebring C, Svensson G, Gustafsson L-G. The impacts of climate change and urbanisation on drainage in Helsingborg, Sweden: Combined sewer system. Journal of Hydrology. 2008;350:100-113.
16. Nie L, Lindholm O, Lindholm G, Syversen E. 2009. Impacts of climate change on urban drainage systems – A case study in Fredrikstad, Norway, Urban Water Journal. 2009;6(4):323-332.
17. McDonald RI, Green P, Balk D, Fekete BM, Revenga C, Todd M, Montgomery M. Urban growth, climate change, and freshwater availability. Proceedings of the National Academy of Sciences. 2011;108(15):6312-6317.
18. Shrestha A, Chaosaku IT, Priyankara DPMP, Chuyen LH, Myat SS, Syne NK, Irvine KN, Koottatep T, Babel MS. Application of PCSWMM to explore possible climate change impacts on surface flooding in a peri-urban area of Pathumthani, Thailand. Journal of Water Management Modeling; 2014. DOI:10.14796/JWMM.C377.
19. Huong HTL, Pathirana A. Urbanization and climate change impacts on future urban flooding in Can Tho city, Vietnam. Hydrol. Earth Syst. Sci. 2013;17:379-394.
20. Tait SJ, Ashley RM, Cashman A, Blanksby J, Saul AJ. Sewer system operation into the 21st century, study of selected responses from a UK perspective. Urban Water Journal. 2008;5(1):77-86.
21. Hirschman DJ, Caraco D, Drescher SR. Linking stormwater and climate change: Retooling for adaptation. Watershed Science Bulletin. 2011;11-18.
22. Food and Agriculture Organization (FAO). Climate Change, Water and Food Security, Technical Background Document; 2008. HLC/08/BAK/2.
23. Grigg NS. Hydrology and management of drought in the United States. In. New Directions for Surface Water Modeling, IAHS Publ. No. 181. 1989;303-312.
24. Showstack R. U.S. federal government tries to get ahead of the curve with drought planning. EOS. 1999;80(33):365-367.
25. Bruins HJ. Proactive contingency planning vis-a-vis declining water security in the 21st century. Journal of Contingencies and Crisis Management. 2000;8(2):63-72.
26. Estrela T , Vargas E. Drought management plans in the European Union. The case of Spain. Water Resour Manage. 2012;26:1537-1553.
27. Fontaine MM, Steinemann AC, Hayes MJ. State drought programs and plans: survey of the western U.S. Natural Hazards Review. 2014;15(1):95-99.
28. Fu X, Tang Z. Planning for drought-resilient communities: an evaluation of local comprehensive plans in the fastest growing counties in the US. Cities. 2013;32:60-69.
29. Isler PL, Merson J, Roser D. 'Drought proofing' Australian cities: implications for climate change adaptation and sustainability. International Journal of Social, Management, Economics and Business Engineering. 2010;4(10):353-361.

30. Klein RJT, Nicholls RJ,Thomalla F. Resilience to natural hazards: How useful is this concept? Environmental Hazards. 2003;5:35-45.
31. Fiksel J. Sustainability and resilience: Towards a systems approach. Sustainability: Science, Practice, & Policy. 2006;2(2):4-21.
32. Cutter SL, Barnes L, Berry M, Burton C, Evans E, Tate E, Webb J. A place-based model for understanding community resilience to natural disasters. Global Environmental Change. 2008;18:598-606.
33. Sahely HR, Kennedy CA, Adams BJ. Developing sustainability criteria for urban infrastructure systems. Canadian Journal of Civil Engineering. 2005;32:72-85.
34. Tanguay GA, Rojaoson J, Lefebvre JF, Lanoie P. Measuring the sustainability of cities: an analysis of the use of local indicators. Ecological Indicators. 2010;10:407-418.
35. Shen LY, Ochoa JJ, Shah MN, Zhang X. The application of urban sustainability indicators – a comparison between various practices. Habitat International. 2011;35:17-29.
36. Mori K and Christodoulou A. Review of sustainability indices and indicators: towards a new City Sustainability Index (CSI). Environmental Impact Assessment Review. 2012;32(1):94-106.
37. Ng YXY. A study of urban heat island using "Local Climate Zones" – the case of Singapore. British Journal of Environment and Climate Change. This Volume.
38. Irvine KN, McCorkhill G, Caruso J. Continuous monitoring of conventional parameters to assess receiving water quality in support of CSO abatement plans. Water Environment Research. 2005;77(5):543-552.
39. Irvine KN, Murphy TP. Assessment of eutrophication and phytoplankton community impairment in the Buffalo River Area of Concern. Journal of Great Lakes Research. 2009;35:83-93.
40. Irvine KN, Richey JE, Holtgrieve GW, Sarkkula J, Sampson M. Spatial and temporal variability of turbidity, dissolved oxygen, conductivity, temperature, and fluorescence in the lower Mekong River-Tonle Sap system identified using continuous monitoring. International Journal of River Basin Management. 2011;9(2):151-168.
41. Chen TS, Yen MC, Weng SP. Interaction between the Summer Monsoons in East Asia and the South China Sea: Intraseasonal Monsoon modes. Journal of the Atmospheric Sciences. 2000;57(9):1373–1392
42. Sun Y and Ding Y. Effects of intraseasonal oscillation on the anomalous East Asian summer monsoon during 1999. Advances in Atmospheric Sciences. 2008;25(2):279–296.
43. Yim SY, Yeh SW, Wu R, Jhun JG. The influence of ENSO on decadal variations in the relationship between the East Asian and Western North Pacific summer monsoons. Journal of Climate. 2008;21(13):3165-3179.
44. Zhou W, Chan JCL. Intraseasonal oscillations and the South China Sea summer monsoon onset. International Journal of Climatology. 2005;25:1585-1609.
45. Chang CH. Preparedness and storm hazards in a global warming world: lessons from Southeast Asia. Natural Hazards. 2011;56(3):667-679. DOI: 10.1007/s11069-010-9581-y. Available: http://hdl.handle.net/10497/4545
46. National Oceanic and Atmospheric Administration. Global summary of the day, climatic data online. National Climate Data Center, National Oceanic and Atmospheric Administration. (no date). Accessed on 18 March 2014. Available: http://www7.ncdc.noaa.gov/CDO/cdoselect.cmd?datasetabbv=GSOD
47. Dybas C. Hurricanes at the equator: impossible perfect storm. National Science Foundation; 2003.Accessed on 3 October 2007. Available: http://www.nsf.gov/od/lpa/news/03/tip030512.htm.

48. Padgett G. Worldwide tropical weather summary. 2001. Accessed on 15 September 2007.
Available: http://www.australiasevereweather.com/cyclones/2002/summ0112.htm.
49. Chang CP, Ching HL, Kuo HC. Typhoon Vamei: An equatorial tropical cyclone formation. Geophysical Research Letters. 2003. Accessed on 20 December 2005. Available: http://www.weather.nps.navy.mil/cpchang/papers-pdf/vamei.pdf
50. Malaysian Meteorological Department. Tropical Storm Vamei causes widespread floods in Johor. Accessed on 5 October 2007. Available: http://www.kjc.gov.my/people/research/webpages/sigwx/dis2701_e.htm.
51. Johnson RH, Chang CP. Winter MONEX: A quarter century and beyond. American Meteorological Society; 2007. Accessed on 5 Oct 2007. Available: http://www.wmo.int/pages/prog/arep/tmrp/documents/WinterMONEX25MeetingReport.pdf.
52. The Malaysian National News Agency. Recent floods claimed five lives and caused substantial damage; 2002. Accessed on 14 October 2007. Available: www.bernama.com.
53. Lee LC. Johor Floods: Once in 100 years, twice in three weeks? Malaysiakini; 2007. Accessed on 1 April 2008. Available: http://www.malaysiakini.com/letters/62176.
54. Tangang FT, Juneng L, Salimun E, Vinayachandran PN, Yap KS, Reason CJC, et al. On the roles of the northeast cold surge, the Borneo vortex, the Madden-Julian Oscillation, and the Indian ocean Dipole during the extreme 2006/2007 flood in southern Peninsular Malaysia. Geophysical Research Letters; 2008. DOI: 10.1029/2008GL033429.
55. Chatterjea K. Severe wet spells and vulnerability of urban slopes: case of Singapore. Natural Hazards. 2011;56(1):1-18. DOI: 10.1007/s11069-009-9362-7.
56. Channel News Asia. The driest month since 1869 for Singapore; 2014. Accessed on 18 March 2014. Available: http://www.channelnewsasia.com/news/singapore/the-driest-month-since/1020150.html
57. Feng Z. Reasons for dry spell run hot and cold. Asia One News; 2014. Accessed on 18 March 2014. Available: http://news.asiaone.com/news/world/reasons-dry-spell-run-hot-and-cold.
58. The Guardian. Singapore and Malaysia hit by extreme dry spell; 2014. Accessed on 18 March 2014.
Available: http://www.theguardian.com/environment/2014/feb/26/singapore-malaysia-extreme-dry-spell.
59. Rahman R. More thundery showers expected as 27-day dry spell ends. Channel News Asia; 2014. Accessed on 18 March 2014.
Available:http://www.channelnewsasia.com/news/singapore/more-thundery-showers/1037978.html
60. Global Post. Water crisis as drought dries up Malaysian reservoirs. Global Post; 2014. Accessed on 18 March 2014.
Available: http://www.globalpost.com/dispatch/news/afp/140219/water-crisis-drought-dries-malaysian-reservoirs
61. Australia Network News. Drought forces Malaysia to expand water rationing around Kuala Lumpur; 2014, February 28. Accessed on 18 March 2014. Available: http://www.abc.net.au/news/2014-02-28/an-drought-forces-malaysia-to-expand-water-rationing-around-kua/5292466.
62. Westphal KS, Laramie RL, Borgatti D, Stoops R. Drought management planning with economic and risk factors. Journal of Water Resources Planning and Management. 2007;133(4):351-362.

63. Pagsuyoin SA, Santos JR. Modeling the effects of drought in urban economies using regional input-output analysis. British Journal of Environment and Climate Change. This Volume.
64. Susskind L. Responding to the risks posed by climate change: cities have no choice but to adapt. Town Planning Review. 2010;81(3):217-235.
65. Francisco HA. Adaptation to climate change, needs and opportunities in Southeast Asia.ASEAN Economic Bulletin. 2008;25(1):7-19.
66. Chen CC, Maksimovic C, Voulvoulis N. Institutional capacity and policy options for integrated urban water management: a Singapore case study. Water Policy. 2011;12:53-68.
67. Caballero-Anthony M and Pau K. From water insecurity to niche water diplomacy: The Singapore experience. NTS Insight; 2012. Accessed on 18 June 2014. Available: http://www.rsis.edu.sg/nts/HTML-Newsletter/Insight/NTS-Insight-nov-1201.html
68. Luan IOB. Singapore water management policies and practices.Water Resources Development. 2010;26(1):65-80.
69. Sanchez-Rodriguez R. Learning to adapt to climate change in urban areas. A review of recent contributions.Current Opinion in Environmental Sustainability. 2009;1:201-206.
70. Irvine KN, Chua LHC, Eikass HC. The Four National Taps of Singapore: A holistic approach to water resources management from drainage to drinking water. Journal of Water Management Modeling; 2014. Doi:10.14796/JWMM.C375.
71. Department of Statistics Singapore. Latest Data; 2013. Accessed on 18 June 2014. Available: http://www.singstat.gov.sg/stats/keyind.html#popnarea
72. Tortajada C, Pobre K. The Singapore–Malaysia water relationship: an analysis of the media perspectives. Hydrological Sciences Journal. 2011;56(4):597-614.
73. Public Utilities Board. The Singapore water story (Water: From vulnerability to strength); 2013. Accessed on 20 June 2014. Available:http://www.pub.gov.sg/water/Pages/singaporewaterstory.aspx#sthash.OsbGqliY.dpuf
74. Kamer J, Thomson R, Mackenzie T, Ooi LK, Phillips BC. Investigating sewer rehabilitation in the Marina Reservoir Catchment in Singapore. Edinburgh: 11th International Conference on Urban Drainage; 2008.
75. Bennet A. Potable water: New technology enables use of alternative water sources. Filtration+Separation. 2011;24-27.
76. Chew YCM, Watanabe C, Tou Y. The challenges in Singapore NEWater development: Co-evolutionary development for innovation and industry evolution. Technology in Society. 2011;33:200-211.
77. Bremere I, Kennedy M, Stikker A, Schippers J. How water scarcity will effect the growth in the desalination market in the coming 25 years. Desalination. 2001;138:7-15.
78. Khawajia AD, Kutubkhanaha IK, Wie J-M. Advances in seawater desalination technologies. Desalination. 2008;221:47-69.
79. Greenlee LF, Lawler DF, Freeman BD, Benoit Marrot B, Moulin P. Reverse osmosis desalination: Water sources, technology, and today's challenges. Water Research. 2009;43:2317-2348.
80. Expert Panel on Drainage Design and Flood Protection Measures (EPDDFP). Report on Conclusions and Recommendations of the Expert Panel on Drainage Design and Flood Protection Measures; 2012. Accessed on 8 September 2014. Available:https://app.mewr.gov.sg/data/ImgCont/1524/Expert_Panel_Report_on_Drainage_Design_and_Flood_Protection_Measures.pdf.

81. Chang NB. Hydrological connections between low-impact development, watershed best management practices, and sustainable development. Journal of Hydrologic Engineering. 2010;384-385.
82. Shamsi UM. Low impact development for stormwater quantity and quality. In Dynamic Modeling of Urban Water Systems, Monograph 18, In W James, KN Irvine, JY Li, EA McBean, RE Pitt, SJ Wright eds. 2010;13. CHI, Ontario.
83. Lawson SO, Joksimovic D, Li JY, Walters M. The perceptions of stake-holders in low impact development planning. In James W, Irvine KN, Li JY, McBean EA, Pitt RE, Wright SJ, eds. Cognitive Modeling of Urban Water Sys-tems, Monograph 19. 2011;19. CHI, Ontario.
84. Zhang G, Hamlett JM, Saravanapavan T. Representation of low impact development scenarios in SWMM. In W James, K Irvine, J Li, E McBean, R Pitt, and S Wright eds. Dynamic Modeling of Urban Water Systems, Monograph 18. 2010;12. CHI.
85. Stovin V. The potential of green roofs to manage urban stormwater. Water and Environment Journal. 2010;24:192-199.
86. Wong THF Framework for stormwater quality management in Singapore. In 12th International Conference on Urban Drainage, Proceedings, Porto Alegre, Brazil, International Water Association 2011.
87. Garrison N, Hobbs K. Rooftops to Rivers II: Green Strategies for Controlling Stormwater and Combined Sewer Overflows. Natural Resources Defense Council, New York, NY; 2011.
88. Rossman LA . Modeling low impact development alternatives with SWMM. In Dynamic Modeling of Urban Water Systems, Monograph 18, In W James, KN Irvine, JY Li, EA McBean RE Pitt, and SJ Wright eds. 2010;11. CHI, Ontario.
89. Irvine KN. Climate change and urban hydrology: Research needs in the developed and developing worlds. In: W James, KN Irvine, D Joksimovic, JY Li, EA McBean, RE Pitt, JG Vasconcelos, SJ Wright, JS Wu, eds. Pragmatic Modeling of Urban Water Systems. Monograph 21. 2013;11. Computational Hydraulics International, Ontario.
90. Holman-Dodds JK, Bradley AA, Potter KW. Evaluation of hydrologic benefits of infiltration based urban storm water management. Journal of the American Water Resources Association. 2003;39(1):205-215.
91. Gill SE, Handley JF, Ennos AR, Pauleit S. Adapting cities for climate change: the role of the green infrastructure. Built Environment. 2007;33(1):115-133.
92. Pitt R, Voorhees J. Modeling green infrastructure components in a combined sewer area. In W James, KN Irvine, JY Li, EA McBean, RE Pitt, and SJ Wright, eds. Cognitive Modeling of Urban Water Systems, Monograph 19. 2011;8. CHI, Ontario.
93. Brooks N, Anderson S, Ayers J, Burton I, Tellam I. Tracking adaptation and measuring development (Climate Change Working Paper No 1). London/Edinburgh: International Institute of Environment and Development; 2011.
94. WMO. Guide to Hydrological Practices, 5th ed. WMO-No. 168; 1994.
95. National Public Radio. Why typhoon Haiyan caused so much damage (Report); 2013. Accessed on 18 March 2014. Available: http://www.npr.org/2013/11/11/244572227/why-typhoon-haiyan-caused-so-much-damage
96. National Disaster Risk Reduction and Management Council. NDRRMC Updates re Effects of TY YOLANDA (HAIYAN).National Disaster Risk Reduction and Management Council; 2014. Accessed on 15 June 2014. Available: http://www.ndrrmc.gov.ph/index.php?option=com_content&view=article&id=1177:updates-re-effects-of-ty-yolanda-haiyan

97. Holthaus E. Super Typhoon Haiyan, one of the strongest storms ever seen, hit the Philippines with record force. Quartz; 2013. Accessed on 18 March 2014. Available: http://qz.com/144734/super-typhoon-haiyan-one-of-the-strongest-storms-ever-seen-is-about-to-hit-the-philippines/
98. Pedrasa I. Finding a Filipino word for storm surge: 'Daluyong' or 'humbak'? ABC-CBNnews; 2013. Accessed on 15 June 2014. Available: http://www.abs-cbnnews.com/focus/11/18/13/finding-filipino-word-storm-surge-daluyong-or-humbak
99. Rodolfo K. Call storm surges 'silakbô'. Philippine Daily Inquirer. 2013. Accessed on 15 June 2014. Available: http://opinion.inquirer.net/68333/call-storm-surges-silakbo
100. Travis J. Hurricane Katrina: Scientists' fear come true as Hurricane floods New Orleans. Science. 2005;309(5741):1656–1659.
101. The Straits Times. Flash floods wreak havoc; 2010. Accessed on 23 July 2010. Available:http://www.straitstimes.com/BreakingNews/Singapore/Story/STIStory_541057.html
102. Tan KW. A greenway network for Singapore. Landscape and Urban Planning. 2006;76:45-66.
103. Tortajada C, Joshi YK. Water demand management in Singapore: Involving the public. Water Resources Management. 2013;27(8):2729-2748.
104. Biswas AK. Droughts or floods: What is important for Singapore? Global-is-asian. 2012;40-41.
105. Mitchell T, Tanner T. Adapting to climate change: challenges and opportunities for the development community. Teddington: Tearfund; 2006.
106. Brooks N, Adger W, Kelly M. The determinants of vulnerability and adaptive capacity at the national level and the implications for adaptation. Global Environmental Change. 2005;15(2):151-163.

6

Impact Assessment of Hydrology and Water Quality in the Saugahatchee Creek under Projected Land Use and Climate Change Scenarios Using WARMF

Sushban Shrestha[1], Xing Fang[2*], Rajesh Sawant[3] and Luke J. Marzen[4]

[1]Trimble Navigation Limited, Auburn, Alabama 36830, USA.
[2]Department of Civil Engineering, Auburn University, Alabama 36849, USA.
[3]School of Forestry and Wildlife Sciences, Auburn University, Alabama 36849, USA.
[4]Department of Geology and Geography, Auburn University, Alabama 36849, USA.

Authors' contributions

Authors collaborated for the completion of this work. Author SS designed the study, ran the model for all cases, performed the data analyses, prepared and revised the manuscript. Author XF provided valuable suggestions for the study design, supervised the data analyses, and revised the manuscript. Author RS developed the land use land cover scenarios and revised the manuscript. Author LJM supervised land use land cover study and revised the manuscript. All authors read and approved the final manuscript.

ABSTRACT

The hydrology and water quality of a stream or reservoir can be affected due to rapid urbanization and land use change in its watershed. Climate change, if it occurs, is likely to have additional impacts on hydrology and water quality of the watershed system. In this study, a watershed model WARMF (Watershed Analysis Risk Management Framework) was applied to the Saugahatchee Creek Watershed which includes two stream branches that were listed on State of Alabama's 303(d) list of impaired water for nutrients and organic enrichment/dissolved oxygen. WARMF model for the

**Corresponding author: Email: xing.fang@auburn.edu;*
Note: The peer review process of this paper was handled by the editor-in-chief.

Saugahatchee Creek Watershed was developed and model calibration and validation were performed. The model was then used to investigate hydrologic and water quality response to two different land use scenarios (LU 2009 and LU 2030) and four statistically downscaled future climate scenarios derived from Canadian Global Coupled Model (CGCM3) and Hadley Centre Coupled Climate Model (HadCM3). Temperature, dissolved oxygen, total nitrogen, total phosphorus, and algal concentration were the water quality parameters simulated along with flow. Based on monthly average of daily predicted values, the effect due to land use change was not significant except for nutrient concentration. The monthly average of daily total phosphorus concentration for LU 2030 is predicted to increase up to 72% more than baseline (LU 2009) under past climate conditions (1981–2010). Based on model results, the monthly average of daily surface water temperature is predicted to rise for all future climate scenarios. The monthly average of daily flow is predicted to increase corresponding to CGCM3 (annual average increase of 88%) and decrease corresponding to HadCM3 scenarios (annual average decrease of -49%). Accordingly, nutrient concentration is expected to decrease corresponding to CGCM3 and increase corresponding to HadCM3 scenarios. DO concentration are predicted to fall up to 2.3 mg/l (monthly average), especially in summer for the four climate scenarios. Combined land use and climate change scenarios cause the increase in nutrient concentrations for future land use and climate scenarios (e.g., annual TP from 0.082 mg/l for the baseline to 0.203 mg/l for HadCM3 A2 20s scenario). Chlorophyll-a concentration during the growing season is expected to increase to 25.8 and 26.3 µg/l under HadCM3 A2 and B2 scenarios due to combined effect, respectively, in comparison to 18 µg/l for the baseline (1981–2010 and LU2009). The results of this study can be incorporated into watershed management and planning strategies.

Keywords: Hydrology; water quality; land use change; climate change; Saugahatchee; WARMF.

1. INTRODUCTION

During the recent years, rapid urbanization has led to massive land use changes, pervious forest soils have been reduced and industrial and residential areas have been increased. Similarly, the atmospheric concentration of greenhouse gases is believed to be increasing, thereby leading to climate change. Climate change, if it happens, will cause increase in temperature, evaporation, evapotranspiration, precipitation variability and extremes [1]. These alterations will change the hydrological behavior of watershed ecosystem in physical, chemical, and biological terms. The potential effects of land use and climate change are not limited to quantity; it can have serious impacts on water quality of the streams in the watershed. These changes will further affect water quality in downstream lakes and reservoirs that may alter fish habitat and have an effect on indigenous fish populations [2-5]. In this study, we utilized the Watershed Analysis Risk Management Framework (WARMF) [6,7] to develop a region-specific modeling framework to systematically assess the impact on flow and water quality in streams and reservoirs of the Saugahatchee Creek Watershed using future land use and future climate scenarios as model inputs.

The Intergovernmental Panel on Climate Change (IPCC) Fourth Assessment Report summarized that the linear trend over the last 50 years is warming of 0.13 (0.10–0.16) °C per decade, nearly twice temperature increase for the last 100 years, and it is projected to further increase by 0.2°C per decade for the next two decades [8]. The IPCC [1] summarized the impact of climate change on freshwater systems are mainly due to increases in temperature and precipitation variability. Higher water temperatures, increased precipitation intensity, and longer periods of low flows exacerbate water pollution, with impacts on ecosystems (e.g., fish habitat), human health, water system reliability, and operation cost [1].

The outputs from various General Circulation Models (GCMs) like Couple Global Climate Model (CGCM3) from Canadian Climate Centre Modeling and Analysis (CCCma), and Hadley Centre Couples Climate Model (HadCM3) developed in the United Kingdom are available to generate future climate scenarios. However, the GCMs were not designed to analyze the hydrological impact at the watershed scale and therefore have coarser spatial resolution as compared to what is required for watersheds impact studies. GCMs are inherently unable to represent watershed scale features and dynamics for hydrological impact studies [9,10]. To bridge this gap, the techniques have been developed to downscale GCMs output into local meteorological variables required for hydrological modeling, usually referred to as downscaling techniques. The downscaling techniques can be statistical or dynamic. The Statistical Down Scaling Model (SDSM) was used in this study, considering its advantages over dynamic method because statistical method is computationally undemanding and provides station or local-scale climate information based on GCM-scale output [11].

Hydrological impact of land use change has been investigated in a variety of studies using modeling methods [12-15]. The water quality in a watershed is affected directly by vegetative cover and agricultural and other land management practices [16]. Bhattarai et al. [16] used BASINS-SWAT model to estimate the effect of land use change on nitrogen and phosphorus runoff and sediment deposition in a small watershed in the Alabama Wiregrass Region. Similarly, many researchers have pointed out adverse effect of increasing urban land use on water quality [17-20].

Many previous studies have assessed the impact of climate change on hydrology [21-26]. Dibike and Coulibaly [27] applied statistical downscaling techniques to generate future climate scenarios in the Saguenay watershed in Canada at a local watershed scale and simulated the corresponding flow based on the downscaled future climate data as an input to hydrological models. Rich et al. [28] applied WARMF to assess impacts of extended droughts and increased temperature due to climate change on hydrology of the San Juan Basin. Simulations showed that drought and increased temperature impact water availability and lead to increased frequency of critical storages. The assessment of impact of climate change on water quality in the southeastern United States revealed that watersheds are likely to have higher nitrogen levels and lower dissolved oxygen problems [29]. However, very few have conducted the climate change impact study for water quantity as well as quality [29-32]. Wang investigated the individual and combined impact of future land use and climate change in the Wolf Bay Watershed using SWAT. United States Environmental Protection Agency (USEPA) is evaluating the impacts of land use and climate change of hydrology and water quality in major rivers basins throughout the United States using watershed models, HSPF and SWAT [33,34]. Limited water quality parameters such as total nitrogen, total phosphorus concentration, etc. are considered in previous impact studies.

The objective of this study was to assess the impact of land use change and climate change (downscaled to local watershed scale) based on CGCM3 and HadCM3 future climate scenarios, on flow and five water quality parameters (water temperature, dissolved oxygen, total nitrogen, total phosphorus, and algal concentration) in the streams of the Saugahatchee

Creek Watershedin Alabama (Fig. 1). The physically-based WARMF model for the Saugahatchee Creek Watershed (WARMF-SCW) was set up, calibrated, and validated for observed flow and water quality, then it was run for past and future climate scenarios, and their potential impacts were evaluated. After we have a better understanding on potential changes of stream and reservoir water quality due to climate warming, we can further study its impacts on fish habitat using different oxythermal criteria [35].

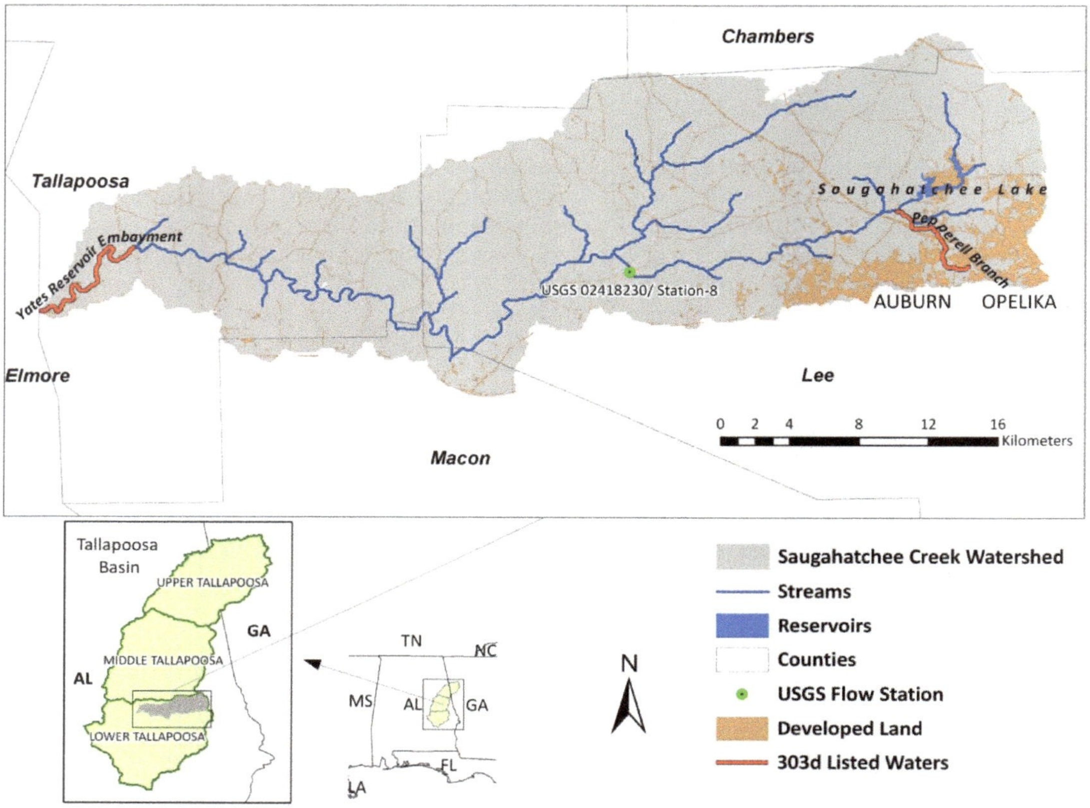

Fig. 1. Geographical location of the Saugahatchee creek watershed in the tallapoosa basin including surrounding counties in Alabama, USA. The monitoring Station-8 is at the same location as the USGS gaging station 02418230

2. MATERIALS AND METHODS

2.1 Study Area

The watershed of concern is the Saugahatchee (also referred as Sougahatchee) Creek Watershed, located mostly in Piedmont region of eastern Alabama with an area of approximately 550 km^2 (Fig. 1). Beginning with its headwaters in Chamber and Lee counties, the Saugahatchee Creek runs westward through parts of Macon and Tallapoosa counties until it enters Yates Reservoir and converges to Tallapoosa River. Two segments in the Saugahatchee Creek Watershed (Fig. 1) were listed on the Alabama Department of Environmental Management (ADEM)'s 303 (d) list of impaired waters under the federal

Clean Water Act [36]. Pepperell branch, a tributary to the Saugahatchee Creek was listed as impaired waters for nutrients and the portion of the Saugahatchee Creek entering Yates Reservoir (Yates Reservoir Embayment) was listed for nutrients and organic enrichment and dissolved oxygen. The headwater portion of the watershed is the Auburn-Opelika metropolitan area in Lee County (Fig. 1) with a 2009 population of 135,883 (140,247 in 2014). It was the 19th fastest growing metro area in the United States between 1990 and 2000<http://en.wikipedia.org/wiki/Auburn,_Alabama_metropolitan_area>. The growth of the cities has resulted in land use change and might have potential changes in stream water quality.

2.2 Models and Data

Complex hydrological models such as AGNPS (Agricultural Nonpoint Source Pollution Model [37]); BASINS (Better Assessment Science Integrating point and Nonpoint Sources [38]); HSPF (Hydrological Simulation Program – Fortran [39]); GWLF (Generalized Watershed Loading Functions [40]); SWAT (Soil and Water Assessment Tool [41]); and WARMF [6] have been frequently applied to study watershed hydrology in the United States (U.S.) and all over the world. A physically based, dynamic watershed model WARMF was applied to this study to the Saugahatchee Creek Watershed for assessing hydrology and water quality impact due to land use and climate change. Although other models, discussed above, would yield similar results, WARMF was applied here for its integration of stream and one-dimensional (1-D) reservoir water quality models, user friendly interface and ability to assess the impact of point and nonpoint sources with varying land use and meteorological scenarios.

WARMF is an integrated watershed model with simulation models and databases under one GIS-based graphical user interface (GUI). The algorithms embedded in WARMF are adapted from many well established codes such as ILWAS (Integrated Lake Watershed Acidification Study [42,43]), ANSWERS (Areal Nonpoint Source Watershed Environmental Response Simulation [44,45]), SWMM (Storm Water Management Model [46]) and WASP (Water Quality Analysis Simulation Program [47]).

WARMF represents a watershed by dividing it into a network of land catchments, river segments, and one- or two-dimensional reservoirs. Land catchment is further divided into a canopy layer, a snowpack, and up to five soil layers. Each compartment is considered as a seamlessly connected continuously stirred tank reactor (CSTR) for flow routing and mass balance calculation. WARMF simulates the process of canopy interception, snowpack accumulation and snowmelt, infiltration through soil layers, evapotranspiration, surface runoff, and groundwater exfiltration to river segments. The water from the upstream river segment is mixed with water in the river segment from previous time step and the point and nonpoint loads entering the river segment during the time step. Heat budget and mass balance are performed to calculate the temperature and concentration of various water quality constituents in each soil layer, river segment, and reservoir [6].

The flowchart of the study involving different models, data, and scenarios is given in (Fig. 2). The delineated watershed map with land catchments, stream segments and reservoirs, is required for WARMF, to which input data can be given and simulation results can be viewed. BASINS [38] provides watershed delineation tool which helps to delineate watershed based on digital elevation model (DEM) and river network. To delineate watershed, BASINS data download tool extracts nationally derived databases like boundaries for States and Cataloging Units, DEM, National Hydrography Dataset (NHD) and National Land Cover

Dataset (NLCD). A BASINS project can be set up for 8-digit hydrologic unit code (HUC) watershed by selecting the geographic area of interest from among the entire 48 contiguous United States. The watershed delineation developed from BASINS can be imported into WARMF [48]. The Saugahatchee Creek Watershed is a subset watershed of Lower Tallapoosa Watershed (Fig. 1) in Alabama (HUC 03150110). BASINS project for HUC 03150110 was built and only those catchment layers that drain into the Saugahatchee Creek were selected for the WARMF-SCW model [49].

WARMF uses readily available data from National Oceanic and Atmospheric Administration (NOAA), USEPA, and United States Geological Survey (USGS) and other online sources to predict hydrology and water quality in streams. To run the simulation, land use data, meteorological data, air quality data, and soil properties are required. Observed hydrology and water quality data are required for model calibration and validation (Fig. 2). Land use represents the characteristics of watershed. Land use data can be directly imported into WARMF as a GIS shapefile. Land use shapefile used for model calibration was obtained for the Saugahatchee Creek Watershed for 2009 (LU 2009) (Fig. 3). The meteorological data required for WARMF includes precipitation, dew point temperature, minimum and maximum air temperature, cloud cover, air pressure and wind speed. National Climatic Data Center (NCDC) Global Summary of Day, online climate dataset was used to download the necessary weather data for station in Montgomery, AL (Montgomery Dannelly Field) except cloud cover data which was calculated using recommended formula [49]. The dry and wet deposition for air quality was obtained from National Atmospheric Deposition Program (NADP) and USEPA Clean Air Status and Trends Network (CASTNET) website for GAS153 and AL10 Station, respectively [49]. There are three major point-source dischargers that contribute to the Saugahatchee Creek. The average flow discharge from these point dischargers and constituent loadings were obtained from data collected by ADEM and Auburn University [36] in previous studies.

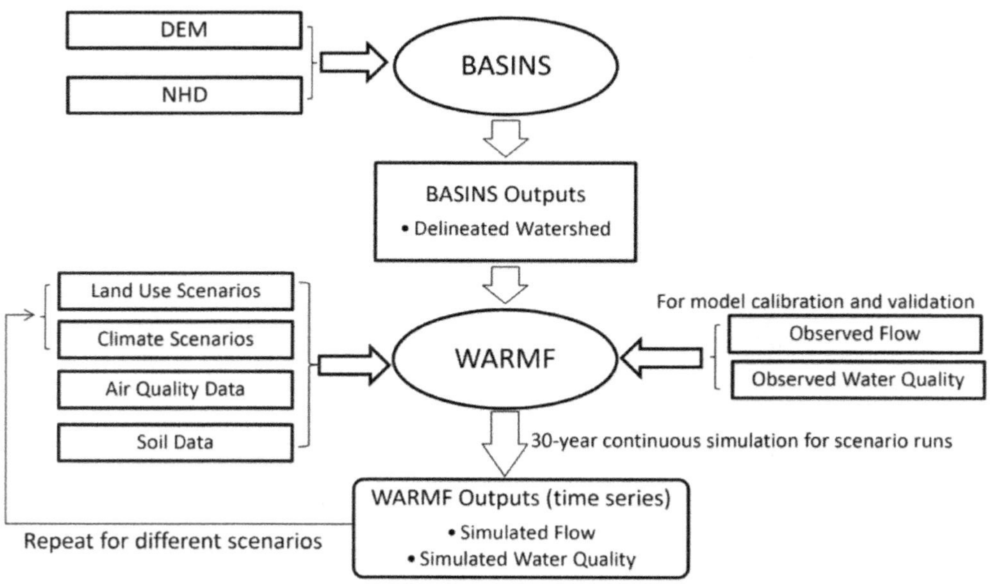

Fig. 2. Flowchart of the study involving different models, data and scenarios

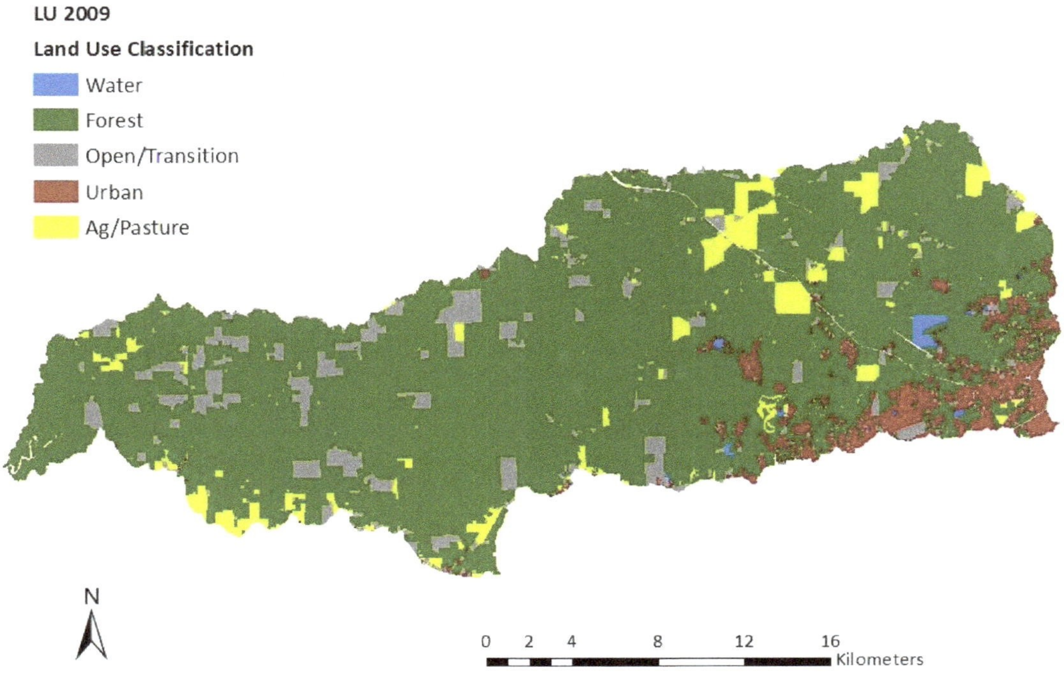

Fig. 3. 2009 land use map of the Saugahatchee creek watershed (LU 2009)

2.3 Land Use Change Scenarios

There have been significant changes in land use pattern of the Saugahatchee Creek Watershed over the last two centuries, particularly in recent years [50]. The upper watershed is undergoing rapid transition from forest to urban/developed land. Forest area has declined and urban and transitional area has increased. The changes in land use distribution are expected to bring changes in water quality, including surface flow, nutrient runoff, and sedimentation levels [16]. Multi-temporal datasets of Landsat5 Thematic Mapper imagery, aerial photographs, and other vector datasets were used to determine land use land cover (LULC) changes over the study periodfor year 1991, 2001 and 2009. Historical analyses of changes in the LULC over the study area were done to model changes in the LULC and to develop and validate logistic regression model. Landsat data was processed utilizing unsupervised classification in ERDAS Imagine with ISODATA clustering algorithm [51]. The ISODATA clustering method uses the minimum spectral distance formula to form clusters by iteration. It was followed by cluster busting method to improve accuracy of land cover classification. The classified land cover maps were produced by employing five types of land cover categories: Water, Forest, Open/Transition, Urban, and Ag/Pasture. The accuracy assessment was carried out to evaluate overall accuracy of LULC classification results. The overall accuracies for the pixel based unsupervised classification with cluster busting for year 1991, 2001, and 2009 were 90%, 90%, and 90.4%, respectively [52].

Urbanization patterns in the Saugahatchee Creek Watershed were modeled using GIS and remote sensing imagery coupled with logistic regression analysis. For modeling purposes, the spatial unit utilized was the Lee County tax parcels. Parcel data is most commonly used in city planning and unlike per pixel classification, which is limited to land cover, using parcels as spatial unit of classification it is possible to determine land use associated with it and provides better spatial configuration of land cover. The raster dataset of LULC classification for year 1991, 2001, and 2009 was then transferred from the image pixels to vector dataset of Lee County parcels using a "Majority" algorithm in ERDAS Imagine. By employing various drivers of LULC change, such as distance of a parcel from major road, school, commercial and industrial areas, provision of utility, and population density as variables, a land use model based on multiple logistic regression analysis was developed and validated over the period 1991 to 2009 [52].

The land use conversion model developed using logistic regression model in ArcGIS was then used to predict future land use of 2030 (LU 2030).For this study, LU 2009 was used as baseline scenario in logistic regression model to predict development for the year 2030. For each parcel, probability of change is computed with the fitted model for the 2030 scenario. The projected LU 2030 (Fig. 4) was then used as future land use scenario (Table 1) [52].

The change detection analysis of the study area done by Sawant [52] for the period from 1991 to 2009 shows conversion of forest to municipal land use in the Saugahatchee Creek Watershed has been the most dominant LULC change in the past two decades. Since 1991 footprint of urban area within the Saugahatchee Creek Watershed has increased by 36.5% to 7,035 acres in year 2009. Although the area of urbanization is only about 5%of the total area of the Saugahatchee Creek Watershed, the urbanization has likely played a role in impairment of the Pepperell branch of the Saugahatchee Creek. Most of the urban changes occurred within and immediate vicinity of the existing urban land use areas. The land change detection shows spread of urban land use in the surrounding areas [52].

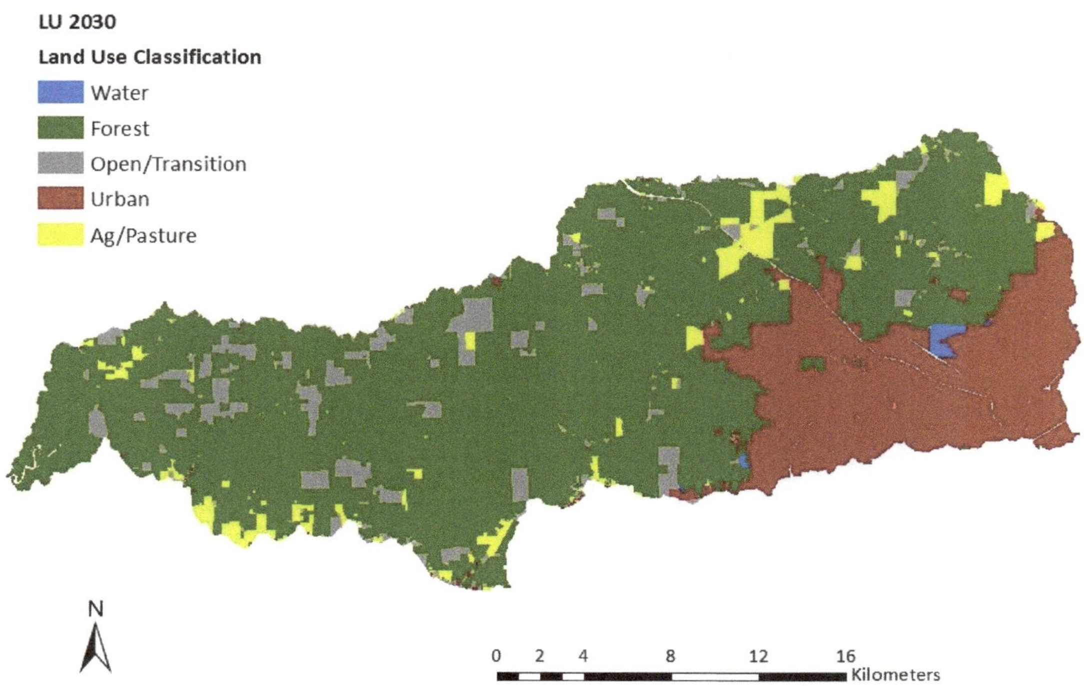

Fig. 4. Projected 2030 land use map of the Saugahatchee Creek Watershed (LU 2030)

Table 1. Land use change in the Saugahatchee creek watershed

Land use categories	Area in km^2 (percentage of total area)	
	LU 2009	**LU 2030**
Water	2 (0.4%)	2 (0.3%)
Forest	449 (83.3%)	385 (71.4%)
Open/Transition	43 (8.0%)	37 (6.9%)
Urban	18 (3.4%)	94 (17.5%)
Ag/Pasture	26 (4.8%)	21 (3.9%)

2.4 Climate Change Scenarios and Downscaling

Future climate projected using GCMs have coarser spatial resolutions than what is required for a watershed scale study. For hydrologic and water quality impact studies, local or station scale meteorological variables are required, which can be derived using large-scale atmospheric variables available from GCM outputs. Future climate change scenarios were downscaled for the Saugahatchee Creek Watershed at daily time scales, using four GCM outputs (CGCM3 A1B, CGCM3 A2, HadCM3 A2 and HadCM3 B2). The software used for downscaling is SDSM [11].

In statistical downscaling techniques, the quantitative relationships are established between large-scale atmospheric variables (predictors) and local or station surface variables (predictands). The National Centers for Environmental Prediction (NCEP) and National Center for Atmospheric Research (NCAR) worked together in a reanalysis project to produce a physically consistent retroactive record of more than 50 years of global analyses of atmospheric fields to support the needs of research and climate monitoring communities [53]. Reanalysis project involved the recovery of data from many observed and measurement systems, quality controlled and assimilated with a data assimilation system kept unchanged over the reanalysis period. The main objective of reanalysis is to eliminate perceived climate jumps associated with changes in data assimilation system and provide consistent records of temperature, precipitation, winds and many other variables that describe climatic conditions from the past to the present. The daily NCEP/NCAR reanalysis data were selected to represent the large-scale predictors in the SDSM model. NCEP/NCAR data have been used in several downscaling studies in different regions over the world [27,54].

Using SDSM, the appropriate large-scale predictor variables were selected from the list of predictors obtained from NCEP/NCAR reanalysis data for the period of 1961–1990, based on regression techniques, to downscale predictands (such as station precipitation, maximum and minimum temperature). Table 2 lists the predictor variables (from NCEP/NCAR) screened for downscaling and the predictands. SDSM constructs a downscaling model with parameters of the model based on multiple regression equations, given observed daily weather data (predictand) and the selected large-scale NCEP predictors for the same time period. The observed daily weather data at Montgomery Dannely Field station from 1961 to 1990 were downloaded from NOAA's National Climatic Data Center (NCDC) web site. The climate data from 1961–1975 were used to develop the regression model and the model regression weights (as parameter file) produced were then used to validate the climate conditions for the period from 1976–1990. The detailed process of selecting predictors, calibration results (such as R^2), parameter file generation, and validation is presented elsewhere by Shrestha [49].

Table 2. List of predictands (station climate parameters) and corresponding predictors used in SDSM model to downscale GCMs outputs

Station parameters	Predictors from NCEP/NCAR and CGM[1]
Precipitation	500 hPa divergence (p5zh), Relative humidity at 500 hPa (r500), Specific humidity at 500 hPa (s500), Relative humidity at 850 hPa (r850), and Specific humidity at 850 hPa (s850)
Maximum temperature	Mean temperature (temp), and 500 hPa geopotential height (p500)
Minimum temperature	Mean temperature (temp), and Near surface specific humidity (shum)

[1]*variable name used in SDSM is given inside parenthesis*

The parameter file was then used to downscale future climate to local watershed scale based on predictor derived from GCMs. SDSM future climate predictors may be obtained for any global land area through data portal maintained by the Canadian Climate Impacts Scenarios Group (CCIS) <http://www.cics.uvic.ca/scenarios/index.cgi?Scenarios>. The predictors are available for CGCM3A1B, CGCM3 A2, HadCM3 A2, and HadCM3 B2scenarios.

Given the latitude and longitude of the Saugahatchee Creek Watershed, CGCM3 A1B, CGCM3 A2, HadCM3 A2, and HadCM3 B2 predictors were extracted for the nearest grid along with NCEP/NCAR predictors interpolated to the same grid as GCMs. The future climate projection scenarios were separated into three time frames and termed as 20s (2011–2040), 50s (2041–2170), and 80s (2071–2100). Figs. 5 and 6 show the patterns of downscaled monthly average precipitation, maximum (T_{max}) and minimum (T_{min}) temperatures compared to baseline (1961-1990) for the study watershed along with standard deviation corresponding to CGCM3 A1B, CGCM3 A2, HadCM3 A2, and HadCM3 B2 scenarios, respectively. The mean monthly maximum and minimum temperatures are projected to increase over time up to 5.98°C in July from CGCM3 A2 80s (Fig. 5), especially similar increases in summer months for all four scenarios, e.g., 4.82°C average increase from the baseline in July–September for CGCM3 A2 80s (Fig. 5). The standard deviations of T_{max} and T_{min} over 30 year periods are higher in the winter months (up to 7.3°C) and lower in summer (~2°C). The CGCM3 downscaled results projected slight changes in precipitation during early 21st century (20s), but precipitation is projected to increase during the later century (Fig. 5) whereas for HadCM3 scenarios, precipitation decreased from the baseline, especially during summer (Fig. 6). For CGCM3 A2 80s, monthly average precipitations in summer months (June–August) are projected to increase 21.4% to 68.1% (Fig. 5); while those are projected to decrease -16.7% to -76.1% for HadCM3 A2 80s (Fig. 6). Two scenarios (A1B and A2) of CGCM3 (Fig. 5) projected more increases in the winter months (November and December) and early spring (February).

The possible uncertainties that statistical model can carry is due to its assumption that the statistical relationships between the past predictors and predictands will remain valid in the future climate scenarios [11].

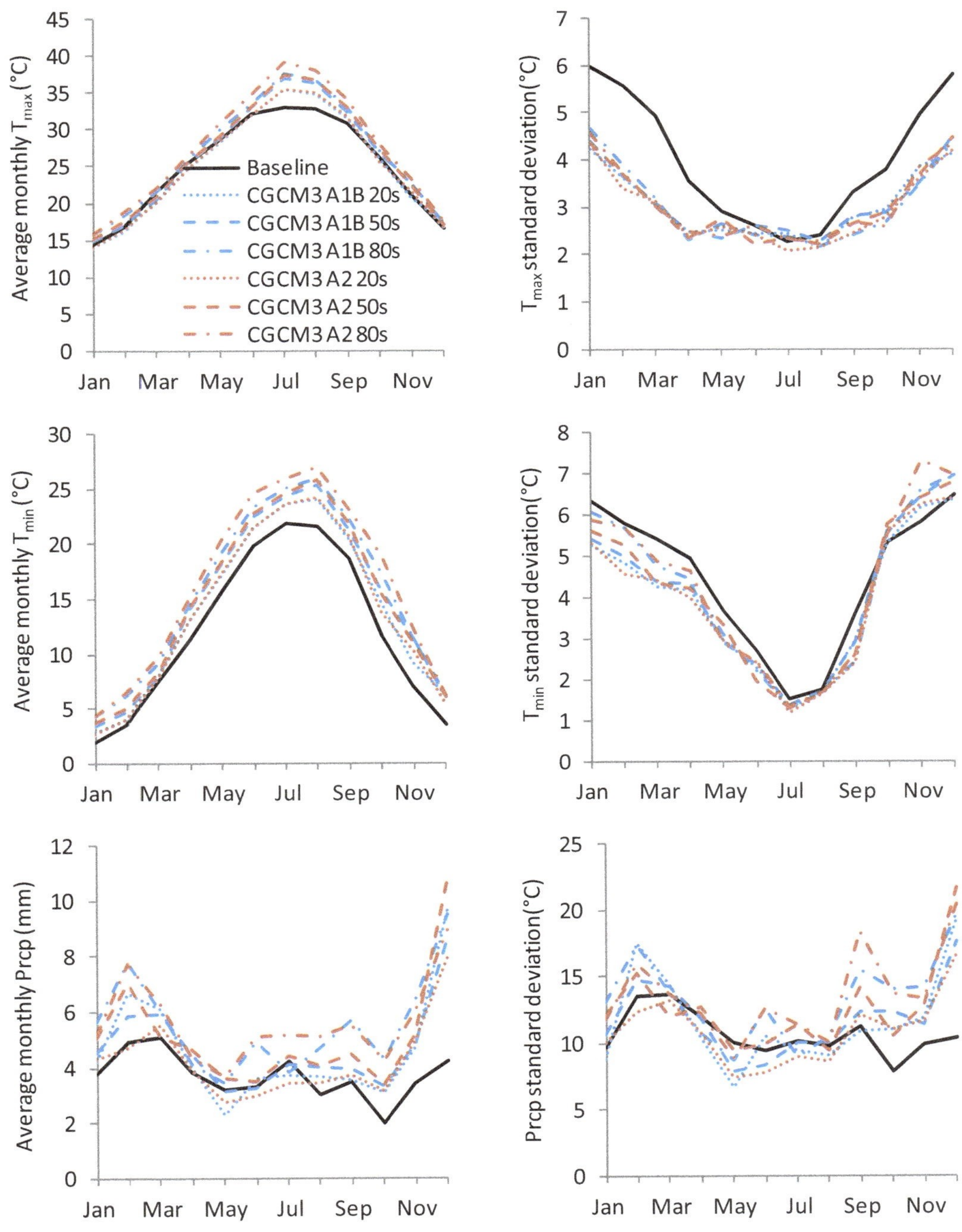

Fig. 5. General trends in average monthly maximum temperature (*T_{max}*), minimum temperature (*T_{min}*), and precipitation (*Prcp*) corresponding to downscaled climate change scenarios based on CGCM3 A1B and CGCM3 A2 for three 30-year time periods (20s for 2011-2040, 50s for 2041-2070, and 80s for 2071-2100)

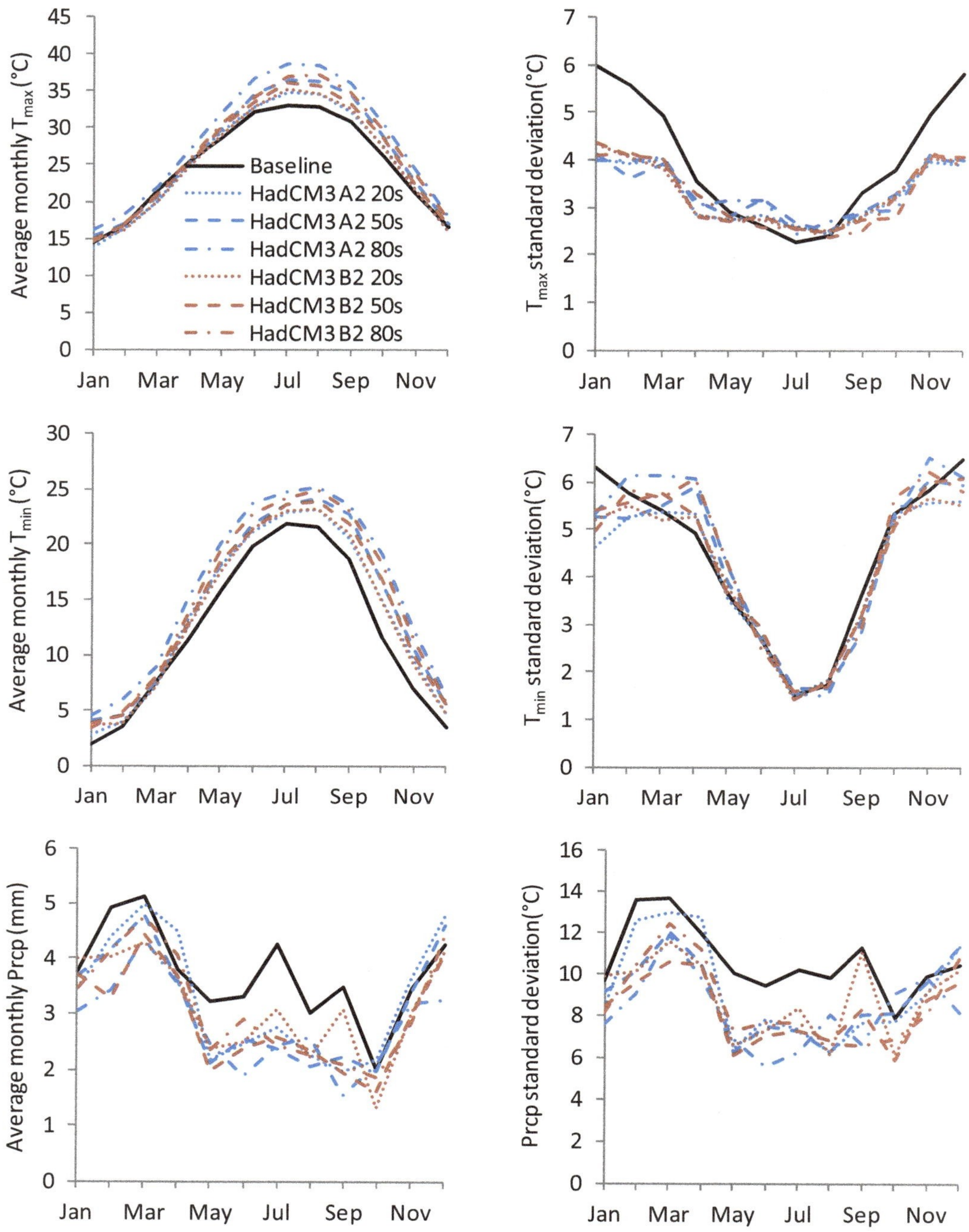

Fig. 6. General trends in average monthly maximum temperature (T_{max}), minimum temperature (T_{min}), and precipitation (*Prcp*) corresponding to downscaled climate change scenarios based on HadCM3 A2 and HadCM3 B2 for three 30-year time periods (20s for 2011-2040, 50s for 2041-2070, and 80s for 2071-2100)

3. RESULTS AND DISCUSSION

3.1 Model Calibration and Validation

The Saugahatchee Creek Watershed was divided into 44 subcatchments, 40 stream segments, and 2reservoirs (Fig. 1) [49]. The hydrologic and water quality parameters can be assigned to these segments individually, referred as catchment coefficients, river coefficients, and reservoir coefficients in WARMF manual [7]. There are another set of parameters known as system coefficients that apply to all catchments in the watershed [7]. During the calibration, some of catchment, river, reservoir, and system coefficients were manually adjusted to obtain a best fit (Table 3) based on guidance from the model parameter study [55].

Table 3. Keyparameter values after WARMF calibration

Parameters	Units	Literature range	Calibrated value
Precipitation weighting factor	-	0.5 - 1.5	0.74
Evaporation magnitude	-	0.6 - 1.4	0.91
Evaporation skewness	-	0.6 - 1.4	0.9
Number of soil layers	-	1 - 5	3
Thickness of soil layers	cm	> 0	8 – 79
Saturation moisture	-	0.2 - 0.6	0.35 – 0.45
Field capacity	-	0 - 0.4	0.18 – 0.31
Initial moisture	-	0 - 0.6	0.25
Horizontal conductivity	cm/day	> 0	3600 – 5600
Vertical conductivity	cm/day	> 0	1800 – 2800
Aeration factor	/day	0.2 - 1	0.5
Sediment oxygen demand	$g/m^2/day$	0.1 – 2	0.8

The simulation of watershed model can be judged satisfactory if Nash-Sutcliffe efficiency [56] (NSE)>0.5, if the ratio of the root mean square error to the standard deviation of measured data (RSR) ≤0.7, and if the percent bias (PBIAS) is within ±25% for flow [57].

Nash-Sutcliffe model efficiency coefficient is defined as:

$$NSE = 1 - \left[\frac{\sum_{i=1}^{n}\left(Y_i^{obs}-Y_i^{sim}\right)^2}{\sum_{i=1}^{n}\left(Y_i^{obs}-Y^{mean}\right)^2}\right] \quad (1)$$

Where Y_i^{obs} stands for the ith observed data, Y_i^{sim} stands for the corresponding simulated value, and Y^{mean} is mean value of all observed data. The ratio RSR is calculated as:

$$RSR = \frac{\sqrt{\sum_{i=1}^{n}\left(Y_i^{obs}-Y_i^{sim}\right)^2}}{\sqrt{\sum_{i=1}^{n}\left(Y_i^{obs}-Y^{mean}\right)^2}} \quad (2)$$

Percent bias is calculated using Eq. (3):

$$PBIAS = \left[\frac{\sum_{i=1}^{n}\left(Y_i^{obs}-Y_i^{sim}\right)*100}{\sum_{i=1}^{n}\left(Y_i^{obs}\right)}\right] \quad (3)$$

The observed flow daily data is available at USGS Gage Station 02418230 (Fig. 1) in the Saugahatchee Creek near Loachapoka since 2000. The flow was calibrated for the period 2000–05 and validated for the period 2006–09 (Fig. 7). The model simulation was run three years (model spin-up period) prior to the calibration period starting from 1997 to minimize the effect of initial conditions used in the model. The simulation of average daily flow resulted in NSE, RSR and PBIAS values for calibration (validation) as 0.64, 0.60 and -5.72% (0.56, 0.66 and -9.48%), respectively. These parameter values indicate overall satisfactory model calibration and validation of the WARMF-SCW.

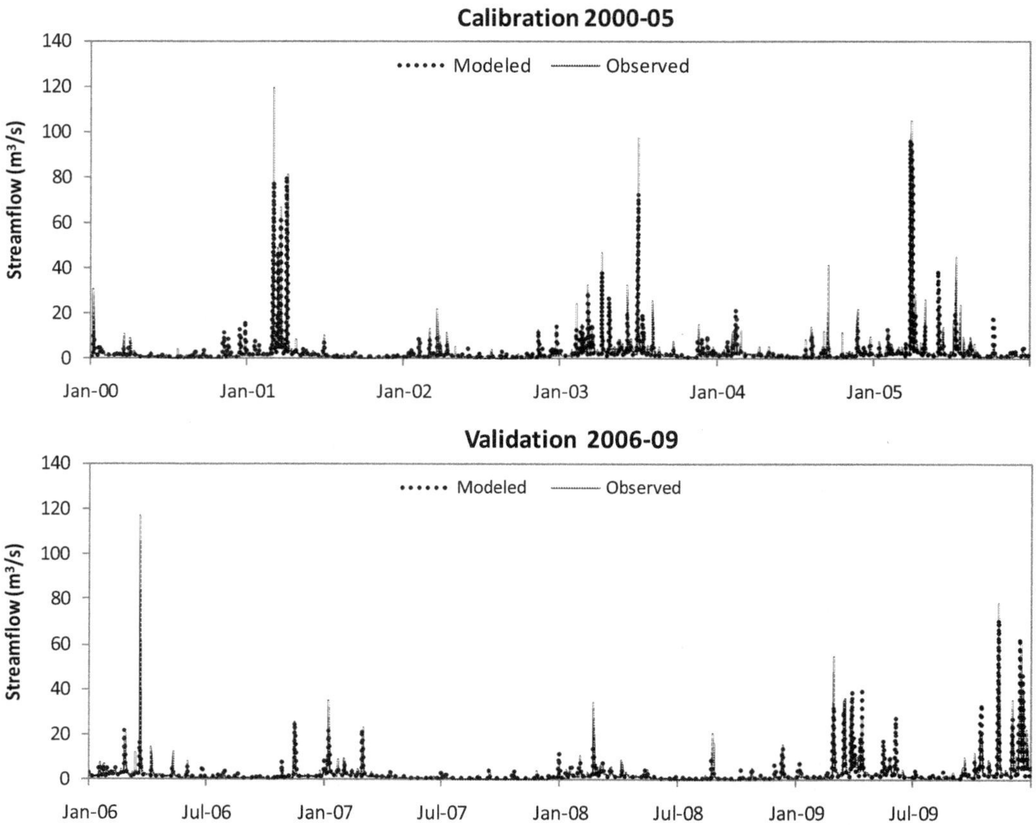

Fig. 7. Flow calibration (2000–2005) and validation (2006–2009) using WARMF-SCW at the USGS 02418230 Station in the Saugahatchee creek near Loachapoka, Alabama

ADEM and Auburn University collected water quality data in the Saugahatchee Creek Watershed in 2000–2002 [36]. The observed water quality data such as temperature, dissolved oxygen (DO), total phosphorus (TP), and nutrients were available for Station-8 (Fig. **1**) at the same location as USGS Gage Station 02418230 but not in daily basis[35].Therefore, water quality calibration was performed based on visual comparison of simulated and observed data. Time series plots shown on (Fig. 8) indicate the WARMF-SCW model simulates these water quality parameters reasonably well.

3.2 Impact of Land Use and Climate Change on Hydrology and Water Quality

The impact study was performed on the Saugahatchee Creek Watershed to evaluate the effects on hydrology and water quality due to future land use and future climate scenarios. The impact was analyzed under three categories: Impact due to land use change only, impact due to climate change only, and impact due to combined land use and climate change. Flow and water quality parameters were reported at the watershed outlet (Yates Reservoir Embayment) and their monthly average of daily values for baseline and various land use and climate change scenarios were calculated and compared for impact analysis.

3.2.1 Impact due to land use change only

The baseline scenario corresponding to land use of 2009 and future land use scenario corresponding to projected land use of 2030 were analyzed using 30 years (1981-2010) simulation results to investigate the effects of land use change only. (Fig. 9) shows monthly average of daily values for flow and water quality parameters under the baseline (LU 2009) and future land use (LU 2030) scenarios simulated using same 30-year (1981-2010) meteorological data.

The monthly averages of daily flow for LU 2009 ranged from4.14 (September) to 14.22 (March) m^3/s and are projected to rangefrom4.62 to 14.29 m^3/s for LU 2030. The increase or decrease in the monthly average flow from LU 2009to LU 2030is projected to be not more than 12.6% (Fig. 9). Similarly, for surface water temperature, there is very little or no change projected between two land use scenarios. The increase or decrease in the monthly average surface water temperature corresponding to LU 2030is less than 0.5% (Fig. 9). Therefore, land use change from 2009 to 2030 will not have significant impact on flow and surface water temperature in the Yates Reservoir Embayment for the simulation period.

Surface dissolved oxygen concentration in the Yates Reservoir Embayment, in terms of monthly average of daily values, will not experience much change due to land use change, partly because the simulation was performed on daily time step. The water body with predominance of algae shows a larger fluctuations in dissolved oxygen than less productive water with low algal concentration [58]. The water bodies with algal concentration higher than 15 µg/l are categorized as eutrophic and higher than 40 µg/l as hypereutrophic water bodies [59]. Higher eutrophic level implies algal abundance and hence exhibit higher rate of photosynthesis, respiration, and decomposition. For eutrophic and hypereutrophic systems, the dissolved oxygen concentration tends to increase during day time due to algal photosynthesis dominating over respiration and decomposition; whereas the system will have less dissolved oxygen concentration during night as there is no sunlight for photosynthesis [60,61]. Due to limited data availability, the simulation was run on daily time steps in the model, which doesn't model diurnal phenomenon [49]. Fig. 9 shows the relative change in monthly average of daily DO concentration to the baseline corresponding to LU 2030 is 2.0% at most.

In the Saugahatchee Creek Watershed, forest land has been transformed into urban land (Table 1). Pesticides, industrial waste, urban stormwater and sewage are the major source of nutrients. Therefore, the major impact that land use change has on the watershed is to the nutrient concentration in the streams and reservoirs. LU 2030 projects to produce monthly average TP concentration to be from 55.8 to 71.7% greater than baseline scenario (Fig. 9). The relative change in monthly average of daily TN concentration to the baseline corresponding to LU 2009is projected to range from1.0 to 7.7% (Fig. 9).

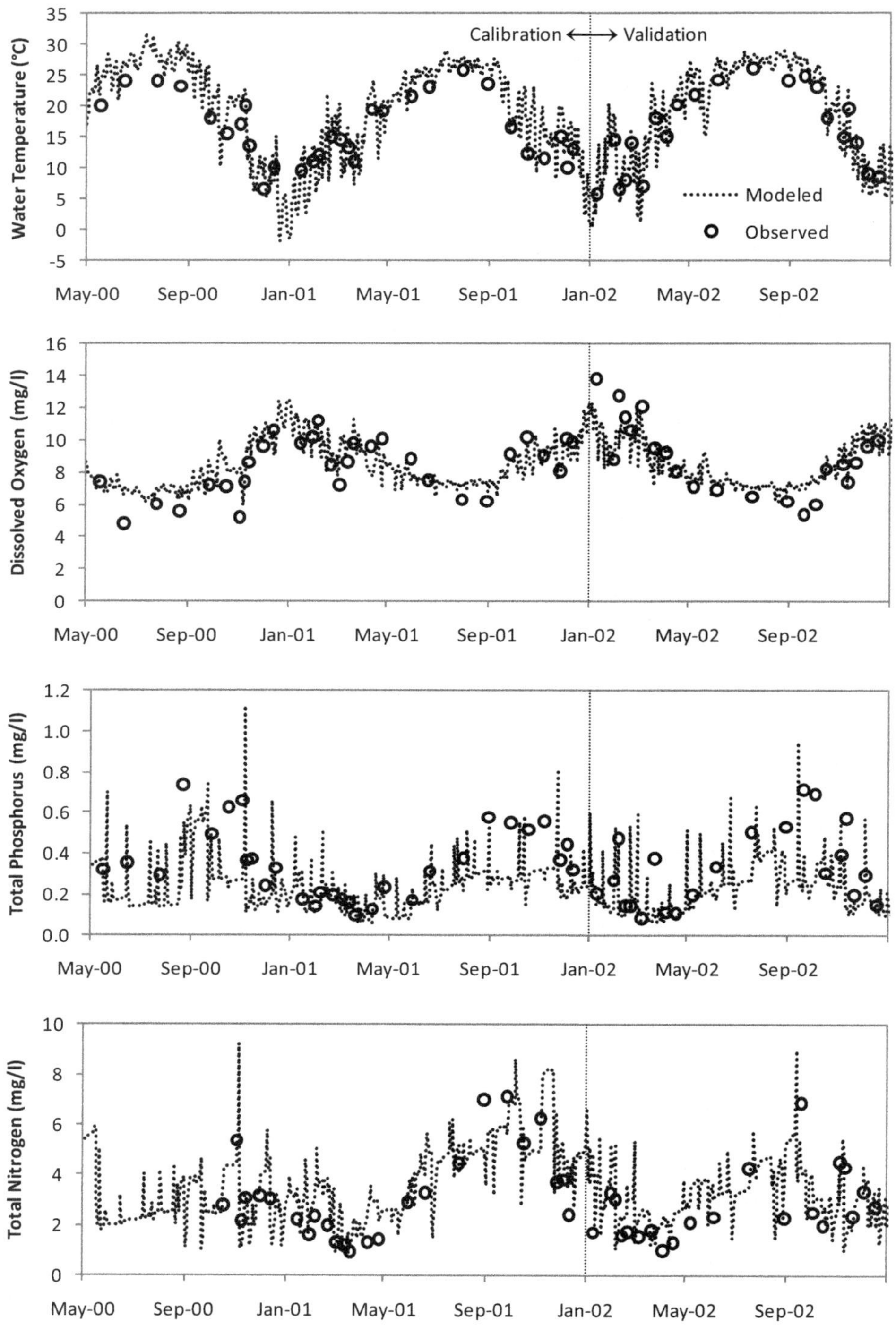

Fig. 8. Water quality calibration (2000–2001) and validation (2002) using WARMF-SCW at the Station-8 near Loachapoka, Alabama

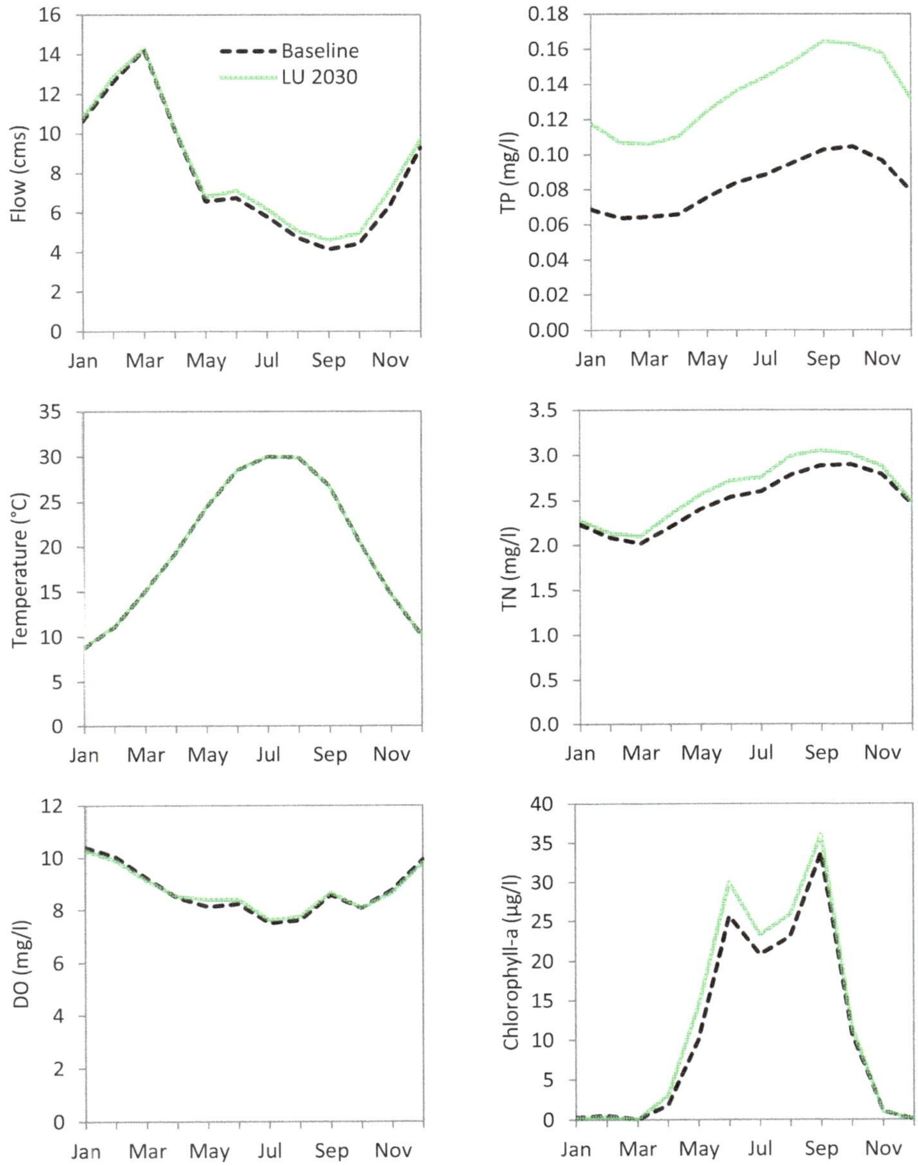

Fig. 9. Simulated monthly average flow and water quality parameters under the baseline (LU 2009) and future land use scenario (LU 2030)

The algal growth in streams and reservoirs are seasonal. The growing season for the Saugahatchee Creek Watershed has been identified as April through October [36]. Monthly average chlorophyll-a concentration under 2009 land use varied from 1.8 to 33.7 µg/L during growing season at Yates Reservoir Embayment, which is a highly eutrophic reservoir. Fig. 9 shows the relative change in the monthly average of daily algal concentration to the baseline corresponding to LU 2009 rangesfrom-32.8 to 72.1% or -0.2 to 4.4 µg/L (Fig. 9).

3.2.2 Impact due to climate change only

The impact on flow and other water quality parameters due to future climate scenarios was analyzed under four different GCMs output scenarios, namely CGCM3 A1B, CGCM3 A2, HadCM3 A2, HadCM3 B2 scenarios. Each GCMs output was further separated into three 30-year time frames: 20s (2011-2040), 50s (2041-2070), and 80s (2071-2100). For comparing results of climate change scenarios to the baseline, the monthly average of daily projected values over the period of 30-year time frame of 20s, 50s, and 80s were calculated and compared with the monthly average of daily simulated values over the baseline period (1981-2010). Land use (LU 2009) was kept constant for the baseline and climate change scenarios to analyze the effect of climate change only. Figs.10 to 13 show the monthly average of daily values of projected flow and water quality parameters under different future climate scenarios.

The monthly average of daily flow is projected to increase corresponding to CGCM3 A1B and B2 scenarios, whereas projection corresponding to HadCM3 A2 and B2 scenarios produces lower flows (
Fig. **10**–13), especially towards the end of 21st century (2080s). These projections are reflective of input precipitation pattern for the future climate scenarios (Fig. 5 and 6). For example, under CGCM3 A280s scenario (Fig. 11), projected flow is increased by 17% (July) to 256% (December) with average increase of 88%, while under HadCM3 A2 80s scenario (Fig. 12), projected flow is decreased by -31% (May) to -62% (December) with average decrease of -49%.

The surface water temperature is projected to increase, especially in summer. The increase in monthly average of daily surface water temperature reaches as high as 2.8°C and3.6°C for CGCM3 A1B and A2 scenarios (Figs. 10 and 11); and 5.4°C and 4.6°C for HadCM3 A2 and B2 scenarios, respectively (Figs. 12 and 13). The monthly average of daily surface DO will decrease by 1.8 mg/l and2.3 mg/l at most for CGCM3 A1B and A2 scenarios (Figs. 10 and 11); and 2.0 mg/l and 1.4 mg/l at most for HadCM3 A2 and B2 scenarios (Figs. 12 and 13), respectively. The decrease in DO concentration is partly because of projected increases in water temperature as dissolved oxygen solubility is inversely proportional to water temperature [62].

For CGCM3 A1B and A2 scenarios, nutrients are projected to decrease in the stream, whereas for HadCM3 A2 and B2 scenarios, nutrient concentrations are projected to increase. It corresponds to the flow decrease in HadCM3 A2 and B2 scenarios (Fig. 6) as low flow tends to have higher concentration of nutrients. The monthly average TP is projected to decrease by 41.7% and 38.4% at most for CGCM3 A1B and A2 scenarios (Figs. 10 and 11); and increase as high as 112% and 96% for HadCM3 A2 and B2 (Figs. 12 and 13), respectively. The monthly average TN will decrease by 40.6% and 36.8% at most for CGCM3 A1B and A2 scenarios (Figs. 10 and 11); and increase as high as 84.5% and 59.7% for HadCM3 A2 and B2 (Figs. 12 and 13), respectively. Precipitation variability projected by GCMs may result in higher nutrient runoff during heavy storms and nutrient concentration tends to increase during low flow condition as well [63].

The algal concentration is significant only during the growing season (April through October). For CGCM3 A1B and A2 scenarios, algal concentrations are projected to decrease in the whole summer because of projected decrease in nutrients (Figs. 10 and 11). Under the baseline scenario, algal concentration during the growing season is 18 μg/l (eutrophic) ranging from 1.8 (April) to 33.7 μg/l (September), and under CGCM3 A2 80s scenario

(Fig. 11), algal concentration during the growing season is projected to be 2.8 µg/l (oligotrophic) ranging from 0.2 to 5.5 µg/l. For HadCM3 A2 and B2 scenarios, algal concentrations are projected to increase most of the time except during late-summer of 2080s when it decreases (Figs. 12 and 13). Under HadCM3 A2 20s, 50s, and 80s scenarios, algal concentrations during the growing season are projected to be 24.1, 20.5, and 16.1 µg/l (Fig. 12), respectively. The algal growth rate increases with the increase in temperature up to the optimum level and then decreases with further increase in temperature [6].

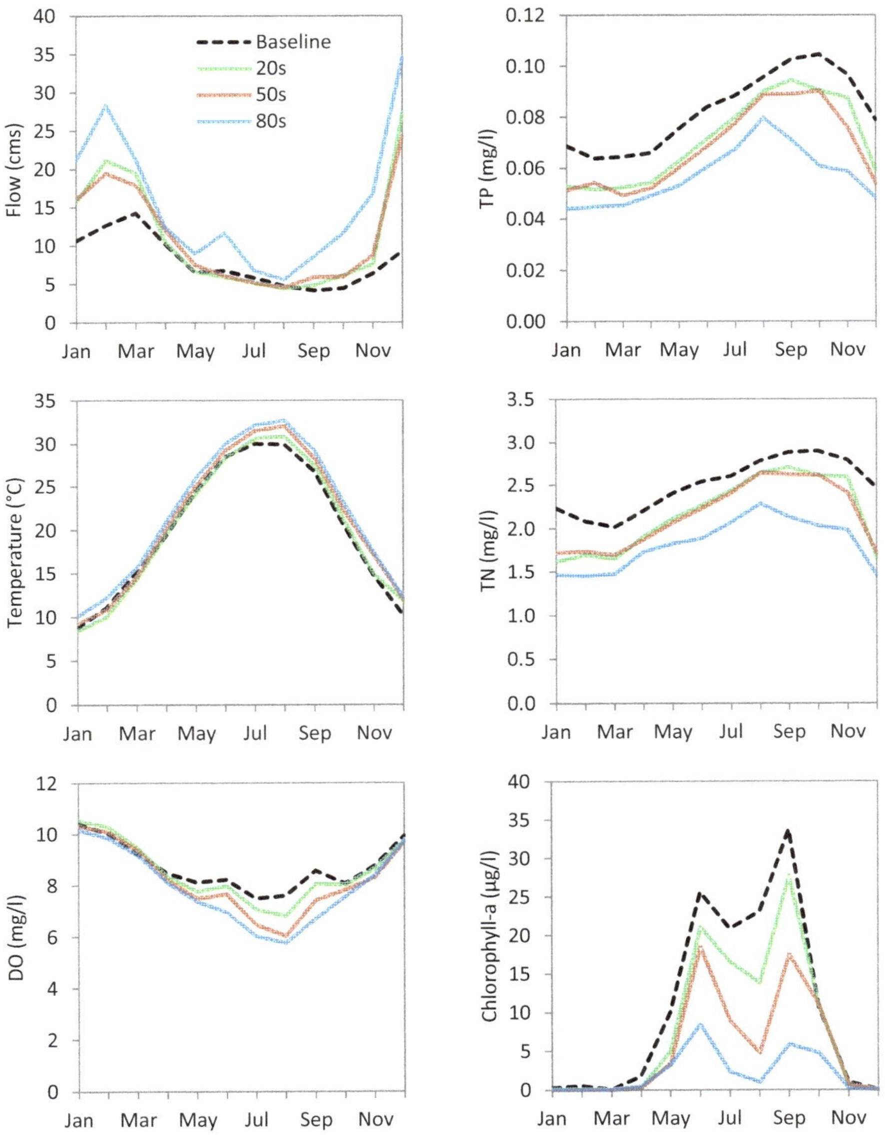

Fig. 10. Projected monthly average flow and water quality parameters under the baseline (1981–2010) and three climate change scenarios (20s, 50s, and 80s) downscaled with CGCM3 A1B

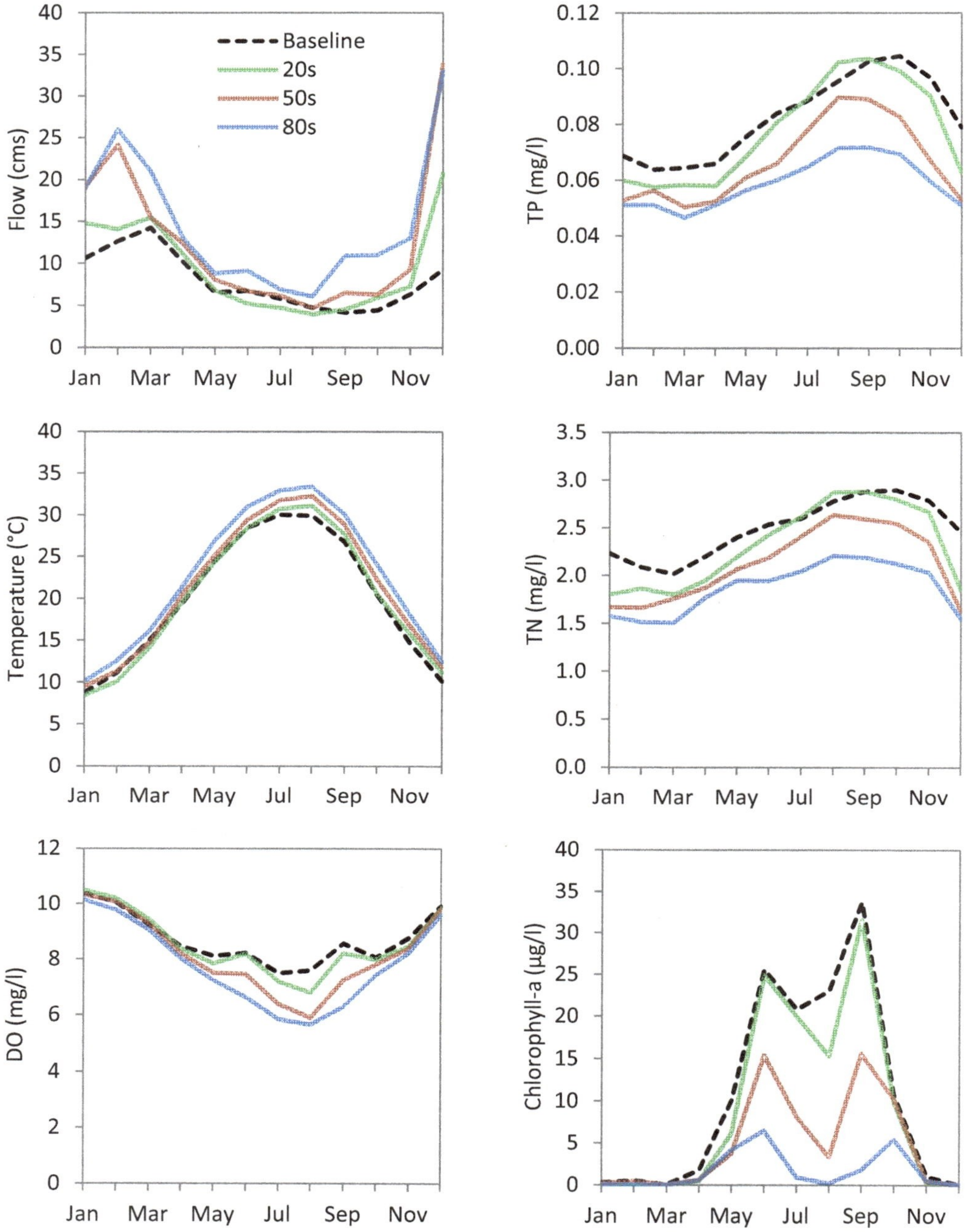

Fig. 11. Projected monthly average flow and water quality parameters under the baseline (1981–2010) and three climate change scenarios (20s, 50s, and 80s) downscaled with CGCM3 A2

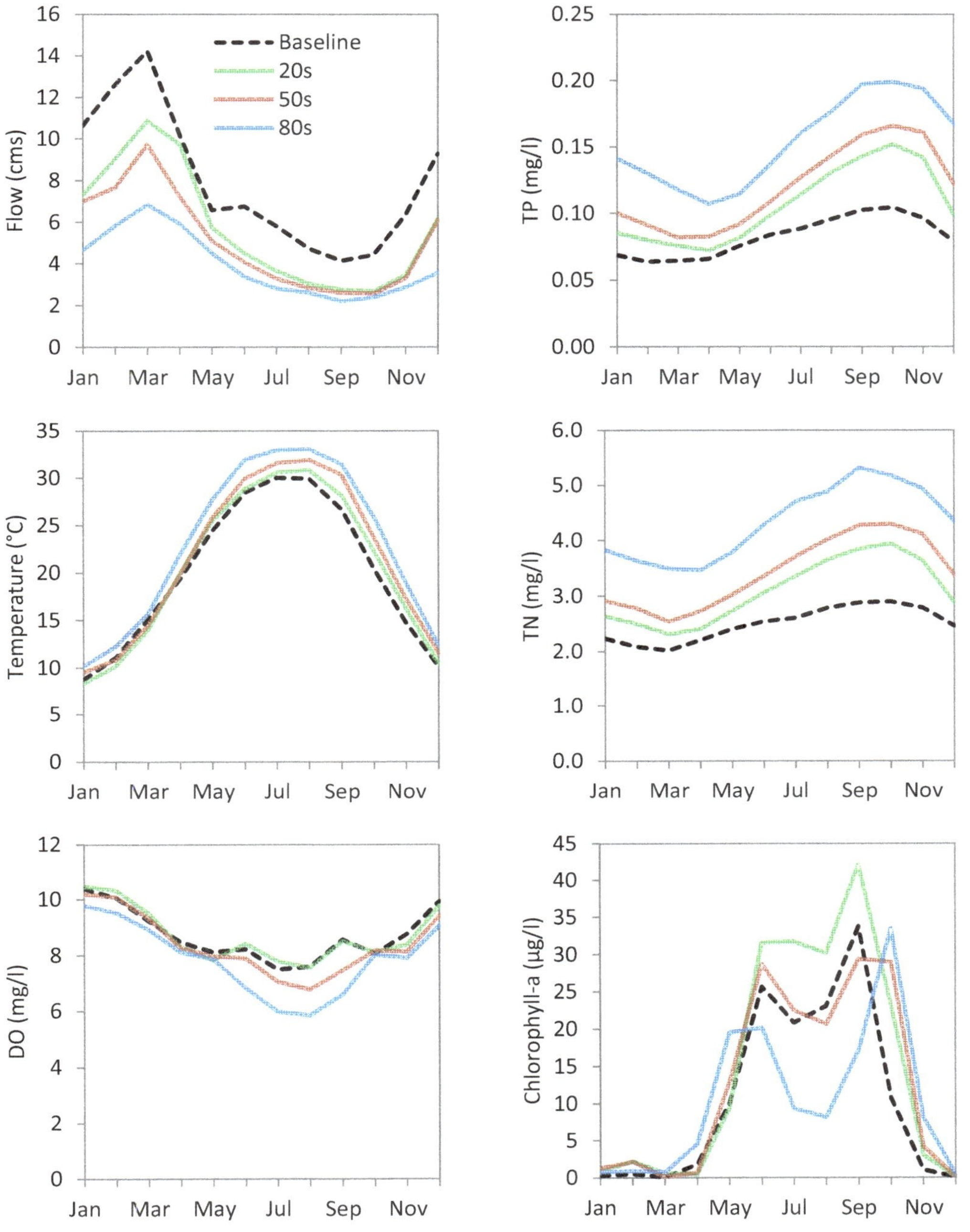

Fig. 12. Projected monthly average flow and water quality parameter under the baseline (1981–2010) and three climate change scenarios (20s, 50s, and 80s) downscaled with HadCM3 A2

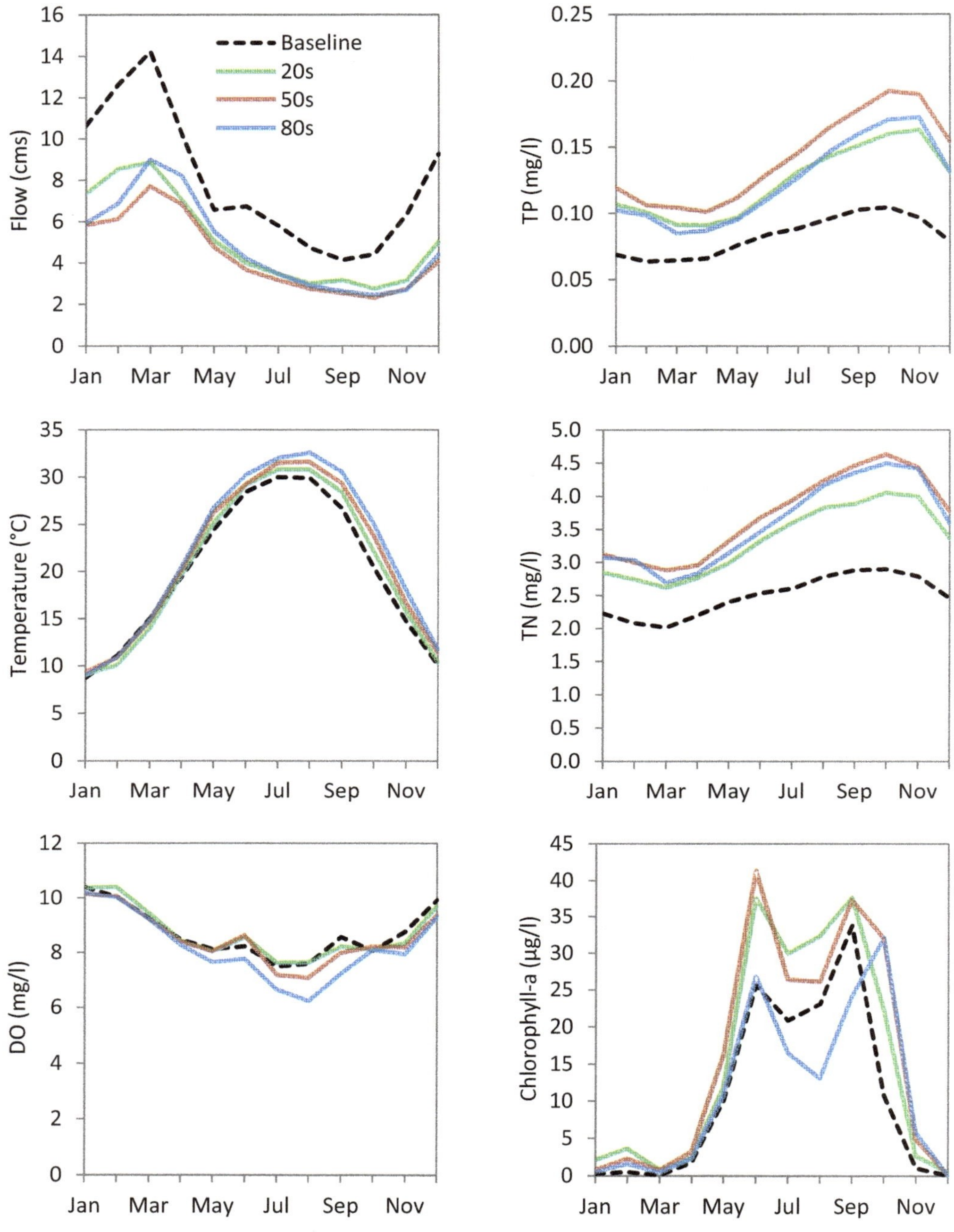

Fig. 13. Projected monthly average flow and water quality parameters under the baseline (1981–2010) and three climate change scenarios (20s, 50s, and 80s) downscaled with HadCM3 B2

3.2.3 Impact due to combined land use and climate change

For combined effect, the monthly average of daily projected values under the combined land use and climate change scenario (LU 2030 and 20s) were computed and compared with the monthly average of daily simulated values under the baseline scenario (LU 2009 and 1981-2010). Four different GCM output scenarios (CGCM3 A1B and A2; HadCM3 A2 and B2) were used for the analysis (Fig. 14).

The monthly average of daily flow is projected to increase by as high as 194.2% and 125.7% under CGCM3 A1B and A2 scenarios; and decrease by 35% and 40% at most under HadCM3 A2 and B2 scenarios, respectively (Fig. 14). The surface water temperature, in terms of monthly average of daily values, is expected to increase by as high as 17.5% and 11.4% under CGCM3 A1B and A2 scenarios; and increase as high as 9.8% and 8.5% under HadCM3 A2 and B2 scenarios; respectively (Fig. 14). The combined effect deteriorates oxygen levels in the streams and reservoirs. The average monthly DO is projected to decrease slightly under all four scenarios, with 9.6% or 0.7 mg/l at most for CGCM3 A2 scenario (Fig. 14).

The monthly average TP and TN concentration is projected to increase due to combined effect. The annual average TP for the baseline and four scenarios (CGCM3 A1B, CGCM3 A2, HadCM3 A2, and HadCM3 B2) are 0.082, 0.108, 0.123, 0.177, and 0.203 mg/l, respectively. The increase in TP and TN concentration is higher under HadCM3 compared to CGCM3 scenarios. Chlorophyll-a concentration during the growing season is expected to increase to 25.8 and 26.3 µg/l under HadCM3 A2 and B2 scenarios due to combined effect, respectively (Fig. 14).

4. SUMMARY AND CONCLUSION

In this paper, a physically-based dynamic watershed model, WARMF, was applied to assess the impact on hydrology and water quality in streams and reservoirs of the Saugahatchee Creek Watershed in Alabama, USA, due to land use change and climate change. The model was calibrated for flow using six years (2000–2005) of stream flow data from USGS gage station and the flow validation was conducted for the period 2006–2009. Water quality calibration and validation was performed for the period 2000–2002 based on available data. By employing various drivers of LULC change, future land use scenario of 2030 is projected using multiple logistic regression analysis after the land use model was validated over the period 1991 to 2009. Future climate scenarios derived from CGCM3 and HadCM3 for 30-year future period of 20s (2010–2040), 50s (2040–2070), and 80s (2070–2100), were downscaled to local scale using SDSM downscaling technique and compared to baseline scenarios of 1981-2010. The response of watershed model to these different land use and climate change scenarios was analyzed by comparing the monthly average of daily projected values of flow and water quality for 30-year period.

Land use change scenario for 2030projects forest areas will be reduced to 71.4% from 83.3% in 2009 and urban areas will be increased to 17.5% from 3.4% in 2009 in the watershed (Table 1). For the effects of land use change only, the simulation over the period of 30 years (1981–2010) predicts increases in flow to be 12.6% or less whereas increase in TP concentration to be up to 71.7% for LU 2030 in comparison to the baseline scenario LU 2009. These results demonstrate that land use change has major impact on nutrient concentration but less significant impact on flow. The algal increase up to 4.4 µg/lis expected during the growing seasons in the streams and reservoirs due to increased level of nutrients.

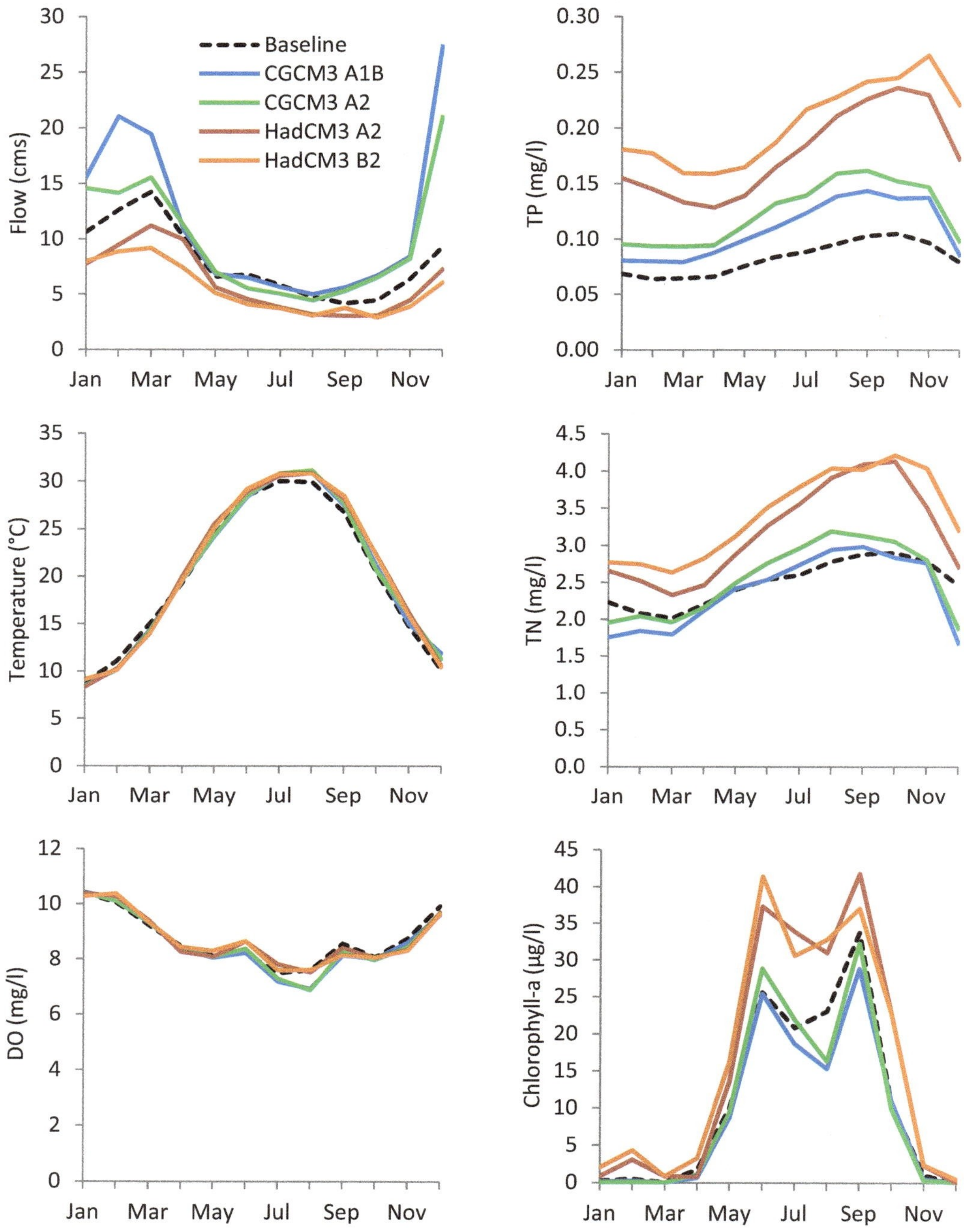

Fig. 14. Projected monthly average flow and water quality parameters under the baseline (1981–2010 and LU 2009) and combined land use change and climate change scenario (20s and LU 2030) downscaled with CGCM3 A1B and A2, and HadCM3 A2 and B2

Future climate scenarios were obtained by statistically down scaling the derived outputs from CGCM3 and HadCM3 models. Climate change scenarios CGCM3 A1B, CGCM3 A2, HadCM3 A2, and HadCM3 B2 show increasing pattern of maximum and minimum temperatures in 20s, 50s, and 80s. The anomalies of air temperature rise in these future climate scenarios to the baseline scenario (1981–2010) are higher especially in summer (up to 6.0°C in July from CGCM3 A2 80s). The watershed response to these changes without considering land use change, based on monthly average of daily values, will result rise in water temperature ranging from 2.8 to 5.4°C, especially during summer which is obvious corresponding to the rise in air temperatures. The monthly average of daily flow is predicted to increase corresponding to CGCM3 (annual average increase of 88%) and decrease corresponding to HadCM3 scenarios (annual average decrease of -49%). Accordingly, nutrient concentration (TP and TN) is expected to decrease corresponding to CGCM3 (Figs. 10 and 11) and increase corresponding to HadCM3 (Figs. 12 and 13) scenarios. The results show contrasting watershed response on flow and nutrient concentration based on which GCM is selected for future climate scenarios. DO concentration are predicted to fall up to 2.3 mg/l (monthly average), especially in summer for the four climate scenarios.

Combined effect due to land use change and climate change adds more to increases in nutrient concentrations underHadCM3 A2 and B2scenarios as both land use change and climate change cause nutrient concentration to increase. For CGCM3 A1B and A2 scenarios, nutrient concentrations are shifted from relatively decreasing under climate change only effect to increasing under combined effect. The annual average TP for four scenarios (CGCM3 A1B, CGCM3 A2, HadCM3 A2, and HadCM3 B2) are projected to be 0.108, 0.123, 0.177, and 0.203 mg/l, respectively, comparing 0.082 mg/l for the baseline. The average monthly DO is projected to decrease slightly under all four scenarios (up to 9.6% or 0.7 mg/l decrease for CGCM3 A2 scenario, Fig. 14).

Flow, temperature, nutrients concentration, algae, and dissolved oxygen concentration, all interact with each other in a complex watershed system. The impact study under different scenarios of land use and climate change using advanced watershed model gives us better understanding about how management alternatives can be launched during watershed planning.

CONSENT

Not applicable.

ETHICAL APPROVAL

Not applicable.

ACKNOWLEDGEMENTS

This study was partially funded through the Auburn Water Research Center. We would like to acknowledge anonymous reviewers for providing constructive comments that helped us to improve the manuscript.

COMPETING INTERESTS

Authors have declared that no competing interests exist.

REFERENCES

1. Kundzewicz ZW, Mata LJ, Arnell NW, Doll P, Kabat P, Jimenez B, et al. Freshwater resources and their management, in Climate change: Impacts, adaptation and vulnerability: contribution of working group II to the fourth assessment report of the intergovernmental panel on climate change, Parry ML, Canziani OF, Palutikof JP, Linden PJvd, and Hanson CE, Eds. Cambridge, UK: Cambridge Univ Press. 2007;173-210.
2. Chang LH, Railsback SF, Brown RT. Use of reservoir water quality model to simulate global climate change effects on fish habitat. Climatic Change. 1992;20:277-296.
3. Fang X, Stefan HG, Eaton JG, McCormick JH, Alam SR. Simulation of thermal/dissolved oxygen habitat for fishes in lakes under different climate scenarios: Part 3. Warm-water fish in the contiguous US. Ecological Modelling. 2004;172(1):55-68.
4. Stefan HG, Fang X, Eaton JG. Simulated fish habitat changes in North American lakes in response to projected climate warming. Transactions of the American Fisheries Society. 2001;130:459-477.
5. Sinokrot B, Stefan HG, McCormick MJ. Modeling of climate change effects on stream temperatures and fish habitats below dams and near groundwater inputs. Climatic Change. 1995;30:181-200.
6. Chen CW, Herr JW, Weintraub L. Watershed Analysis Risk Management Framework (WARMF): Update One–A decision support system for watershed analysis and total maximum daily load calculation, allocation and implementation. Publication No. 1005181. Electric Power Research Institute, Palo Alto, California; 2001.
7. Herr J, Weintraub L, Chen C. User's guide to WARMF: Documentation of graphical user interface. EPRI. Palo Alto, CA. Report EP-P2346/ C. 2000;1054.
8. IPCC. Summary for policymakers, in Climate change 2007: The physical science basis. Contribution of working group I to the fourth assessment report of the intergovernmental panle on climate change, Solomon S, Qin D, Manning M, Chen Z, Marquis M, Averyt KB, Tignor M, Miller HL, Eds., 1st ed. Cambridge, United Kingdom and New York, NY, USA: Cambridge University Press; 2007.
9. Samuels R, Rimmer A, Hartmann A, Krichak S, Alpert P. Climate change impacts on Jordan River flow: Downscaling application from a regional climate model. Journal of Hydrometeorology. 2010;11(4):860-879.
10. Wigley T, Jones P, Briffa K, Smith G. Obtaining sub-grid-scale information from coarse-resolution general circulation model output. Journal of Geophysical Research. 1990;95(2):1943-1953.
11. Wilby RL, Dawson CW. SDSM 4.2–A decision support tool for the assessment of regional climate change impacts. Lancaster University, UK; 2007.
12. Breuer L, Huisman JA, Willems P, Bormann H, Bronstert A, Croke BFW, et al. Assessing the impact of land use change on hydrology by ensemble modeling (LUCHEM). I: Model intercomparison with current land use. Advances in Water Resources. 2009;32(2):129-146.
13. Choi JY, Engel BA, Muthukrishnan S, Harbor J. GIS based long term hydrologic impact evaluation for watershed urbanization. Journal of the American Water Resources Association. 2003;39(3):623-635.
14. Choi W, Deal BM. Assessing hydrological impact of potential land use change through hydrological and land use change modeling for the Kishwaukee River basin (USA). Journal of Environmental Management. 2008;88(4):1119-1130.

15. Kim Y, Engel BA, Lim KJ, Larson V, Duncan B. Runoff impacts of land-use change in Indian River Lagoon watershed. Journal of Hydrologic Engineering. 2002;7(3):245-251.
16. Bhattarai G, Srivastava P, Marzen L, Hite D, Hatch U. Assessment of economic and water quality impacts of land use change using a simple bioeconomic model. Environmental Management. 2008;42(1):122-131.
17. Ouyang T, Zhu Z, Kuang Y. Assessing impact of urbanization on river water quality in the Pearl River Delta Economic Zone, China. Environmental Monitoring and Assessment. 2006;120(1):313-325.
18. Schoonover JE, Lockaby BG, Pan S. Changes in chemical and physical propertiesof stream water across an urban-rural gradient in western Georgia. Urban Ecosystems. 2005;8(1):107-124.
19. Sliva L, Williams DD. Buffer zone versus whole catchment approaches to studying land use impact on river water quality. Water Research. 2001;35(14):3462-3472.
20. Tu J, Xia ZG, Clarke KC, Frei A. Impact of urban sprawl on water quality in eastern Massachusetts, USA. Environmental Management. 2007;40(2):183-200.
21. Chang H, Knight CG, Staneva MP, Kostov D. Water resource impacts of climate change in southwestern Bulgaria. Geo Journal. 2002;57(3):159-168.
22. Gleick PH, Chalecki EL. The impacts of climatic changes for water resources of the Colorado and Sacramento San Joaquin River Basins. Journal of the American Water Resources Association. 1999;35(6):1429-1441.
23. Liu L, Liu Z, Ren X, Fischer T, Xu Y. Hydrological impacts of climate change in the Yellow River Basin for the 21st century using hydrological model and statistical downscaling model. Quaternary International. 2011;244:211-220.
24. Novotny EV, Stefan HG. Stream flow in Minnesota: Indicator of climate change. Journal of Hydrology. 2007;334(3-4):319-333.
25. Xu ZX, Zhao FF, Li JY. Response of streamflow to climate change in the headwater catchment of the Yellow River basin. Quaternary International. 2009;208(1-2):62-75.
26. Zhang X, Srinivasan R, Hao F. Predicting hydrologic response to climate change in the Luohe River basin using the SWAT model. Transactions of the ASABE. 2007;50(3):901-910.
27. Dibike YB, Coulibaly P. Hydrologic impact of climate change in the Saguenay watershed: comparison of downscaling methods and hydrologic models. Journal of Hydrology. 2005;307(1-4):145-163.
28. Rich PM, Weintraub LH, Chen L, Herr J. Climate change impacts on hydrology and water management of the San Juan Basin. American Geophysical Union, Fall Meeting. 2005;1054.
29. Cruise JF, Limaye AS, Al-Abed N. Assessment of impacts of climate change on water quality in the southeastern United States. Journal of the American Water Resources Association. 1999;35(6):1539-1550.
30. Bouraoui F, Galbiati L, Bidoglio G. Climate change impacts on nutrient loads in the Yorkshire Ouse catchment (UK). Hydrology and Earth System Sciences. 2002;6(2):197-209.
31. Neff R, Chang H, Knight CG, Najjar RG, Yarnal B, Walker HA. Impact of climate variation and change on Mid-Atlantic Region hydrology and water resources. Climate Research. 2000;14(3):207-218.
32. Tu J. Combined impact of climate and land use changes on streamflow and water quality in eastern Massachusetts, USA. Journal of Hydrology. 2009;379(3-4):268-283.
33. Wang R. Modeling hydrologic and water quality responses to changing climate and land use/cover in the Wolf Bay Watershed, South Alabama. M.S. thesis, Auburn University, Auburn, AL; 2010.

34. Butcher JB, Parker A, Johnson T, Weaver CP. Nationwide watershed modeling to evaluate potential impacts of climate and land use change on hydrology and water quality. Presented at the Proceedings of the 2010 Watershed Management Conference, Madison, WI; 2010.
35. Fang X, Jiang L, Jacobson PC, Stefan HG, Alam SR, Pereira DL. Identifying cisco refuge lakes in Minnesota under future climate scenarios. Transactions of the American Fisheries Society. 2012;141(6):1608-1621.
36. ADEM. Final total maximum daily load: Nutrients & OE/DO. Pepperell branch AL03150110-0201-700 Nutrients; Sougahatchee Creek Embayment (Yates Reservoir) AL03150110-0204-101 Nutrients & OE/DO. Alabama Department of Environmental Management (ADEM), Montgomery, AL; 2008.
37. Young RA. AGNPS, agricultural non-point-source pollution model: A watershed analysis tool. Department of Agriculture, Agricultural Research Service, National Agricultural Library . Document Delivery Services Branch; 1987.
38. USEPA. Better Assessment Science Integrating point and Nonpoint Sources (BASINS) Version 3.1: User's Manual. United States Environmental Protection Agency EPA-823-C-04-004; 2004.
39. Johanson RC, Imhoff JD, Davis HH, Jr. Users manual for hydrological simulation program - Fortran (HSPF). Environmental Research Laboratory, US Environmental Protection Agency, Athens, GA; 1980.
40. Haith DA, Shoenaker LL. Generalized watershed loading functions for stream flow nutrients. JAWRA Journal of the American Water Resources Association. 1987;23(3):471-478.
41. Arnold JG, Soil G. SWAT (Soil and Water Assessment Tool). Grassland, Soil and Water Research Laboratory, USDA, Agricultural Research Service; 1994.
42. Chen CW, Gherini SA, Hudson R, Dean J. The integrated lake-watershed acidification study. Model principles and application procedures. Electric Power Research Institute report EA-3221, Palo Alto, California, USA. 1983;1.
43. Gherini S, Mok L, Hudson R, Davis G, Chen C, Goldstein R. The ILWAS model: formulation and application. Water, Air & Soil Pollution. 1985;26(4):425-459.
44. Beasley DB, Huggins LF, Monke EJ. Answers: A model for watershed planning. Trans of the ASAE. 1980;23(4):938-944.
45. Beasley DB, Huggins LF. ANSWERS, areal nonpoint source watershed environment response simulation: User's manual. Department of Agricultural Engineering, Purdue University, West Lafayette, IN; 1981.
46. Huber WC, Dickinson RE, Barnwell Jr TO. Storm water management model; version 4. Environmental Protection Agency, United States; 1988.
47. Ambrose R, Wool T, Martin J. WASP 5, The water quality analysis simulation program version 5.00. ASCI Corporation, Athens, Georgia; 1993.
48. Systech Engineering Inc. Creating a WARMF 6.1 application using a BASINS 3.1 delineation. A User's Guide; 2005.
49. Shrestha S. Watershed response to land use and climate change in the Saugahatchee Creek Watershed using the WARMF model. Thesis for the Master of Science, Department of Civil Engineering, Auburn University, Auburn, Alabama; 2011.
50. AWW. Saugahatchee Creek Watershed: past, present, and future. Saugahatchee Creek Watershed management plan. Alabama Water Watch (AWW), Auburn University, Alabama 36849; 2005.
51. ERDAS. ERDAS Field Guide, Sixth edition. Erdas LLC, Atlanta GA. 2002;657.
52. Sawant R. Land use land cover change projection for use in municipal water resource planning in the saugahatchee watershed. Thesis for the Master of Science, Department of Geology and Geography, Auburn Unviersity; 2012.

53. Kalnay E, Kanamitsu M, Kistler R, Collins W, Deaven D, Gandin L, et al. The NCEP/NCAR 40-year reanalysis project. Bulletin of the American Meteorological Society. 1996;77(3):437-471.
54. Tatli H, Dalfes N, Mentes S. A statistical downscaling method for monthly total precipitation over Turkey. International Journal of Climatology. 2004;24(2):161-180.
55. Keller A. Peer review of the Watershed Analysis Risk Management Framework (WARMF) – An evaluation of WARMF for TMDL applications by independent experts using USEPA guidelines; 2000.
56. Nash JE, Sutcliffe JV. River flow forecasting through conceptual models part I - a discussion of principles. Journal of Hydrology. 1970;10(3):282-290.
57. Moriasi DN, Arnold JG, Van Liew MW, Bingner RL, Harmel RD, Veith TL. Model evaluation guidelines for systematic quantification of accuracy in watershed simulations. Transactions of the ASABE. 2007;50(3):885-900.
58. Ansa-Asare OD, Marr IL, Cresser MS. Evaluation of modelled and measured patterns of dissolved oxygen in a freshwater lake as an indicator of the presence of biodegradable organic pollution. Water Research. 2000;34(4):1079-1088.
59. Carlson RE. A trophic state index for lakes. Limnology Oceanography. 1977;22(2):361-369.
60. O'Connor DJ, Di Toro DM. Photosynthesis and oxygen balance in streams. Journal of the Sanitary Engineering Division. 1970;96(2):547-571.
61. Odum HT. Primary production in flowing waters. Limnology and Oceanography. 1956;1(2):102-117.
62. APHA. Standard methods for the examination of water and wastewater, 18 ed. American Public Health Association (APHA), Washington, DC; 1992.
63. Weaver LT, Fuller ML. Stream-water quality during storm-runoff events and low-flow periods in the St. Clair River/Lake St. Clair Basin, Michigan. U.S. Geological Survey, U.S. Department of the Interior, Open-File Report. 2007;1201.

7

Climate Impact on Freshwater Biodiversity: General Patterns in Extreme Environments of North-Eastern Siberia (Russia)

Sophia Barinova[1*], Viktor Gabyshev[2] and Olga Gabysheva[2]

[1]Institute of Evolution, University of Haifa, 199 Aba Khoushy Ave., Mount Carmel, Haifa, 3498838, Israel.
[2]Institute for Biological Problems of Cryolithozone Sb Ras, 41 Lenin Avenue, Yakutsk 677980, Russia.

Authors' contributions

This work was carried out in collaboration between all authors. Author SB designed the study, performed the statistical analysis, wrote the protocol and wrote the first draft of the manuscript. Authors VG and OG managed the analyses for the study. All authors read and approved the final manuscript.

ABSTRACT

Aims: The aims of the current study are to reveal the response of high latitude riverine planktonic algal communities in northeastern Siberia to extreme climatic conditions of its habitats.
Study Design: We implemented diverse statistical methods, which represent some new approaches in freshwater algal diversity analysis.
Place and Duration of Study: Institute of Evolution, University of Haifa, Israel, Institute for Biological Problems of Cryolithozone SB RAS, Russia, between June 2008 and January 2014.
Methodology: We collected 800 samples of phytoplankton from 400 sites of 12 northeastern Siberian rivers in gradients of climatic and chemical variables that we analyzed. New indices - Geo-associated and Dynamic Habitat Index were included in this

**Corresponding author: Email: barinova@research.haifa.ac.il;*

analysis. Statistical methods for comparative floristic analyses were used for calculating the similarity of algal communities among the sampling stations. Multiple regression stepwise statistical analysis on phytoplankton including chemical and climatic variables data was performed. Species diversity in algal communities and their environmental variables relationships were calculated.

Results: As a result, 1283 species (1637 taxa of species and infraspecies) from six taxonomic divisions were identified in phytoplankton communities. Species richness as a whole increased to the north. Abundance and biomass were highly correlated. Two types of phytoplankton communities were identified: a southern community with increasing diatoms and a northern group with decreasing diatoms to the north. Diatoms prevailed but were replaced by green algae in high mountains or by green and *Chrysophyta* algae and *Cyanobacteria* in the Arctic. We revealed major variables that considered stimulating or stress factors with helps of statistical prorgams.

Conclusion: Statistical analyses of phytoplankton in 12 large rivers revealed an increase in species richness to the north with community structure changing under stimulation of air temperature, ice-free periods, humidity, and trophic variables were stimulants and water transparency and speed flow were considered stress factors.

Keywords: Freshwater algae; diversity; ecology; climate adaptation; Northern Siberia.

1. INTRODUCTION

Relationships between freshwater algal biodiversity and environmental conditions can be determined as adaptation levels between the species and the community as a whole. Bio-indication and related integrative methods for aquatic environment sustainable assessment are based on the principle of congruence between community composition and the complexity of environmental factors. In any case, freshwater biodiversity is still a much underestimated component of global biodiversity [1] but can be estimated using diverse methods to measure major climatic variables [2,3]. However, there are still many problems in defining the role of climatic factors in predicting the community's response to environmental changes. In an analysis of freshwater algae diversity we encountered certain difficulties. First, it is difficult to determine the scope of communities involved, as well as to define the scope of research tasks and the relevance of operative approaches [4]. The effects of major climatic variables, such as temperature and altitude, on freshwater algae distribution is widely discussed in recent literature [5-13] but still remains a problem. We planned this study because there is a lack of detailed distributional information for most freshwater organism groups and an absence of distribution-climate models; therefore, studies should be aimed at increasing our knowledge about these aspects of the ecology of freshwater organisms [14]. We know only one relevant study for bio-diversity assessment over large gradient of latitude in South America [15] but it has been do on the lake secosystems. Many studies about planktonic algal communities in boreal lakes has been do for revealing gamma-diversity response to climatic variables [16] and importance of phytoplankton studies [17] with conclusion that diversity is the best predictor for resource use efficiency of phytoplankton communities across considerable environmental and climatic [18] gradients. Riverine communities studies under Far East climatic gradient up to now has been in initial stage [19].

The freshwater algal communities' comparisons experience allows us to use a productive approach when there are visible climatic responses in the ecoregional, riverine basin floras, and infra-specific generalization levels for comparison.

The aims of the current study are to reveal the response of high latitude riverine planktonic algal communities in northeastern Siberia to extreme climatic conditions of its habitats. Thus, we tried to implement diverse statistical methods, which represent some new approaches in freshwater algal diversity analysis.

2. MATERIALS AND METHODS

2.1 Sampling and Laboratory Studies

Samples for the study were collected in the summer low water season of June-August 2008 – 2011 from 12 large East Siberian Rivers: Anabar, Olenek, Indigirka, Yana, Kolyma, Aldan, Amga, Vilyui, Lena, Vitim, Olekma, and Chara (Fig. 1). The study areas are located on territories of Yakutiya, Buryatiya, Zabaykalskiy Kray and Magadanskaya, Amurskaya, and Irkutskaya Oblast.

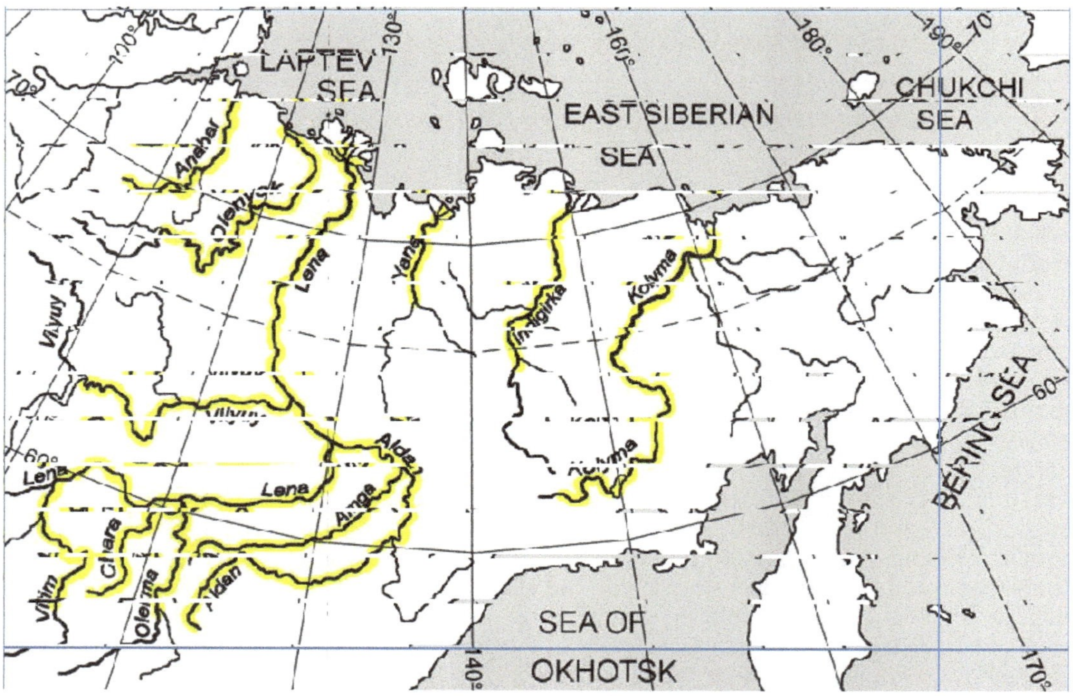

Fig. 1. Map of the studied (yellow) rivers in north-eastern Siberia

A total of 303 water samples and 800 phytoplankton samples were collected for hydrochemical and algological analyses, respectively. Some samples were collected from the littoral and others from the fairway of the rivers (from the 0 – 0.3 m depth in both cases).

Samples for chemical analysis were taken in 2.5 L dark glass bottles and transported to the laboratory during two days in an ice box with 4°C. Water chemical analyses followed standard methods set out in [20,21]. Gas regime components (CO_2, O_2, and BOD_5) and Secchi disk depth, were determined in situ. Concentrations of other chemicals were determined in the laboratory.

One and a half liter water samples for phytoplankton quantitative analyses were concentrated on Sartorius membrane filters (pore size 1.2 µm) by pressure filtration, using the phytoplankton concentrator of our own design [22]. Samples for qualitative analyses were collected with an Apstein plankton net (SEFAR NITEX filter fabric, mesh size 30µm) and fixed in 4% neutral formaldehyde solution. Microscopy was performed with an Olympus BH2 microscope under magnification 600 – 1000. Algal cells were counted in a Nageotte counting chamber (volume 0.01 cm^3). Biovolumes of different taxa were calculated by approximating the cell shape to simple geometrical shapes, using our own linear measurements of cells. The volume to specific weight conversion factor was assumed to be a unity. Diatom shells were studied in 2500 permanent slides with Bio mount media. No less than 100 cells of each abundant species were calculated from each quantitative sample.

2.2 Taxonomic Analysis and Functional Classification

For taxonomic identification a common international handbook series were used. Modern species' names in our work come from Algaebase [23], employing the common system nomenclature derived from Cavalier-Smith [24].

The mean phytoplankton community cell size was estimated from the ratio between total biovolume and total abundance according to [25].

The Shannon's diversity index [26], which reflects the degree of abundance equality among the species in the community and is correlated to the entropy of the ecosystem [27], was calculated as:

$$\bar{H} = -\sum_{i=1}^{s} \frac{n_i}{N} \log_2 \frac{n_i}{N} \qquad (1)$$

Where: H – Shannon species diversity index, bit, N – common individuals abundance, s – species number; n_i – the number of individuals in each species.

2.3 Statistical Analysis

Statistical methods for comparative floristic analysis were used for calculating the similarity of algal communities among the sampling stations on the basis of Sørensen's similarity indices that automatically calculated with GRAPHS Program [28].

Statistical analysis of correlation between species diversity and major climatic condition variables was calculated by distance-weighted least squares using the Statistica 7.1 Program.

Multiple regression Stepwise statistical analysis on phytoplankton, including chemical and climatic variables' data, was performed via the Statistica 7.0 software in order to determine the variables with strongest influence on the algal communities in studied rivers.

2.4 Description of Study Site

Eastern Siberia is the Asian territory of the Russian North-East, from the Yenisei basin to the Chukotsky peninsula [29]. Our study area with a large amplitude of latitude, altitude, temperature, and precipitation range [30] (Fig. 1) is bounded on the north by seas Laptev and East Siberian, on the west by the watershed of the Central Siberian Plateau, from the south by the mountain plateau of Stanovoy Ridge, from the southeast by the Yudoma-Maisky highlands and southern spurs of the Verkhoyansk Range, and from the east by the Kolyma Range [31]. Studied rivers are from the Lena 608 and Kolyma 609 ecoregions in the Freshwater Ecoregions of the World (FEOW) [32].

The territory of the region is represented by all forms of relief: mountains, plateaus, intermountain valleys (basins) and lowlands, and are distributed over two natural geographic zones – arctic and subarctic, and three natural zones - tundra, forest tundra and taiga, and mountain landscapes. Eastern Siberia is characterized by almost universal distribution of permafrost - continuous (integral) in the northern latitude of Vilyui River and intermittent in the southern half of the area. Power permafrost in the central part of the region is 350 – 450 m, the maximum depth found in the river basin Olenyok and reach 1,500 m [33]. In the south of the region power permafrost decreases; there are more or less significant areas devoid of permafrost (taliks) [34]. A climatic condition of Eastern Siberia is largely determined by its geographical location within the Asian continent and is protected by mountains from the Atlantic and Pacific Oceans. The main features of the climate of Eastern Siberia – a clear, severe, with little snow, stable and long winters and fairly dry, short, and hot summers. The studied water bodies represent twelve of the largest rivers of Eastern Siberia and belong to the basin of the Arctic Ocean. Our study area extends in a longitudinal direction of 106°53'E to 160°58'E, and in latitude – from 56°13' to 73°10'N. The total length of all studied rivers sites reaches more than 17,000 km.

2.5 Floristic Materials

The regional freshwater algal flora is represented in the main monographs of Vassilyeva-Kralina et al. [35,36] and Vasilyeva et al. [37]. Our papers are focused on the floristic diversity and community structure of phytoplankton in twelve major rivers of Eastern Siberia [38-48]. Freshwater algal diversity and ecology under anthropogenic impact study are represented in [49-52].

2.6 Environmental Data

Environmental and Dynamic Habitat Index, DHI [53] data we compiled from the Institute for Biological Problems of Cryolithozone SB RAS, Yakutsk meteorological data base (see Appendix as supplementary file). The DHI index is an important variable which correlated with the latitude of the sampling sites is calculated as a result of landscape variation and sensitivity to climate change for the North America from central part of U.S.A., Canada and to the Polar. Climatic parameters have been taken from http://worldclim.org/. Geo-association classes of the river basin placement are created by us as:

1 – Large transitional rivers;
2 – Rivers flows mainly in the latitudinal direction;
3 – River located in the south of the region;
4 – Rivers of the central part of the region;
5 – Most of the river basin is located north of the Arctic Circle;
6 – River basin is located north of the Arctic Circle.

3. RESULTS

We revealed 1,283 species (1,637 taxa of species and infraspecies) of algae and cyanobacteria in 800 samples collected during 2008 – 2011 from 12 rivers and 400 sampling stations (See Appendixas supplementary file). The *Bacillariophyta* prevail in six taxonomic divisions. Species richness varied from 9 – 146 taxa in different stations and represent diverse algal communities even in each studied river.

Studied rivers have mostly flow in a northerly direction. They are covered with ice most of the year (See Appendix as supplementary file) with ice free days from 112 (Anabar) to 149 (Amga), and the average water temperature in summer ranges from 3.8 to 23°C. Speed of the flow varies from 0.28 to 3.61 m sec^{-1}, and water transparency from 0.04 to 4.50 m. Waters are mostly alkaline with pH from 6.68 to 9.28, fresh (TDS 13–452 mg L^{-1}), well oxygenated (6.0 – 13.6 mg O_2 L^{-1}) with CO_2 saturation of 0-8.8 mg L^{-1}. Nutrients concentration varied in nitric nitrogen (0 – 2.16 mg L^{-1}), phosphorous (0-0.36 mg L^{-1}), Si (0.4 – 4.63 mg L^{-1}), and Fe (0 – 3.0 mg L^{-1}) in quantity enough for algal biomass development. The elements that defined water quality such as ammonia (0.020 – 1.88 mg L^{-1}), nitrites (0 – 0.034 mg L^{-1}), and BOD (0.04 – 4.94 mg L^{-1}) reflect wide range of water quality but cannot be interpreted as unfavorable for survival of algae.

Therefore, we can analyze major biological variables such as species richness, abundance, and biomass changing over stations in a northern direction to reveal algal preferences in extreme climatic zones.

Climatic characteristics of the region form the system of complex interactions. Precipitation correlated with DHI (Pearson 0.60, *P*<.01), decreasing to the north, but their abundance is also associated with the number of days of open water, that is, with the temperature and light intensity (Fig. 2). Maximal values of rainfall are in the southern region, where the ice-free period is longer as well as in the northern sites where there are more days of open water. The DHI index is an important variable which correlated with the latitude of the sampling sites. It can be seen that DHI is lower for the northern habitats (Fig. 3).

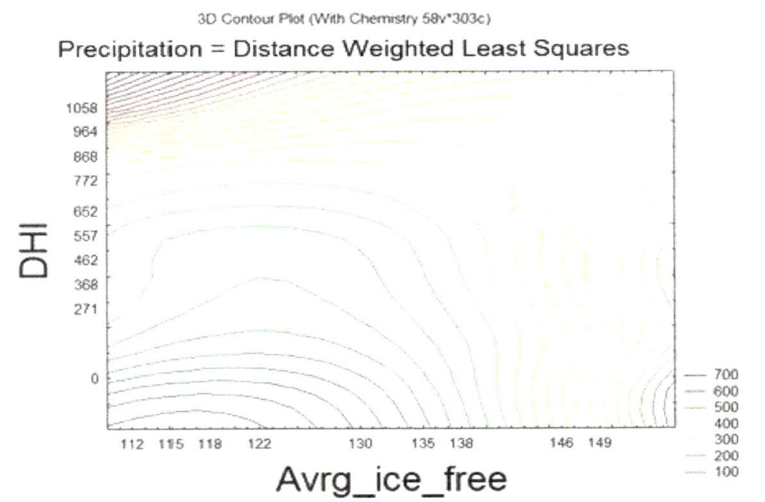

Fig. 2. Plot of precipitation distribution over the ice-free periods (Avrg_ice_free) and Dynamic Habitat Index (DHI)

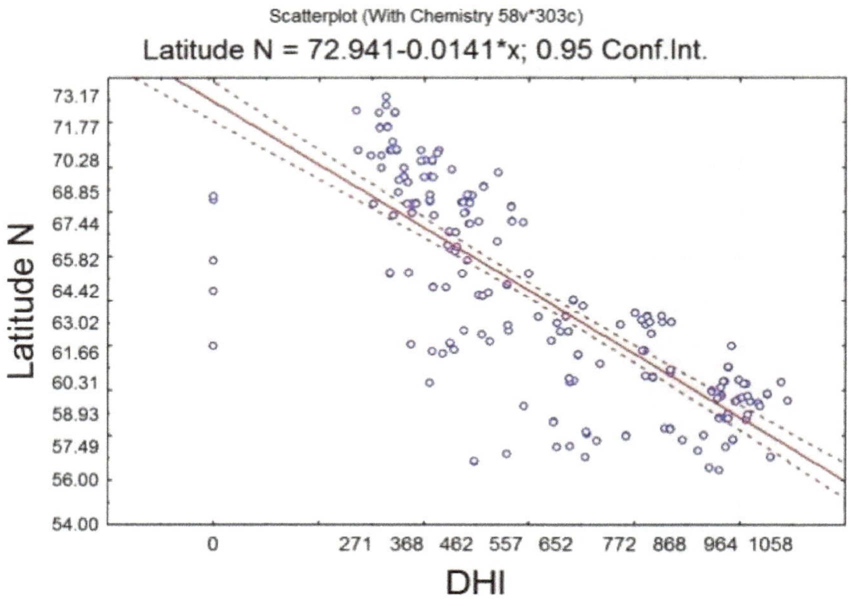

Fig. 3. Distribution of DHI over latitude of studied habitats

In rivers, species richness increased to the north excluding the southernmost river Vitim. Diatoms usually increased to the north, but in the middle of the studied area, they were replaced by *Chlorophytes* (Amga, Vilyui, Kolyma). *Chrysophyta* species number increased to the river mouth in the Indigirka River and changed of *Cyanobacteria* in the northernmost area of the river Anabar. Abundance and biomass of phytoplankton varied widely in the lowermost part of the northern Anabar River (58.9 cell L^{-1}; 0.0002 mg L^{-1}) and were higher in the largest rivers: Lena, Kolyma, Aldan, and Vilyui (more than one million cells per liter and about one mg L^{-1}). That means that the strongest climatic impact can be seen in the northern area rivers where the environment is unfavorable for algal communities. This expectation can be confirmed by data about community complexity. The average cell volume is the lowest in the northern stations but in the largest rivers it is highest, which correlated with the Shannon index of complexity: up to 5.76 in the rivers Lena, Vilyui, Vitim, and Kolyma. The average cell volume has a negative correlation with Shannon index in the studied communities and therefore both variables reflect the community's structure complexity fluctuation under climatic impact.

Abundance and biomass of phytoplankton usually have positive correlation (Pearson coefficient 0.30 – 0.94) with statistical significance (P = .04 – 2 10^{-8}). Each river can be characterized by its own productivity variables, but we revealed some tendencies in community structure over latitude. So, in the southernmost rivers Vitim, Olekma, Chara, Amga, and Aldan, which belong to the Lena River tributaries, diatoms prevail in abundance and biomass (Fig. 4). Northern rivers of the Lena Basin – Lena and Vilyui, changed abundance from diatom to diatom-*Chlorophyta* but in biomass diatoms still prevail. Other studied rivers in the north have more structural complexity of its communities. Usually diatom replaced to diatom-*Chrysophyte-Cyanobacteria* communities to the north. It is remarkable that in the Olenek and Anabar rivers whole basins of which are located above the Arctic Circle, the role of *Chlorophyta* and *Chrysophyta* algae is increased in communities to the mouth (Fig. 5). In the Olenek communities increased also cyanobacteria in abundance and biomass (See Appendixas supplementary file). The major tendency in the communities increase in abundance and biomass of phytoplankton in the north direction in most rivers, except in the Vilyui, Olekma and Aldan where these parameters were slightly decreased, and the Anabar River where there are significant decreases in abundance and biomass. Usually this tendency is correlated with species number except in the Anabar River.

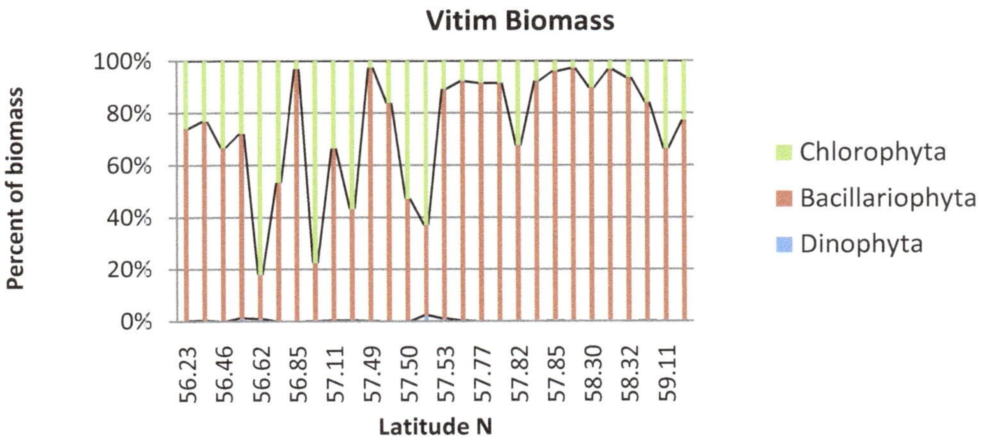

Fig. 4. Distribution of species richness in communities over latitude for typical southern rivers – the Vitim River

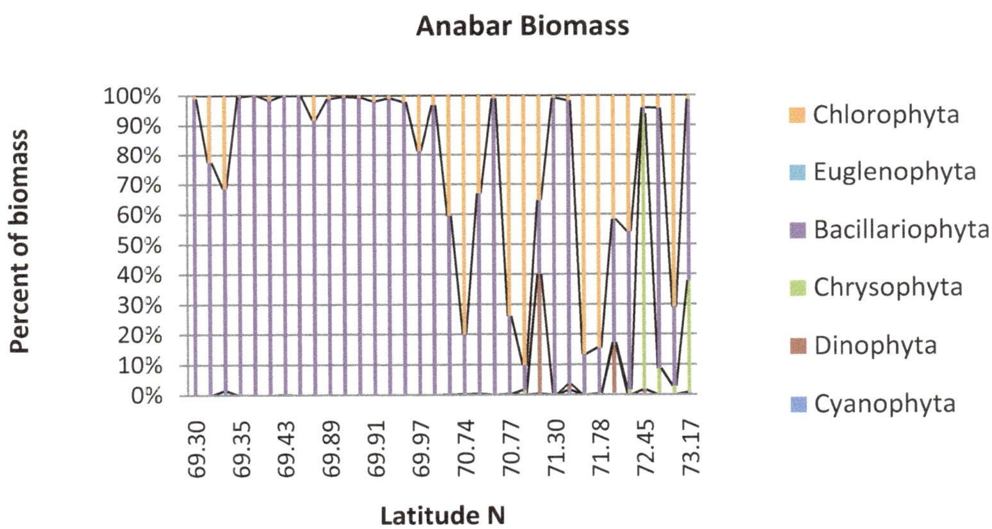

Fig. 5. Distribution of species richness in communities over latitude for typical northern rivers – the Anabar River

Therefore, we chose major community variables (species richness, abundance, biomass, average cell volume) to calculate the relationships with environmental data from (See Appendixas supplementary file). In the first step of calculation we implemented the multiple regression stepwise statistical analysis. (Table 1) show that the most important variables are (in Step 1 column) temperature, ice-free days, trophic variables, and flow speed. Species richness has been influenced by climatic variables – air temperature, precipitation, and ice-free period. Temperature and sunlight stimulate algal abundance and biomass whereas of water transparency and silica suppress algal development. Communities have been enriched by large-celled species and have more complexity in the low mineralized water with low transparency as can be seen in calculating the Biovolume and Shannon index H. Diatom and *Chrysophyte* algae were stimulated by temperature. Ammonia was important for *Cyanobacteria, Chlorophyta*, *Dinophyta*, and *Euglenophyta* whereas water transparency usually suppressed algal communities. Light stimulated cell abundance in diatoms; euglenoids were most abundant in temperate, mineralized, and polluted waters with low transparency, but phosphorous concentrations were important for green algae abundance. Algal biomass was stimulated by the light in green algae and by temperature and water pollution in euglenoids, whereas in diatoms high air temperature had negative influence.

Species richness as a whole increases with increases of water temperature (Fig. 6). Distance-weighted least squares correlation shows that species richness as a whole increases with decreasing DHI to the north as well as with the number of the open-water days (Fig. 7).

Fig. 8 shows the trend of increasing the number of species over the site position. Species richness is higher in the river basin in the north compared to the habitat coordinates. Precipitation plays an important role in increasing species richness in plankton communities, and in how the multi-specific communities form under high rainfall at low latitudes and under lower precipitation at high latitudes (Fig. 9).

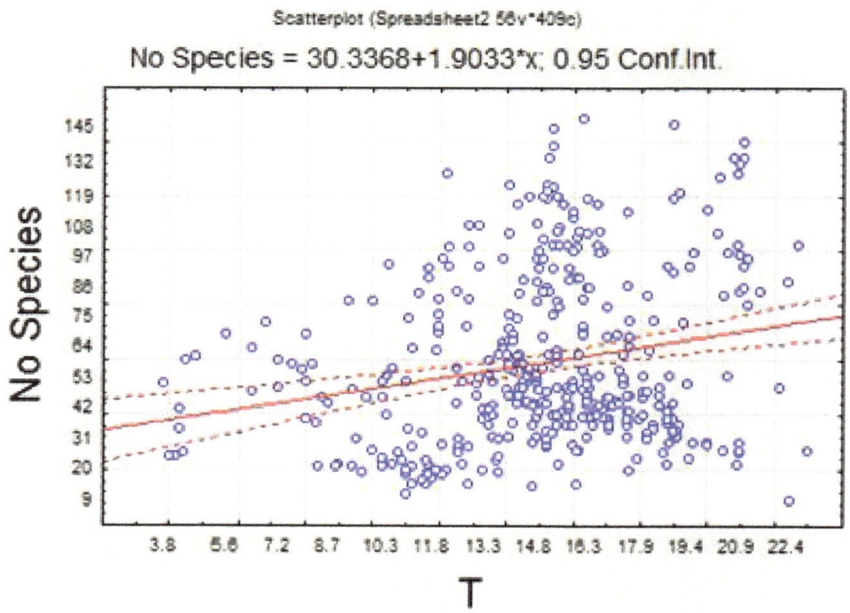

Fig. 6. Distribution of species richness in phytoplankton communities of studied rivers over temperature gradient

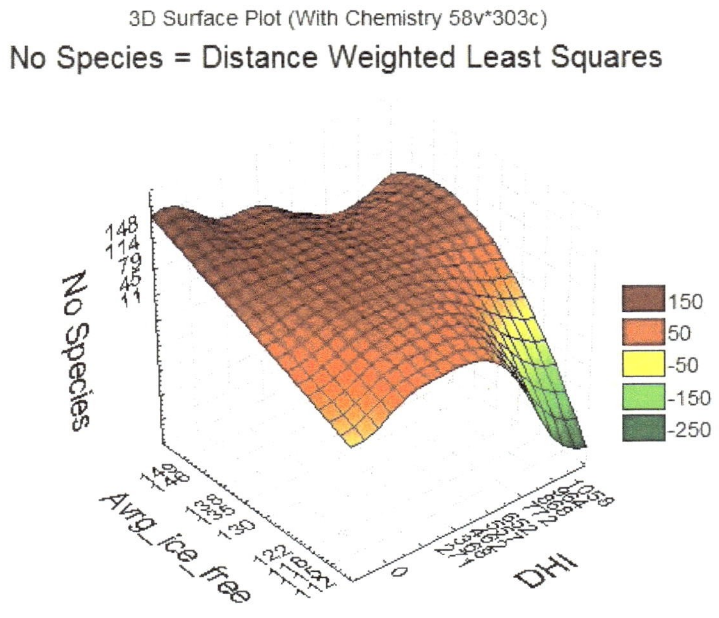

Fig. 7. Distribution of species richness in phytoplankton communities of studied rivers over gradient of DHI and ice-free period

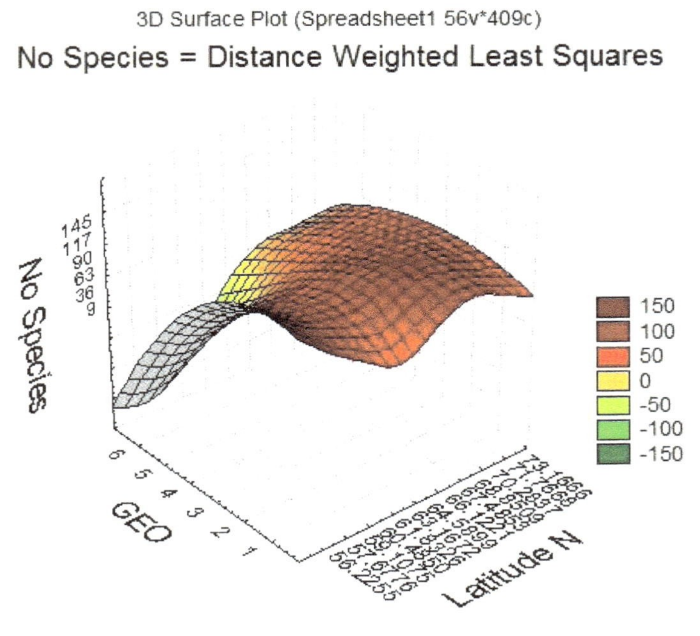

Fig. 8. Distribution of species richness in phytoplankton communities of studied rivers over latitude and gradient of the river basin placement

Table 1. Multiple regression stepwise statistical analysis results for the phytoplankton communities of the north-eastern Siberia Rivers. Abbreviations are gives as in appendix

Depen-dent variable	Step 1	Step 2	Step 3	Step 4	Step 5	Step 6	Step 7	Step 8
No. of Species	T_dry; Ice_free; **T_max**; Precip 0.434***	T_dry; **T_max**; Precip0.372***	T_dry; **T_max** 0.304***	T_dry 0.244***	-	-	-	-
Abundance	**Secchi**; **Si;**T_wat 0.267**	T_wat; **Si** 0.220**	**Si** 0.199*	T_wat 0.165*	T_wat 0.141*	-	-	-
Biomass	**Secchi**; Ice_free 0.264***	**T_max** 0.160**	**T_max** 0.138**	**T_max** 0.111**	-	-	-	-
Bio-volume	**Ice_free**; **TDS**; Secchi 0.161**	-	-	-	-	-	-	-
Cyano-phyta	Ptot; NH4 0.394***	Ptot; PO4; NH4; **Si** 0.372***	Ptot; **PO4**; NH4; **Alt** 0.348***	Ptot; **PO4**; T_dry 0.298***	Ptot; T_dry; **Alt** 0.260***	T_dry; **Alt** 0.217***	T_dry; **Alt** 0.184***	T_dry 0.091***
Bacilla-riophyta	T_mean_D; T_dry; **Secchi** 0.488 ****	T_mean_D; T_dry 0.362****	T_mean_D 0.234****	-	-	-	-	-
Chloro-phyta	**Secchi**;Ice_free; **Si**; T_wat; **T_dry**; NH4 0.712*****	T_wat; NH4; **Si**; **Secchi** 0.446****	T_dry; T_wat; NH4; **Si** 0.382****	T_dry; T_wat 0.318****	T_dry; T_wat 0.269****	T_dry 0.152***	-	-
SpChlo/ SpCyano	T_wat 0.199***	T_wat; T_min 0.175***	T_wat; T_min 0.098***	-	-	-	-	-
Dinophyta	NH4; **Si**; **pH** .329***	**Si**; **pH** 0.297***	CO2; **Si**; **pH** 0.254***	CO2; **Si** 0.169***	CO2 0.096**	-	-	-
Chryso-phyta	T_dry 0.399***	**T_wet**; T_dry **T_max** 0.322***	T_dry 0.288***	T_dry 0.252***	**T_wet** 0.211***	**T_wet** 0.152***	-	-
Eugleno-phyta	NH4 0.181**	NH4 0.157**	NH4 0.126**	NH4 0.091**	-	-	-	-

Table 1 continued.....................

N-Cyano-phyta	-	-	-	-	-	-	-	-
N-Dino-phyta	-	-	-	-	-	-	-	-
N-Chryso-phyta	-	-	-	-	-	-	-	-
N-Bacilla-riophyta	Ice_free; **Secchi** 0.248***	-	-	-	-	-	-	-
N-Eugleno-phyta	BOD; T_wat; Flow; TDS; Secchi0.277***	BOD; Flow 0.191**	BOD; T_wat; Flow 0.162**	BOD 0.102**	-	-	-	-
N-Chloro-phyta	Ptot 0.164**	Ptot; T_wat 0.147**	Ptot 0.078**	-	-	-	-	-
B-Cyano-phyta	**Si**; T_wat 0.192*	T_wat 0.159*	-	-	-	-	-	-
B-Dino-phyta	-	-	-	-	-	-	-	-
B-Chry-sophyta	-	-	-	-	-	-	-	-
B-Bacilla-riophyta	**T_max** 0.152 ***	-	-	-	-	-	-	-
B-Eugle-nophyta	BOD; T_wat; Flow 0.238**	BOD; T_wat 0.218**	BOD; T_wat 0.197**	BOD; T_wat 0.149**	BOD 0.079**	-	-	-
B-Chlo-rophyta	**Flow**; **Secchi**; Ice_free 0.203**	-	-	-	-	-	-	-
Shannon H	Ice_free; **Secchi** 0.804****	Ice_free 0.798****	-	-	-	-	-	-

*Note: Negatively influenced variables are bold; **, ***, **** = statistically significant at P < .01, P < .001; P < .0001 respectively*

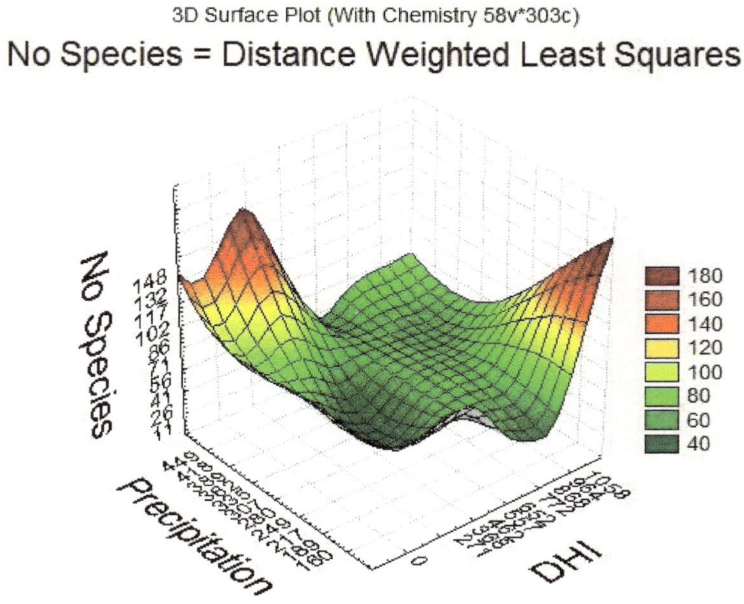

Fig. 9. Distribution of species richness in phytoplankton communities of studied rivers over gradient of DHI and precipitation

Abundance and biomass of phytoplankton in studied rivers correlate, but are not connected directly to each other as seen in (Fig. 10) where an upward trend in both variables in higher latitudes are seen with the change in plankton type. (Fig. 11) reflects the barely noticeable impact of the illumination on the productive variables of phytoplankton, but it is clear that a change in sunlight intensity is a function of ice-free periods; therefore we can see three types of communities.

Diatom biomass as the main part of the studied communities, formed in multi-specific communities, thrived in low-light conditions associated with fewer days in the open ice period (Fig. 12). It is remarkable that the revealed diatom biomass distribution is comparable to precipitation distribution (Fig. 2) in studied regions. This means that climatic variables, such as precipitation, and associated play major roles in increasing species richness and biomass of phytoplankton. At the same time, there is a tendency for the number of species to increase with the growth of the biomass of diatoms under well-lit conditions (Fig. 13).

As can be seen from the distribution of species composition in the typical south and typical north rivers (Figs. 4, 5), except the dynamics of diatoms, an important distinctive role in the northern habitats were made up of green algae and cyanobacteria; we therefore calculated this ratio for the studied communities. Precipitation increases with altitudes of habitats and (Fig. 14) shows that the number of cyanobacteria in the mountainous communities decreases, giving more benefits to the green algae. With increasing latitude and rainfall, the role of blue-green algae also decreased in comparison to the greens (Fig. 15).

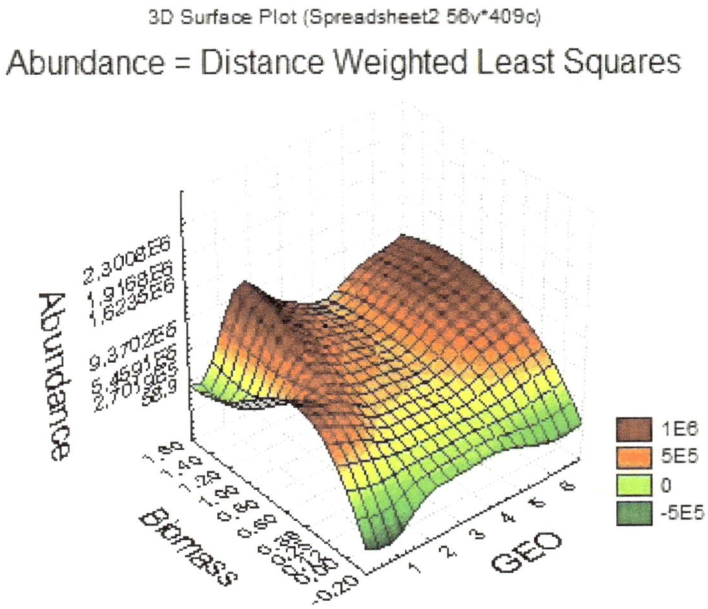

Fig. 10. Distribution of cell abundance in phytoplankton communities of studied rivers over the gradient of its biomass and the river basin placement

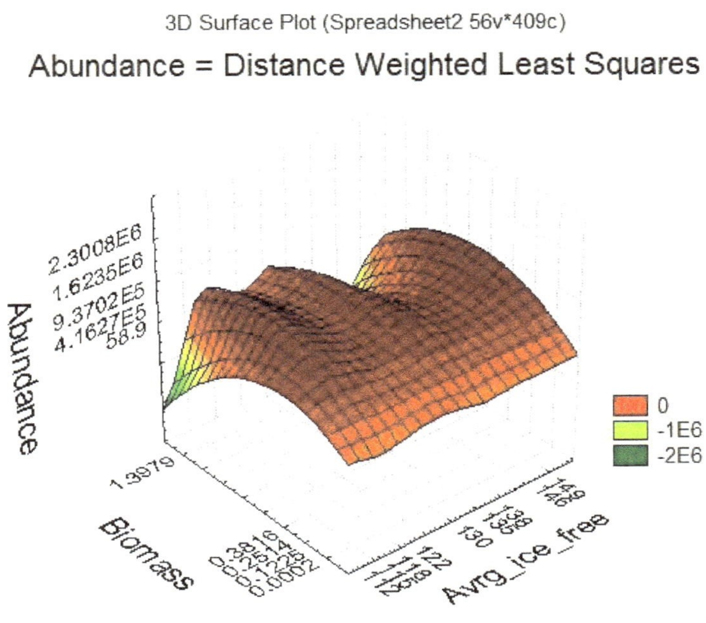

Fig. 11. Distribution of species richness in phytoplankton communities of studied rivers over the gradient of its biomass and ice-free period

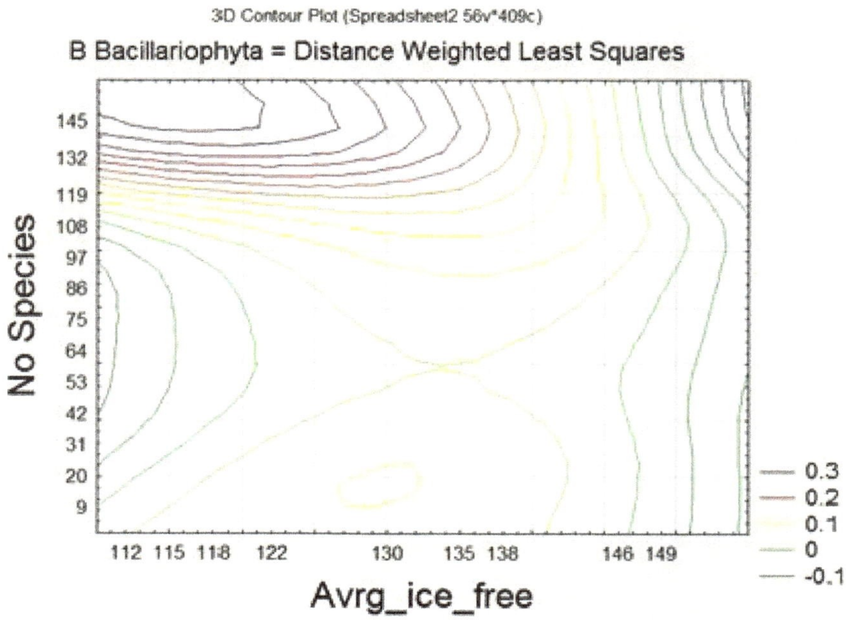

Fig. 12. Distribution of diatom biomass in phytoplankton communities of studied rivers over gradients of species richness and ice-free period

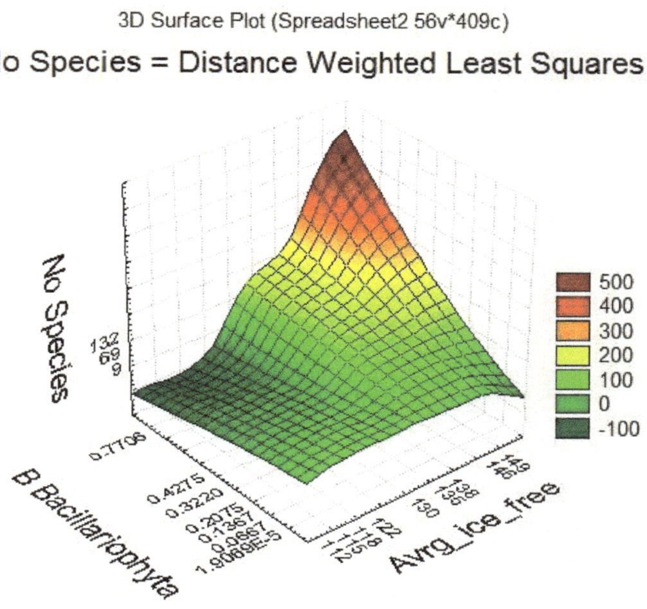

Fig. 13. Distribution of species richness in phytoplankton communities of studied rivers over diatom biomass and ice-free periodgradient

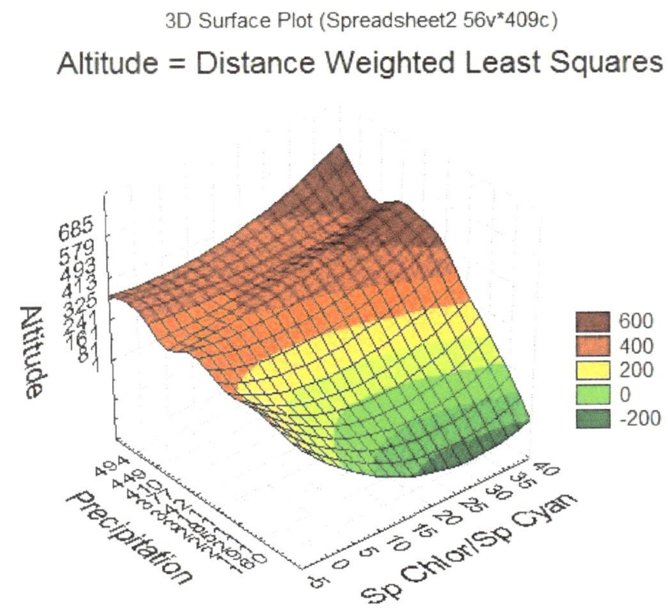

Fig. 14. Altitude, precipitation and *Chlorophyta/Cyanobacteria* ratio relationships in the studied rivers

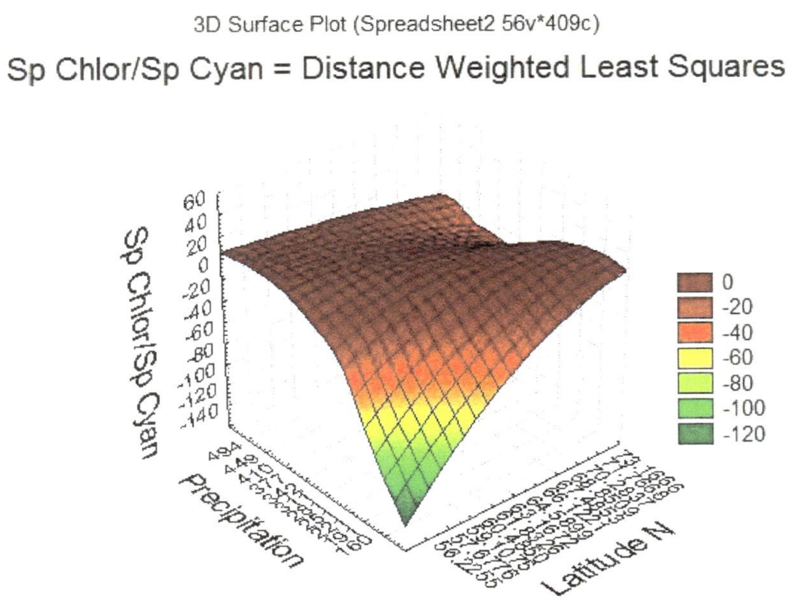

Fig. 15. *Chlorophyta/Cyanobacteria* ratio and precipitation relationships over the latitude of the studied rivers

4. DISCUSSION

Earlier we discussed the system of organizing the data to analyze the diversity of freshwater algal communities [4], which was hierarchical and limited at the top level of the river basin. Here, we used a new approach in which the data are combined at the regional level, thus increasing the level of generalization. Another new approach in our materials was to use not only hydrochemical but also regional climate data. We have also expanded the range of statistical analysis software with Program Graphs and modern version of Statistica.

We chose the same type of community that is only a summer phytoplankton of major rivers; the current of these major rivers predominantly ran from a south-to-north direction, which eliminated non-climate-related fluctuations. We analyzed a large number of samples from 400 sites resulting in the identification of 1283 species (1637 taxa of species and infraspecies) in phytoplankton communities, which is consistent with the Willis pattern [54], pointing to the completeness of the identified material. The studied rivers are mostly alkaline, ice-covered about half of the year with an average water temperature in summer from 3.8 to 23°C and have no limitation for algal biomass development. We included environmental variables from two new indices: The DHI index [48] which strongly decreased with latitude, and the Geo-association index that we created for the river basin placement characteristic.

As a result of statistical programs applied to such large data numbers, our analysis shows the relationships between biological and climate-related variables. We revealed air temperature, ice-free periods, trophic variables, and flow speed as major variables that influenced phytoplankton communities. In contrast, water temperature was proposed as the major factor that influenced freshwater diversity [55] and therefore could be used as an indicator of global climate change [14], which wasn't included as an important variable in the studied Arctic rivers. Rivers in the north-eastern Siberia have two types of phytoplankton communities: southern with increasing diatoms to the north and northern with decreasing diatoms to the north. Species richness as a whole increased to the north and correlated with air temperature, precipitation, and ice-free periods. It is very interesting and non-trivial result which is invisible when only diatom communities have been studied in the climatic gradient [12] where diversity represents only one taxonomic group, such as diatoms [56]. In the Caucasian Mountains and Golan Heights freshwater communities we revealed a similar dynamic in community structure with decreases in diatoms in high-altitude habitats [9,10], which is similar to the Himalayas [56] with increases of *Chlorophyta* species [12]. At the same time anthropogenic factors also regulate the freshwater algal community structure and the distribution over the climatic gradient [11,12]. But our approach in this study suggested the removal of influence of human activities because the selected region is poorly populated and the rivers are large. Studies like this used some new approach that compared data in respect of sampling point latitude and was doing in first time. Whereas general patterns of diversity and productivity in aquatic communities of the lakes are rather developed [52], the riverine phytoplankton is poorly studied in hard climatic gradient. Our results reflect the same diversity-productivity pattern as for a large sequence of lakes [15,57] in climatic gradient.

5. CONCLUSION

Our study conclusions reveal strong environmental stress in high altitude and high latitude climatic conditions. Diatoms in both cases are replaced by green and red algae (in high mountains) or by green and *Chrysophyta* algae and *Cyanobacteria* (in the Arctic). That allows us to conclude that for climatic impact assessments on freshwater algal diversity can be studied the multidivisional communities under large amplitude climatic gradients. A multidivisional approach helps identified factors that regulated freshwater community dynamics over a climatic gradient, such as altitude [10] or latitude (present study), especially in the northern regions [14]. Usually, the assessment and prediction of freshwater diversity dynamics indicate water temperature as a major regulating factor [14,15,57]. But our analysis of large based algal communities and environmental variable relationships revealed the importance of other climatic related factors such as latitude, altitude, ice-free periods, air temperature, and humidity. These factors are stimulated in phytoplankton communities, while water transparency and flow speed are stress factors. Water temperature, on the contrary, is not revealed as an important regulating factor in the distribution of freshwater phytoplankton communities.

ACKNOWLEDGEMENTS

This work has been partly funded by the Ministry of Absorption of Israel.

COMPETING INTERESTS

Authors have declared that no competing interests exist.

REFERENCES

1. Segers H, Martens K, editors. Aquatic biodiversity II: The Diversity of Aquatic Ecosystems. Springer; 2006.
2. Chapin FS (III.), Sala OE, Huber-Sannwald E, editors. Global biodiversity in a changing environment: Scenarios for the 21st Century. New-York, Berlin, Heidelberg: Springer-Verlag; 2001. ISBN: 978-0-387-95286-4 (Print) 978-1-4613-0157-8 (Online)
3. Sun G, Segura C. Interactions of forests, climate, water resources, and humans in a changing environment: Research Needs. British Journal of Environment & Climate Change. 2013;3(2):119-126. DOI: 10.9734/BJECC/2013/6212.
4. Barinova S. Multilevel approach in biodiversity analysis of freshwater algae. Expert opin on Environ Biol. 2013;2(2):1–2. DOI: http://dx.doi.org/10.4172/2325-9655.1000e106.
5. Aboal M, Puig M, Prefasi M. Diatom assemblages in springs in Castellón province (Eastern Spain). Algol Stud. 1998;90:79–95.
6. Sabater S, Roca JR. Ecological and biogeographical aspects of diatom distribution in Pyrenean springs. British Phycol J. 1992;27:203–213. DOI: 10.1080/00071619200650201.
7. Cantonati M, Bertuzzi E, Spitale D, editors. The spring habitat: Biota and sampling methods. Trento: MuseoTridentino di Scienze Naturali; 2007.
8. Barinova S. Algal diversity dynamics, ecological assessment, and monitoring in the river ecosystems of the eastern Mediterranean. Nova Science Publishers, New York, USA; 2011.
9. Barinova S. The effect of altitude on distribution of freshwater algae in continental Israel. Current Topics in Plant Biol. 2011;12:89–95.
10. Barinova SS, Kukhaleishvili L, Nevo E, Janelidze Z. Diversity and ecology of algae in the Algeti National Park as a part of the Georgian system of protected areas. Turk J Bot. 2011;35:729–774. DOI:10.3906/bot-1009-83.

11. Barinova S, Nevo E. Climatic and pollution impact on algal diversity of the freshwater ecosystems in Eurasia. In: Climate Change and Impacts. WY, USA: Academy Publish; 2012.
12. Barinova S, Ali N, Barkatullah, Sarim FM. Ecological adaptation to altitude of algal communities in the Swat Valley (Hindu Cush Mountains, Pakistan). Expert Opin Environ Biol. 2013;2(2):1–15. DOI: http://dx.doi.org/10.4172/2325-9655.1000104.
13. Barinova S, Stenina A. Diatom diversity and ecological variables in the Arctic lakes of the Kostyanoi Nos Cape (Nenetsky Natural Reserve, Russian North). Plant Biosyst. 2013;147(2):397-410. DOI:10.1080/11263504.2012.749956.
14. Heino J, Virkkala R, Toivonen H. Climate change and freshwater biodiversity: detected patterns, future trends and adaptations in northern regions. Biol Rev. 2009;84:39–54. DOI: 10.1111/j.1469-185X.2008.00060.x. Epub 2008 Nov 11.
15. Schiaffino RM, Unrein F, Gasol JM, Massana R, Balague V, Izaguirre I. Bacterial community structure in a latitudinal gradient of lakes: the roles of spatial versus environmental factors. Freshwater Biol. 2011;56:1973–1991. DOI: 10.1111/j.1365-2427.2011.02628.x.
16. Angeler DG, Drakare S. Tracing alpha, beta, and gamma diversity responses to environmental change in boreal lakes. Oecologia. 2013;172:1191–1200. DOI: 10.1007/s00442-012-2554-y.
17. Ptacnik RA, Solimini G, Andersen T, Tamminen T, Brettum P, Lepisto L, Willen E, Rekolainen S. Diversity predicts stability and resource use efficiency in natural phytoplankton communities. PNAS. 2008; 105:5134–5138. DOI: 10.1073/pnas.0708328105.
18. Krupa E, Slyvinskiy G, Barinova S. The effect of climatic factors on long-term dynamics of phytoplankton, zooplankton and macrozoobenthos of the Balkhash Lake (Kazakhstan, Central Asia). Adv Stud Biol. 2014;6(3):115–136. Available: http://dx.doi.org/10.12988/asb.2014.4523.
19. Medvedeva LA, Barinova S, Semenchenko AA. Use of Algae for Monitoring Rivers in the Monsoon Climate Areas (Russian Part of Asian Pacific Region). Int J Envir Res. 2012;1(1):39–44.
20. Alekin OA, Semenov AD, Skopintsev BA. Rukovodstvo po khimicheskomu analizu poverkhnostnykh vod sushy (Manual on chemical analysis of land surface waters). Gidrometeoizdat, Leningrad; 1973. Russian.
21. Semenov AD. Rukovodstvo po khimicheskomu analizu poverkhnostnykh vod sushi (Manual on chemical analysis of land surface waters). Leningrad: Gidrometeoizdat; 1977. Russian.
22. Gabyshev VA. Ustroistvo dlya kontsentrirovaniya fitoplanktona pod davleniyem (Sampling device for concentration of phytoplankton under pressure). Algologia. 2009;19(3):318–320. Russian.
23. Guiry MD, Guiry GM. Algae Base. World-wide electronic publication, National University of Ireland, Galway. Accessed 16 July 2013. Available: http://www.algaebase.org.
24. Cavalier-Smith T. Only six kingdoms of life. Proc R Soc Lond B. 2004;271:1251–1262. DOI: 10.1098/rspb.2004.2705.
25. Pugnetti A, Acri F, Alberighi L, Barletta D, Bastianini M, Bernardi-Aubry F, et al. Phytoplankton photosynthetic activity and growth rates in the NW Adriatic Sea. Chem Ecol. 2004;20(6):399–409. DOI:10.1080/02757540412331294902.
26. Odum EP. The strategy of ecosystem development. Science. 1969;164:262–270. DOI: 10.1126/science.164.3877.262.

27. Good IJ. The population frequencies of species and the estimation of population parameters. Biometrika. 1953;40:237-264. URL: http://links.jstor.org/sici?sici=0006-3444%28195312%2940%3A3%2F4%3C237%3ATPFOSA%3E2.0.CO%3B2-K.
28. Novakovsky AB. Abilities and base principles of program module "GRAPHS". Proceedings of Komi Scientific Center, Ural'sky Branch Russian Academy of Science, Syktyvkar. 2004;27:1–28. Russian.
29. Gvozdetsky NA, Mikhailov NI. Physical geography of the USSR. Asian Part. Moscow: Mysl'; 1978. Russian.
30. Hijmans RJ, Cameron SE, Parra JL, Jones PG, Jarvis A. Very high resolution interpolated climate surfaces for global land areas. Int J Climatol. 2005;25:1965–1978. DOI: 10.1002/joc.1276.
31. Kolosova LN, editor. Geographical atlas. Moscow, main administration of geodesy and cartography under the council of ministers of the USSR; 1982. Russian
32. Freshwater ecoregions of the world (FEOW). Accessed 03 January 2014. Available: http://feow.org/.
33. Desyatkin RV, Okoneshnikova MV, Desyatkin AR. Yakutia soils. Yakutsk: Bichik Publisher; 2009. Russian.
34. Sokolov AA. Hydrography of the USSR. Leningrad: Hydrometeoizdat; 1952. Russian
35. Vassilieva-Kralina II, Remigailo PA, Gabyshev VA, Ivanova AP, Kopyrina LI, Pshennikova EV. Algae. In: Flora Yakutia: Geographical and environmental aspects. Novosibirsk: Nauka; 2010. Russian.
36. Vassilieva-Kralina II, Remigailo PA, Gabyshev VA, Pshennikova EV, Ivanova AP, Kopyrina LI. Algae. In: Diversity of flora Yakutia. Novosibirsk: Siberian Branch of the Russian Academy Publisher; 2005. Russian.
37. Vasilyeva II, Remigailo PA, Gabyshev VA, Ivanova AP, Kopyrina LI. The far north: Plant biodiversity and ecology of Yakutia 2. Flora of Yakutia: Composition and Ecological Structure 2.6. Algae. Dordrecht Heidelberg London New York: Springer; 2010. Russian.
38. Gabyshev VA, Gabysheva OI. Water quality of the Anabar river on the base of phytoplankton structure and hydrochemical data. Siberian ecological journal. 2010;17(4):563–570. Russian.
39. Gabyshev VA, Gabysheva OI. Features phytoplankton growth and physico-chemical properties of water in the Indigirka River. Vestnik NESC FEB RAS. 2010b;3:42–50. Russian.
40. Gabyshev VA, Gabysheva OI. Features phytoplankton growth and physico-chemical properties of the waters of the Yana River in summer. Proceedings of the Irkutsk State University Series "Biology. Ecology". 2010;3(4):82–94. Russian.
41. Gabyshev VA, Gabysheva OI. The study of phytoplankton and physico-chemical parameters of water in the Olenyok River. Vestnik NESC FEB RAS. 2010d;3:51–55. Russian.
42. Gabyshev VA, Gabysheva OI. Features phytoplankton growth and physico-chemical properties of water in middle and lower Vilyuy River and Svetlinsky reservoir. Problems of regional ecology. 2010e;3:45–54. Russian.
43. Gabyshev VA, Gabysheva OI. Current status of phytoplankton and water chemistry in the Mayya River. Siberian Journal of Ecology. 2011;18(1):23–31. Russian.
44. Gabyshev VA, Gabysheva OI. The structure of phytoplankton in the Chara River (Eastern Siberia) and its habitat in the early summer (June). Proceedings of Penza State Pedagogical University n. V.G. Belinsky. Natural sciences. 2012;29:144–151. Russian.

45. Gabyshev VA, Gabysheva OI. Structure of summer phytoplankton in the Olekma River (Eastern Siberia) and its habitat. Proceedings of the Komi Scientific Center, RAS. 2013;1(13):25–31. Russian.
46. Gabyshev VA, Gabysheva OI. Structure of the summer (July) phytoplankton in the Vitim River and its habitat. Vestnik of St. Petersburg. Univ Ser. 2013;3(1):16–27. Russian.
47. Gabyshev VA, Gabysheva OI. The phytoplankton structure and physico-chemical parameters of the waters of the Kolyma River (North-Eastern Siberia) in summer. Siberian Journal of Ecology. 2013;20(3):341–351. Russian.
48. Gabyshev VA, Remigaylo PA, Gabysheva OI. Spatial structure and habitat of phytoplankton in the Aldan River. Proceedings of the Irkutsk State University Series "Biology. Ecology". 2012;5(2):61–69. Russian.
49. Bondarenko NA, Tomberg IV, Logacheva NF, Timoshkin OA. Phytoplankton and hydrochemistry of rivers Vitim, Mama, and Tshuya (Trans-Baikal, the Lena River basin). Proceedings of the Irkutsk State University. Series "Biology. Ecology". 2010;3(4):70–81. Russian.
50. Kuzmin GV. Species composition of phytoplankton in the reservoirs flood zone of the Kolyma Power Station. Preprint. Magadan: Institute of Biological Problems of the North, Far Eastern Branch of AN SSSR; 1985. Russian.
51. Kuzmin GV. Biomass and structure of phytoplanktonic communities in wetlands of the flood zone of the Kolyma Power Station. In: Ecology, distribution and life forms of plants in Magadan region. Vladivostok: Far Eastern Branch of USSR Academy of Sciences; 1987. Russian.
52. Nickanorov AM, Bryzgalo VA, Kosmenko LS, Reshetnyak OS. Mouth area of the Kolyma River in modern conditions of anthropogenic impact. Meteorology and Hydrology. 2011;8:74–88. Russian.
53. Coops NC, Wulder MA, Duro DC, Han T, Berry S. The development of a Canadian dynamic habitat index using multi-temporal satellite estimates of canopy light absorbance. Ecol Indicators. 2008;8:754–766. DOI: 10.1016/j.ecolind.2008.01.007.
54. Barinova SS, Medvedeva LA, Anissimova OV. Diversity of algal indicators in environmental assessment. Tel Aviv: Pilies Studio; 2006. Russian.
55. George DG. The influence of global warming on freshwater plankton communities in Britain. Freshwater Forum. 1991;1(3):204-214. Available: http://aquaticcommons.org/id/eprint/4527.
56. Verma J, Nautiyal P. Longitudinal patterns of distribution of epilithic diatoms in a lesser Himalayan stream. Journal of Hill Research. 2009;22(2):105-109.
57. Cardinale BJ, Hillebrand H, Harpole WS, Gross K, Ptacnik R. Separating the influence of resource 'availability' from resource 'imbalance' on productivity-diversity relationships. Ecology Letters. 2009;12:475-487.

Climate Change and the Risk of Highly Pathogenic Avian Influenza Outbreaks in Birds

Jianhong E. Mu[1*], Bruce A. McCarl[2], Ximing Wu[2] and Michael P. Ward[3]

[1]Department of Applied Economics, Oregon State University, Corvallis, OR, 97330, USA.
[2]Department of Agricultural Economics, Texas A&M University, College Station, TX, 77843, USA.
[3]Faculty of Veterinary Science, University of Sydney, Camden, NSW, 2570, Australia.

Authors' contributions

This work was carried out in collaboration between all authors. Author JEM was a postdoc scholar and was involved in the data collection, statistical analyses, results interpretation and manuscript preparation. Author BAM secured funding and was involved in study design, study execution, results interpretation and manuscript preparation. Author XW was involved in statistical analyses and manuscript preparation. Author MPW was involved in results interpretation and manuscript preparation. All authors read and approved the manuscript.

ABSTRACT

In this paper, we examine the association between climate change and outbreak probability of Highly Pathogenic Avian Influenza A virus (HPAI H5N1) in birds. Climate change is a potential factor for the recent spread of H5N1 outbreaks because it can directly alter the conditions involved in persistence of the virus and disease transmission. Also it can contribute indirectly by changing wild bird migration patterns. Econometric analyses using a dynamic Probit model over monthly data from January 2004 to December 2008 found that a 1% rise in winter total precipitation increases the risk of HPAI H5N1 outbreaks by 0.26%. Spring mean temperature was also found to have positive and significant impacts. Our findings are robust across different model specifications and under out-of-sample tests. Using historical data we find the realized climate change of the last 20 years partly explains the recent expansion in outbreaks. Under future climate change projections, we find that countries having higher projected spring temperature or more winter precipitation or both, such as Japan and Romania, will have large increases in outbreak probabilities. This suggests that climate change may play an even greater role in

*Corresponding author: Email: mujh1024@gmail.com;

the future, although magnitudes will vary across countries and climate projections. From a policy perspective, future climate conditions may give rise to a need for different disease control and prevention strategies.

Keywords: HPAI H5N1; climate change; general circulation models (GCMs); risk assessment.

1. INTRODUCTION

Avian Influenza (AI) or "bird flu" is a contagious animal viral disease [1]. Infections can be divided into highly pathogenic avian influenza (HPAI) and low pathogenic avian influenza (LPAI). LPAI is less contagious, and infected species may not display clinical signs. The HPAI A (H5N1) virus spreads rapidly among infected birds with a high mortality rate (up to 90-100% within 48 hours) and may spread to humans [2].

Since 2003, epidemics of HPAI H5N1 have occurred in domestic and wild bird populations with unprecedented frequency across the globe. This strain was first identified in Hong Kong in 1997 and subsequent outbreaks were observed in East and Southeast Asia during 2003 and 2004. In 2005, outbreaks occurred in central Asia, Russia and Europe. Since then, HPAI H5N1 has continued to spread through Europe, the Middle East, and Africa [3]. As of 2007, 60 countries had experienced outbreaks [4,5], and the number of countries with outbreaks is expanding [6]. In addition, as the end of 2013, there have been 648 reported cases of human infection, resulting in 384 deaths [7].

Globally outbreaks of HPAI H5N1 have resulted in the destruction of more than 250 million domestic poultry including chickens, ducks, turkeys, quail and ostriches, causing large socioeconomic and livelihood losses, as well as affecting food and protein supplies, wildlife populations and public health [3,4]. Determining factors causing this HPAI H5N1 spread and estimating risk probabilities of further outbreaks could improve surveillance, control measures, and disease prevention planning to ultimately reduce losses [8].

Climate has been argued to be a driver of disease transmission and persistence plus a driver of wild bird migration patterns [9,10-11]. In particular, experimental evidence shows that cold temperature and low relative humidity are favorable to the spread of influenza virus and suggests that these two environmental factors could contribute to the seasonality of epidemic influenza [12], thus regional weather monitoring may serve as an integral part of a bird flu surveillance system [13]. In empirical studies, Si et al. [9] indicate that lower levels of moisture and humidity affect wild bird food availability and thereby their distribution and migration patterns. Fang et al. [15] find that precipitation is an important risk factor affecting H5N1 virus distribution in China and Si et al. [10] find that HPAI H5N1 occurrences in wild birds in Europe are highly correlated with increased minimum temperatures and reduced precipitation in January. Ottaviani et al. [11] and Kilpatrick et al. [16] conclude that alterations in cold weather trigger winter movements of migratory birds and have contributed to H5N1 spread outside the traditional migration period in Europe. In the future, since climate change alters these environment conditions, it would almost certainly influence virus transmission and survival outside the host [14].

In terms of disease spread vectors, there has been considerable effort investigating how the HPAI H5N1 virus enters previously unaffected countries. The main pathways that have been identified are wild bird migration, live bird trade and the transport of poultry and poultry products. In terms of migration, studies suggest that wild birds carry HPAI and can introduce it into new areas during migration [9,10,14,16-17]. Wild ducks were found to be able to carry the H5N1 virus asymptomatically [18]. Circumstantial evidence from Russia and Mongolia indicates that wild birds play a significant role in disease spread, because they can become infected and travel long distances before dying [9,19,20-21].

Some studies have attributed the increased frequency of outbreaks to the expanding poultry industry as well as greater movements of live poultry and poultry products [22]. Analysis of spring 2006 Romanian H5N1 outbreaks indicates that poultry movement might have facilitated the spread of infection [23]. However, quantification of the contribution is difficult due to the combination of local, unregulated movements of poultry plus illicit bird trade [24,25], particularly in developing counties. Therefore, few studies have examined the influence of poultry movements [16].

Studies in Romania, Thailand, Indonesia and China suggest that human infection and poultry outbreaks can be aggravated by agricultural and socio-demographic factors [26]. These factors include agricultural population density, and poultry density plus incidence of rice paddy fields, water sources and transportation. Ward et al. [23] and Farnsworth and Ward [27] indicate that the environment and landscape factors (specifically the Danube River Delta) played a critical role in the introduction and initial spread of H5N1 in Romania. Fang et al. [15] find that the distance to the nearest main city, distance to the nearest body of water and distance to the nearest highway contributed to disease spread in China. Others find that the risk of outbreaks increases with poultry density (of both chickens and ducks), road density, and is more likely in areas located near major cities and highway junctions [8,23,28].

Despite the importance of predicting the global outbreak spread plus the potentially contributing nature of climate change, little effort has been applied to project HPAI H5N1 risk across regions and countries under climate change scenarios [24,25]. In this study, we examine the extent to which climate is associated with outbreaks of HPAI H5N1 in birds, and then we project future consequences for HPAI H5N1 spread under projected changes in temperature and precipitation.

2. MATERIALS AND METHODS

Here we first introduce the conceptual framework of our analyses, the selection of variables, and the data used. Finally we present our econometric model.

2.1 Conceptual Framework

Fig. 1 shows a generalized potential relationship between climate change and HPAI H5N1 outbreaks considering direct and indirect influences. Suppose following [18] the probability of HPAI H5N1 outbreaks denoted by Y, is determined by the virus survival time outside of the host (ST), the movement of wild birds (MB), and the environmental conditions including temperature (T), precipitation (P) and extreme weather (EX) as given by the equation:

$$Y = f(ST, MB, T, P, EX, X) \tag{1}$$

Where in addition to the factors defined above, X includes proximity to wild bird migratory flyways, number of live birds traded, per capita gross domestic product (GDP) and population density.

In equation (1), ST is affected by temperature and humidity of the host environment and MB forms the distribution of wild birds which is also impacted by temperature and precipitation through the availability of food and water and via their migration patterns [9, 10 and 11]. Consequently, we can rewrite equation (1) as follows:

$$Y = f(ST(T,P,EX), MB(T,P,EX), T, P, EX, X) \quad (2)$$

Therefore, we could develop a statistical model that includes climate variables reflecting both direct and indirect effects and non-climate variables as shown in the equation below:

$$Y = f(T,P,EX,X) \quad (3)$$

Equation (3) is a reduced form model showing the probability of HPAI H5N1 outbreaks as a function of temperature, precipitation, extreme weather and social-economic development.

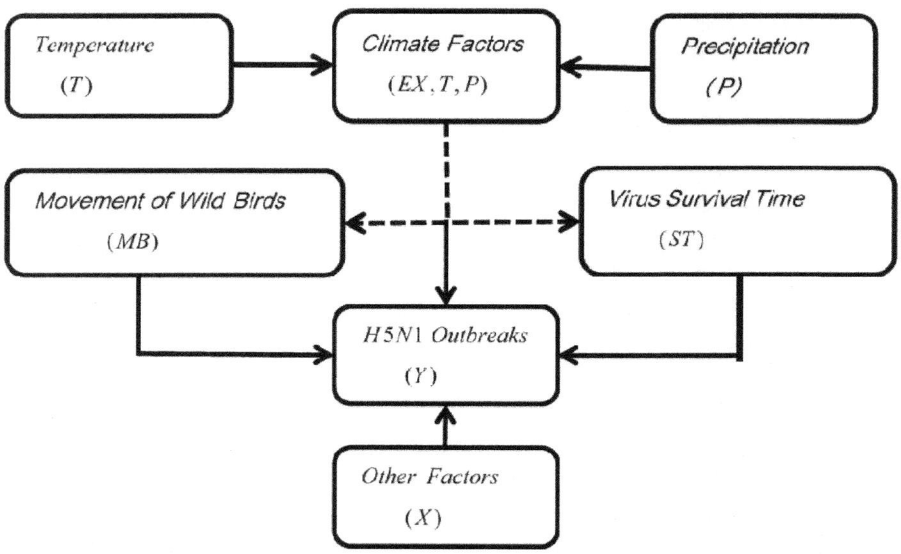

Fig. 1. A conceptual framework regarding climate effects on HPAI H5N1 outbreaks
(Solid lines indicate direct effects of climate on HPAI H5N1 survival and persistence times. Dashed lines indicate the indirect effects of temperature and precipitation on outbreaks via the distribution of wild birds)

2.2 Data and Variables

This study focuses on data regarding HPAI H5N1 outbreaks that occurred between January 2004 and December 2008, the peak period of epidemic activity. The data cover 90 regions in 16 countries in Asia, Africa, Europe and North America. Countries included are Malaysia, South Korea, Cambodia, Indonesia, Thailand, Japan, Vietnam, China, Egypt, Nigeria, Germany, Romania, Turkey, Pakistan, Russia and the United States. Even though there was no HPAI H5N1 outbreak in the U.S., there was a H5N2 outbreak in February 2004 in Texas. Since the U.S. is on migratory flyways, it is vulnerable to H5N1 introduced via wild bird migration. Fig. 2 shows the distribution of H5N1 outbreaks since 2005 that were reported to the World Organization for Animal Health (OIE), suggesting that there exists heterogeneity across regions with 12% having had at least one HPAI H5N1 outbreak in the past 5 years.

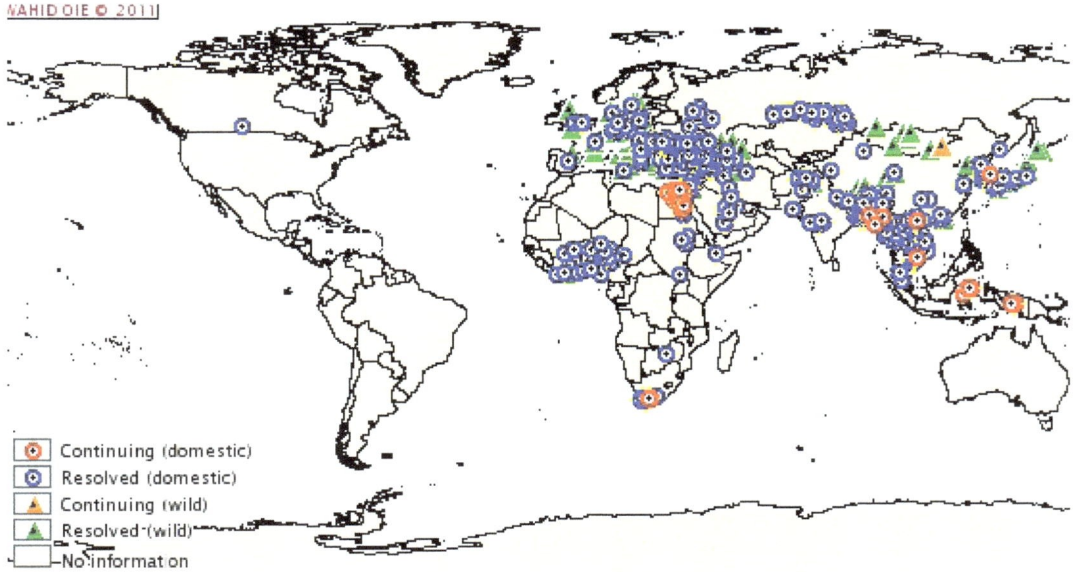

Fig. 2. Outbreaks of HPAI H5N1 in birds since 2005 as portrayed by OIE (2005)

The HPAI H5N1 outbreak incidence data were obtained from the World Animal Health Information Database (WAHID) for 2005–2008. The 2004 data were drawn from the Animal Health Database HANDISTATUS II. The HPAI H5N1 outbreak incidence is a binary variable in which a value of one indicates that a region has at least one reported HPAI H5N1outbreak in domestic poultry or wild birds in a given month while a zero indicates no reported outbreak in that month. We include H5N1 outbreaks in both domestic and wild birds.

Historical monthly climate data were drawn from the National Environmental Satellite Data and Information Service (NESDIS). To construct data on regional climate, we constructed average data across weather stations in each region for temperature and precipitation. Monthly mean temperature was computed in degrees Celsius, and total monthly precipitation including rain and melted snow was computed in millimeters.

Future climate projections were obtained through the Intergovernmental Panel on Climate Change Data Distribution Center (IPCC DDC) for the 2007 IPCC data. Since the projection of future climate change is uncertain, we used multiple General Circulation Models(GCMs) to allow for this uncertainty. Specifically, we choose the following three GCMs:

- The Hadley Center coupled model, version 3 (HAD: CM3), which projects a stable global mean climate [29] and is a mid-sensitivity case [30].
- The Geophysical Fluid Dynamics Laboratory global climate model, version 2.0 (GFDL: CM2.0), which is a model with drastically lower drifts in hydrographic fields

such as temperature and salinity, and display more realistic current climate results that are close to observed values [31,32].

- The Centre National de Recherches Météorologiques coupled atmosphere-ocean climate model, version 3 (CNRM: CM3), which is in close accordance with the present-day climate and predicts a general increase in precipitation throughout the 21st century [33].

We assembled projections of monthly changes in temperature and precipitation of these three GCMs under the A1B SRES [34] emissions scenario for 2030 for each country. We obtained the baseline climate data for the period 1961 to 2000 also from the IPCC DDC. We used the historical climate data from 1971-1980 to 1981-1990 and to 1991-2000 to represent the degree of climate change in those two decades.

Per capita GDP and population density were drawn from the U.S. Department of Agriculture Economic Research Service. Data on country-to-country trade in live birds (which excludes illegal movements) were drawn from the United Nations Food and Agricultural Organization and the U.S. Census Bureau, Foreign Trade Division (for the commodity code H1-0105 - live poultry, domestic fowls, ducks, geese, etc.). We also considered trade of live birds by aggregating the totals from three commodity codes: H2-010632 (live birds including parrots, parakeets, macaws and cockatoos), H1-010631 (live birds of prey) and H1-010639 (live birds excluding H1-010632 and H2-010631). However, the data were incomplete and so were not used in the analysis. For each country, trade data were reported either in dollar values or kilograms. Following Kilpatrick et al. [16], we converted the data to numbers of live birds using the median number of birds per kilogram (obtained from trade data) and 10.61 for poultry (most traded poultry are domestic fowl <185g). Table 1 gives definitions and summary statistics of variables used in this study.

Table 1. Definitions and statistical descriptions of variables

Variable	Definition	Mean	Std.Dev.	Min	Max
AIOtbkProb	Outbreak incidence in a country and month equaling 1 if outbreaks occurred, 0 otherwise	0.12	0.32	0	1
$AIOtbkProb_{t-1}$	Lagged outbreak incidence	0.12	0.33	0	1
sptemp	Spring mean temperature(°C)	5	9	-7	34
ftemp	Fall mean temperature(°C)	5	9	-8	36
wtemp	Winter mean temperature(°C)	2	7	-21	30
spprecp	Spring total precipitation(mm)	18	59	0	2336
fprecp	Fall total precipitation(mm)	19	62	0	1143
wprecp	Winter total precipitation(mm)	11	48	0	2384
cold_month	Dummy variable for whether the month average temperature is <= 4°C	0.12	0.32	0	1
hot_month	Dummy variable for whether the month average temperature is >= 28°C	0.16	0.36	0	1
lnpimport	Log of imported live birds	8.95	5.06	0	16
lnpexport	Log of exported live birds	6.80	5.98	0	17
popden	Population density (people/ square km)	152	115	8	493
gdpden	Per capita GDP($1000)	10	14	0.4	44
EAAFW	East Asia Australian Flyway	0.53	0.50	0	1
CAFW	Central Asian Flyway	0.12	0.33	0	1
BSMFW	Black Sea Mediterranean Flyway	0.30	0.46	0	1
EAFW	East Atlantic Flyway	0.09	0.28	0	1
EFWAFW	East African West Asian Flyway	0.04	0.21	0	1

2.3 Econometric Models

We examine the relationship between the probability of HPAI H5N1 outbreaks and climate plus other social-economic and geographic factors using a dynamic Probit model with unobserved individual effects as described in Wooldridge [35-37]. We use this because we are interested in whether there is state dependence after controlling for unobserved heterogeneity. In other words, we estimate how much past HPAI H5N1 outbreaks affect the probability of current outbreaks after controlling unobserved effects. The dynamic Probit model with unobserved individual effects is written as

$$P(y_{it}=1\mid y_{i,t-1},x_{it},c_i)=G(\rho y_{i,t-1}+\beta x_{it}+c_i) \qquad (4)$$

where y_{it} is a binary variable, indicating whether an outbreak occurred in region i at period t; $y_{i,t-1}$ is the lagged dependent variable allowing the current outbreak probability to be altered by whether the region has suffered previous outbreaks; x_{it} is a $k\times 1$ vector of explanatory variables; c_i is an unobserved individual effect in the region that is allowed to be correlated with some elements of x_{it}; and G is a cumulative distribution function that maps the index into the response probability; ρ and β are parameters to be estimated.

Without loss of generality, we can arrange the observations starting at $t=0$, so y_{i0} is the first observation on y. For $t=1,2,...T$, the density function of y_{it}, under the assumption of normal distribution, can be written as,

$$f(y_{i1},\ldots,y_{iT}\mid y_{i0},x_i,c_i,\beta)=\prod_{t=1}^{T}\Phi(\rho y_{i,t-1}+\beta x_{it}+c_i)^{y_{it}}[1-\Phi(\rho y_{i,t-1}+\beta x_{it}+c_i)]^{1-y_{it}} \qquad (5)$$

Where $\Phi(\bullet)$ is the standard normal cumulative distribution function.

To estimate β and ρ consistently, we need to address the existence of outbreaks before the model starts, which is called the initial conditions problem. By making an additional assumption on c_i as in Wooldridge [35-37], we assume,

$$c_i=\zeta_0+\zeta_1 y_{i0}+\zeta x_i+a_i \qquad (6)$$

Where the conditional distribution of $a_i\mid(y_{i0},x_i)\sim Normal(0,\sigma_a^2)$ and is independent of x_i and y_{i0}.

To avoid a large dimensionality problem in estimation[1], we use the average of x_i over time

---

[1] *Since we have monthly data from January 2004 to December 2008, x_i is a $k\times T$ matrix and x_{it} is a $k\times 1$ vector of independent, contemporaneous explanatory variables, which is too large to be estimated based on our sample size.*

($\overline{x}_i$) to replace x_i, as in Chamberlain [38]. Also to identify time indicators, which do not vary across i, they must be omitted from $\overline{x}_i$ by setting $\zeta = 0$. Therefore the dynamic Probit model with unobserved individual effects becomes

$$f(y_{i1},\dots,y_{iT} \mid y_{i0},x_i) = \prod_{t=1}^{T} \Phi[(\rho y_{i,t-1} + \beta x_{it} + \zeta_0 + \zeta_1 y_{i0} + \zeta \overline{x}_i)\cdot(1+\sigma_a^2)^{-1/2}]^{y_{it}} \times \{1-\Phi[(\rho y_{i,t-1} + \beta x_{it} + \zeta_0 + \zeta_1 y_{i0} + \zeta \overline{x}_i)\cdot(1+\sigma_a^2)^{-1/2}]\}^{1-y_{it}} \quad (7)$$

Although σ_a cannot be identified in Probit models, we can consistently estimate $\zeta_{0a}, \rho_a, \zeta_{1a}, \beta_a$ and ζ_a applying the conditional Maximum Likelihood Method [37] and estimating a Random-Effects model, which is different from the standard random-effects model since we allow for correlation between the error term and independent variables by specifying the conditional distribution of the unobserved effects.

In the estimation function (7), x includes,

- Seasonal climate variables (Temperature and precipitation) and their squared terms permit a nonlinear response and allow the model to reflect previous inconsistent results on disease outbreaks as have arose in the literature [10,15,18].
- Dummy variables reflecting temperature extremes as motivated by the fact that AI viruses have been found to survive for long periods in the environment, especially when temperatures are low [15]. Also two studies showed the virus retained its infectivity at 4°C for more than 100 days but lost its infectivity after 24 hours when kept at room temperature (28°C) [39,40]. Therefore, two temperature indices are included: *cold_month* is 1 when the mean temperature is lower than 4°C, and zero otherwise; *hot_month* is 1 when the mean temperature is higher than 28°C and zero otherwise.
- A set of migratory bird flyway indicators that identify whether the region is on one of five specific wild bird migratory flyways, or zero otherwise. Disease clusters have occurred throughout the East Asia-Australian flyway since 2003. In the Central Asian flyway, disease clusters started emerging in July 2005 and waned in October 2005. In the Black Sea-Mediterranean flyway, clusters lasted from December 2005 to March 2006. Finally, clusters appeared in the East Atlantic and East Africa-West Asian flyway in March and April 2006, respectively [9].
- Economic variables for country, including per capita GDP, population density, and country-to-country trade in live poultry following Kilpatrick et al. [16].

When applying the econometric model to our data, the empirical model for estimation is,

$$\begin{aligned} P(AIOtbkProb_{it} = 1 \mid x_{it}) = \Phi(&\zeta_{0a} + \rho_a AIOtbkProb_{i,t-1} + \beta_{1a} wtemp_{it} + \beta_{2a} wtemp\_sq_{it} \\ &+ \beta_{3a} sptemp_{it} + \beta_{4a} sptemp\_sq_{it} + \beta_{5a} ftemp_{it} + \beta_{6a} ftemp\_sq_{it} \\ &+ \beta_{7a} wprecip_{it} + \beta_{8a} wprecip\_sq_{it} + \beta_{9a} spprecip_{it} \\ &+ \beta_{10a} spprecip\_sq_{it} + \beta_{11a} fprecip_{it} + \beta_{12a} fprecip\_sq_{it} \\ &+ \beta_{13a} cold\_month_{it} + \beta_{14a} hot\_month_{it} + \beta_{15a} \ln pimport_{it} \\ &+ \beta_{16a} \ln pexport_{it} + \beta_{17a} popden_{it} + \beta_{18a} gdpden_{it} \\ &+ \sum_{k=1}^{5} \beta_{19ak} flyway_{it} + \zeta_1 y_{i0} + \zeta \bar{x}_i) \end{aligned} \quad (8)$$

We then use estimated coefficients to derive the average partial effects (APE). If x_{jt} is a continuous variable and $\phi(\bullet)$ is the standard normal probability density function, the APE can be calculated as,

$$\hat{}_{aj} \; (\hat{}_{0a} + \hat{}_{a} y_{i,t-1} + \hat{}_{1a} y_{i0} + \hat{}_{a} x_{it} + \bar{x}_i \, \hat{}_{a}) \quad (9)$$

If x_{jt} is a discrete variable, the APE is computed as,

$$\Phi(\bullet)\,|_{x_{jt}=1} - \Phi(\bullet)\,|_{x_{jt}=0} \quad (10)$$

3. RESULTS AND DISCUSSION

In this section, we present the estimation coefficients, their associated marginal effects and outbreak response elasticities, plus results of the robustness check of using data from 2009 and 2010. We then predict the likelihood of HPAI H5N1 outbreaks under current and future climate scenarios and discuss policy implications.

3.1 Estimation Results

The estimated coefficients from the model are reported in Table 2 as are three model variants included to examine robustness of our results. There the results from the full model given in equation (8) are called model 1, and plus the results from an alternative that excludes the economic variables are labeled model 2 while another variant that excludes the indices of migratory flyways is defined as model 3. The three models give qualitatively similar results for key variables. To save space, we shall focus on model 1, given its greater log-likelihood value and smaller within-sample mean squared error.

Results of Model 1 in Table 2 suggest significant nonlinear effects of climate on HPAI H5N1 outbreaks. For illustration, we plot the marginal effects of spring temperature and winter precipitation in Fig. 3, holding all other variables at their average. Fig. 3 indicates that the probability of disease outbreaks increases as spring temperature increases and then decreases when it reaches a threshold (about 13ºC). The pattern is consistent with the cold_month dummy variable, which is positive and statistically significant (as argued to be the case in Brown et al. [39] and Shahid et al. [40]). Fig. 3 also suggests that higher risks of disease outbreak exist under heavier winter precipitation. It is possible that winter precipitation affects the risk of disease outbreaks through its impacts on water and food resources and thus affects bird population size and virus survival time [18]. With sufficient precipitation, wild birds might stop for water and food on their migration flyways, which would promote virus transmission among birds [11].

Table 2. Estimation of the dynamic Probit model with unobserved effects

	Model1	**Model2**	**Model3**
Variable	**Coefficients**	**Coefficients**	**Coefficients**
AIOtbkProb$_{t-1}$	1.4382***	1.4628***	1.4380***
	(0.0671)	(0.0664)	(0.0671)
Wtemp	0.0108	0.0100	0.0099
	(0.0124)	(0.0124)	(0.0124)
wtemp2	-0.0006	-0.0005	-0.0005
	(0.0005)	(0.0005)	(0.0005)
Sptemp	0.0350**	0.0339**	0.0355**
	(0.0174)	(0.0172)	(0.0174)
sptemp2	-0.0013**	-0.0013**	-0.0013**
	(0.0006)	(0.0006)	(0.0006)
Ftemp	0.0299	0.0281	0.0298
	(0.0207)	(0.0204)	(0.0205)
ftemp2	-0.0013*	-0.0013*	-0.0013*
	(0.0008)	(0.0008)	(0.0008)
wprecp	0.0069***	0.0063***	0.0064***
	(0.0021)	(0.0021)	(0.0021)
wprecp2	0.0000***	0.0000***	0.0000***
	(0.0000)	(0.0000)	(0.0000)
spprecp	0.0004	0.0006	0.0002
	(0.0016)	(0.0016)	(0.0015)
spprecp2	0.0000	0.0000	0.0000
	(0.0000)	(0.0000)	(0.0000)
fprecp	0.0008	0.0008	0.0007
	(0.0012)	(0.0011)	(0.0012)
fprecp2	0.0000	0.0000	0.0000
	(0.0000)	(0.0000)	(0.0000)
cold_month	0.3669**	0.3583**	0.3601**
	(0.1430)	(0.1414)	(0.1422)
hot_month	-0.0524	-0.0614	-0.0493
	(0.0952)	(0.0944)	(0.0950)
lnpimport	0.0894		0.0897
	(0.0921)		(0.0922)
lnpexport	-0.0487**		-0.0476**
	(0.0205)		(0.0203)
popden	0.0411***		0.0395***
	(0.0108)		(0.0107)
gdpden	0.1259		0.1445
	(0.0871)		(0.0891)
EAAFW	0.1376	0.7440***	
	(0.3235)	(0.2137)	
CAFW	0.3129**	0.0510	
	(0.1498)	(0.1188)	
BSMFW	0.0224	0.7805***	
	(0.3640)	(0.2084)	
EAFW	0.9670***	0.9937***	
	(0.2682)	(0.2618)	
EFWAFW	0.0211	0.9209***	
	(0.4535)	(0.2861)	

Table 2 Continued...........

Intercept	0.0043	-1.7493*	0.2971
	(1.2156)	(0.9196)	(1.0692)
LR test of	0.23	0.99	3.38**
rho=0	(0.3160)	(0.1600)	(0.0330)
LR value	-1242.99	-1258.65	-1252.15
MSE	0.0687	0.0690	0.0693
(with-in-sample)	(0.1854)	(0.1874)	(0.1863)
MSE	0.09	0.08	0.09
(out-of-sample)	(0.2187)	(0.2089)	(0.2204)

*Note: * p<0.1, ** p<0.05, *** p<0.01; standard errors are in parenthesis; Model 1 is the full model with all variables; Model 2 is the model without economic variables and Model 3 is the model without indices on wild bird flyways. We omitted estimated results of ζ_1 and ζ because these parameters do not have economic meaning*

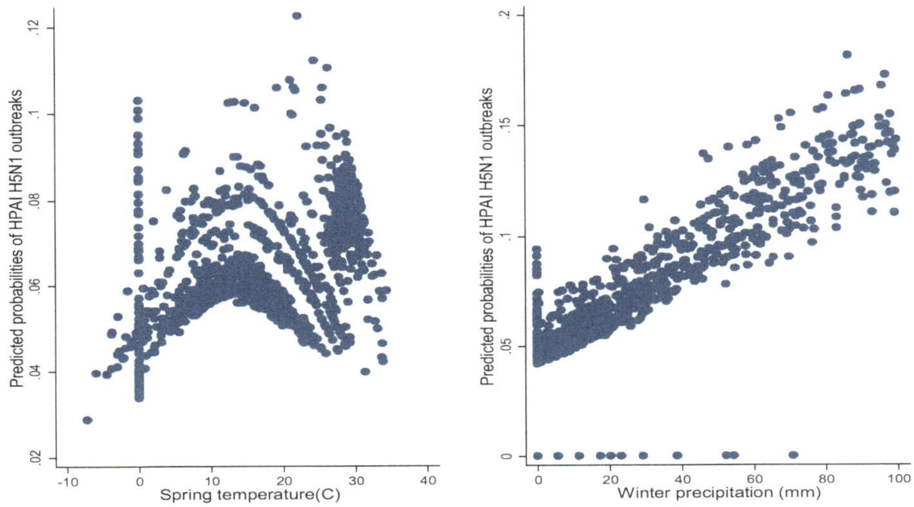

Fig. 3. Marginal effects of spring temperature and winter precipitation on HPAI H5N1 outbreak probability

(*We control all other variables at their sample mean)*

In Table 3, we report the calculated APEs and their associated elasticities. It can be seen that a 1% rise in winter total precipitation increases the risk of HPAI H5N1 outbreaks by 0.26%. Spring mean temperature was also found to have positive and significant impacts. Compared to temperature, precipitation has a larger average partial effect, increasing the risk by 0.12% (model 1).

Previous studies have examined the effect of wild bird movements on HPAI H5N1 spread [9]. Results in both Table 2 and 3 show that countries presence on a migration flyway are statistically significant and positively related to disease outbreaks. The East Atlantic flyway has the largest effects. Previous studies have found that H5N1 outbreaks in some European and African countries were very likely due to earlier seasonal movements of wild birds, particularly from the East Atlantic flyway [9-11]. Therefore, countries on flyways might need to pay more attention to disease control and prevention plans.

As expected, we find that past outbreaks significantly increase the chance of a current outbreak. A number of Asian countries succeeded in eradicating the disease, notably China, Japan, Malaysia, and South Korea, but then all of these countries experienced re-incursions [41]. These repeated outbreaks indicate virus persistence once introduced into an area [42], suggesting that the dynamics of virus survival are also important to a country's decision making on whether to implement disease prevention and control strategies.

Table 3. Average partial effects and the elasticitiy of disease outbreak (change in outbreak probability to change in the independent variable) as derived from regression models

Variable	Model 1		Model 2		Model 3	
	APE	Elasticity	APE	Elasticity	APE	Elasticity
$AIOtbkProb_{t-1}$	0.3292*** (0.0251)	0.7318	0.3468*** (0.0253)	0.7197	0.3345*** (0.0253)	0.7135
Wtemp	0.0020 (0.0022)	0.0875	0.0022 (0.0027)	0.0918	0.0012 (0.0014)	0.0497
Sptemp	0.00005** (0.0000)	0.0050	0.0000** (0.0000)	0.0030	0.0001** (0.0000)	0.0060
Ftemp	0.0002 (0.0001)	0.0171	0.0001 (0.0000)	0.0046	0.0000 (0.0000)	0.0000
Wprecp	0.0012*** (0.0004)	0.2627	0.0014*** (0.0005)	0.2714	0.0007*** (0.0002)	0.1446
Spprecp	0.0000 (0.0000)	0.0000	0.0000 (0.0000)	0.0002	0.0000 (0.0000)	0.0001
Fprecp	0.0000 (0.0000)	0.0017	0.0000 (0.0000)	0.0005	0.0000 (0.0000)	0.0000
cold_month	0.0510** (0.0238)	0.1119	0.0520** (0.0244)	0.1065	0.0517** (0.0242)	0.1088
hot_month	-0.0057 (0.0099)	-0.0164	-0.0071 (0.0103)	-0.0192	-0.0056 (0.0102)	-0.0156
Lnpimport	0.0103 (0.0102)	1.6981			0.0108 (0.0105)	1.7059
Lnpexport	-0.0055** (0.0023)	-0.6954			-0.0056** (0.0023)	-0.6750
Popden	0.0045*** (0.0012)	12.6509			0.0045*** (0.0013)	12.2528
Gdpden	0.0144 (0.0096)	2.5948			0.0164 (0.0101)	2.8406
EAAFW	-0.0124 (0.0364)	-0.1221	0.0871*** (0.0254)	0.8001		
CAFW	0.0306 (0.0223)	0.0689	-0.0085 (0.0137)	-0.0178		
BSMFW	-0.0269 (0.0341)	-0.1491	0.1301*** (0.0421)	0.6722		
EAFW	0.2023** (0.0822)	0.3317	0.2207** (0.0880)	0.3378		
EFWAFW	-0.0268 (0.0314)	-0.0220	0.2180** (0.0981)	0.1668		

*Note: * p<0.1, ** p<0.05, *** p<0.01; standard errors are in parentheses; Model 1 is the full model with all variables; Model 2 is the reduced model without economic variables and Model 3 is the reduced model without indices of wild bird flyways. Please see equation (9) and (10) for the calculation of APE; the elasticities show the effects of a 1% change in the independent variable on the outbreak probability*

We also find the probability of disease outbreaks is positively related to human population density. Apparently a country with a denser population is more likely to spread the disease through contact or trade as found in [14,16]. In addition, we find a statistically significant, negative effect of exporting live birds, showing that trade is another important factor as argued in [16]. Countries with more exports of live birds have less disease outbreaks perhaps because of a combination of a need for more careful management to meet international standards and possible underreporting.

Under alternative model specifications (i.e., model 2 and model 3), we find qualitatively similar results, suggesting that our model exhibits substantial robustness. To further test the model robustness, we use an out of sample evaluation using data from 2009 to 2010. In that case, we predict the probability of disease outbreaks and then calculate the out-of-sample mean squared errors (MSEs). We find that the MSEs across models are very close (see thebottom of Table 2), indicating that our models are stable in both specifications and predictions. Therefore, we continue our analyses to predict the disease outbreak risk using climate data from existing and future climate projections.

3.2 Outbreak Probability under Effects of Climate Change

Using estimated parameters from model 1, we first examine the shift in outbreak probability if we had or did not have the climate change of the last 20 years. Additionally, we examine the effects of projected climate change until 2030.

3.2.1 Outbreaks under past climate change

IPCC (2007) reports that the global average temperature has increased by 0.55ºC per decade from 1970-2006 [43]. Changes in overall precipitation vary by regions and seasons, but globally there has been a statistically significant 2 to 4% increase in the frequency of heavy and extreme precipitation events [44,45].

For countries in this study, Fig. 4 shows the changes in mean temperature (°C) and total precipitation (mm) for recent periods. We use the climate in 1971-1980 as the baseline, and calculate the average difference in temperature and precipitation between 1971-1980 and 1981-1990 for a 10-year comparison and the difference between1971-1980 and 1991-2000 for a 20-year comparison. These data show temperature has increased in all northern hemisphere countries relative to 1971-1980, while changes in precipitation vary across countries and time periods. In most countries, changes in precipitation are close to zero, while in others, for example, Cambodia, Japan, South Korea, Malaysia and Viet Nam, there are large variations of precipitation changes.

To predict the effect of climate change of the last few decades on outbreak probability, we use the changes between decadal average climate data for 1971-1980, 1981-1990 and 1991-2000. We evaluate this using the model 1 parameters, holding all but the climatic variables constant. We compute probability under the climate of 1971-1980, and then compute the change under1981-1990 and 1991-2000 conditions.

Fig. 5 presents these predicted probabilities for each country and shows little change in disease outbreak probability during the 1980s and 1990s. This is because with mixed changes in temperature and precipitation, effects on disease outbreak may trade-off. For example, Romania has a large increase in temperature and a modest decrease in precipitation, but since our statistical results suggest that precipitation plays a more important role in disease outbreaks, the predicted risk decreases. On the other hand, Turkey has increased risk of disease outbreaks because it has increased temperature as well as slightly increased precipitation. Although impacts from past climate change on the risk of disease outbreaks are insignificant for some countries, overall results suggest that temperature and precipitation plays a significant role in outbreaks of HPAI H5N1.

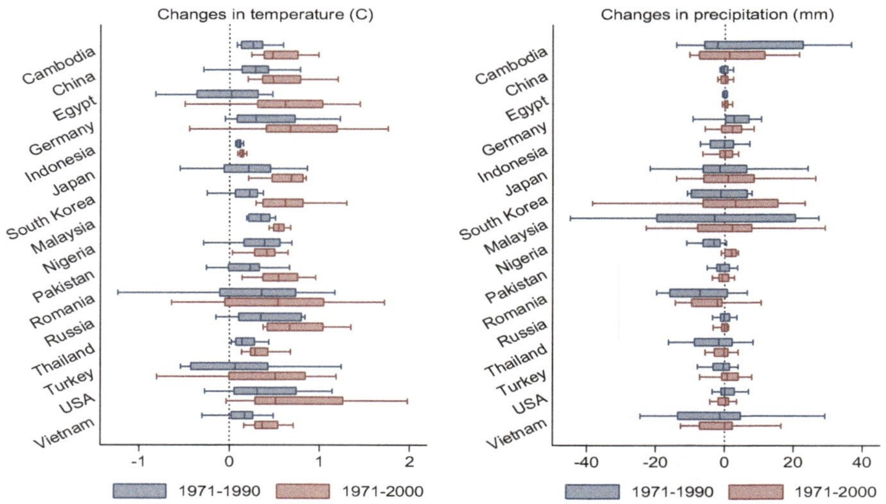

Fig. 4. Changes in winter precipitation and spring temperature between 1971-1980 and recent periods

(we use the climate in 1971-1980 as the baseline to calculate the difference of temperature (left) and precipitation (right) from 1971-1990 and 1971-2000)

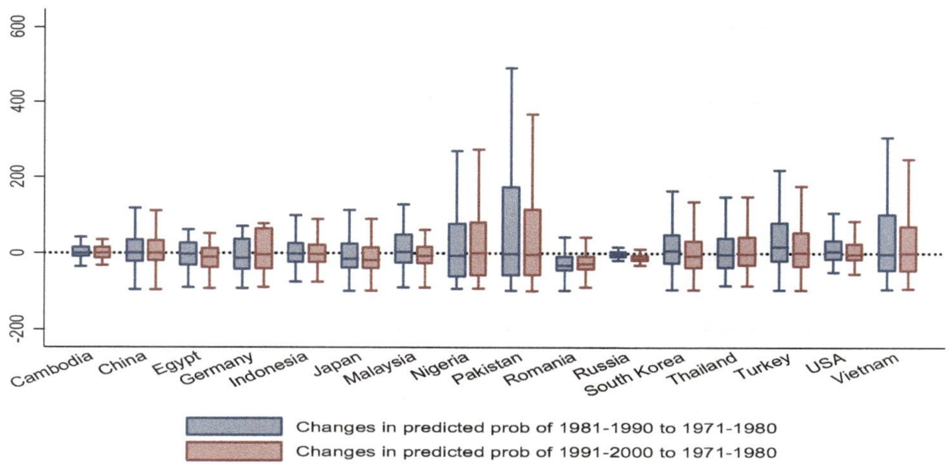

Fig. 5. Percent changes in predicted probabilities of HPAI H5N1 outbreaks under past climate change

(We use the predicted probabilities from 1971-1980 as the baseline and calculate the percent change in predicted probabilities under 1981-1990 and 1991-2000 climate conditions from the baseline, respectively)

3.2.2 Outbreak probability shifts under future climate change

For the effects of future climate change on outbreak risks, we use projected climate data from three GCMs used in IPCC AR4 (2007) under the A1B emission scenario [34] because we note that by 2030, the degree of climate change is not very sensitive to the choice of scenarios across the SRES alternatives [43].

In the climate projections (Fig. 6) all three GCMs project an increase in spring temperature in Indonesia, Japan, Nigeria, Russia, and the US, and an increase in winter precipitation in China, Germany, Japan, Korea, Nigeria, Romania, Russia and the US. There are large variations of changes in temperature and precipitation across countries and GCMs.

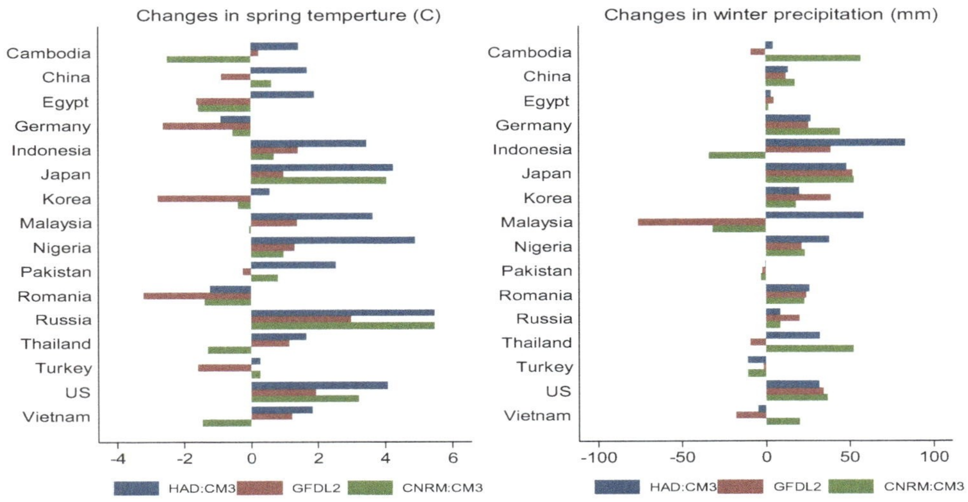

Fig. 6. Changes in spring temperature and winter precipitation under three GCMs

(we use the climate in 1961-2000 as the baseline to calculate the difference of temperature (left)and precipitation (right) from climate models by 2011-2030)

Fig. 7 shows the predicted probabilities of HPAI H5N1 outbreaks across climate scenarios. We find that projected climate change increases the risk of HPAI H5N1 outbreaks in certain countries, mainly depending on the changes in temperature and precipitation or both. In particular, we find increases in disease outbreak probabilities in Japan and Romania, because Japan is projected to have increased temperature and precipitation, and Romania will have increased precipitation and decreased temperature. Additionally, Russia has decreased risk of HPAI H5N1 outbreaks although it has increased temperature and precipitation, but with increased temperature projected in the future, there are fewer cold months with temperature below 4°C. We caution that results in Fig. 7 are not necessarily in accordance with those in Fig. 6 because results in Fig. 7 are generated by changing all climate variables in the model – not only spring temperature and winter precipitation. Note we also include the effects of climate variables that are insignificant in Table 2.

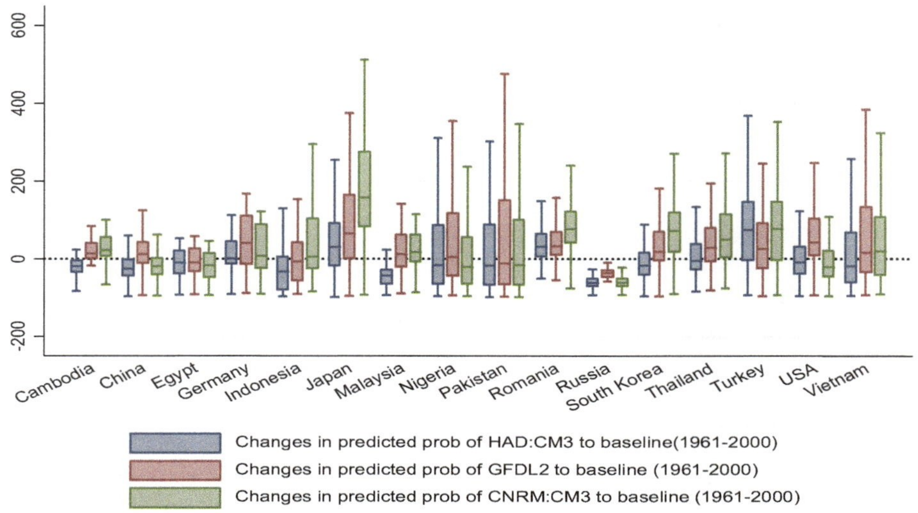

Fig. 7. Percent changes in predicted probabilities of HPAI H5N1 outbreaks under projected climate change

(*We set the predicted probabilities from 1961-2000 as the base and compute the percent changes in predicted probabilities under three GCMs climate conditions from the base, respectively)*

3.3 Discussion

Results in Fig. 7 suggest that climate change is likely to increase the risk of HPAI H5N1 outbreak in a number of countries although different GCMs provide different ranges of increase or decrease. Given this information, those countries with shifts in outbreak levels could use this information to design more aggressive disease prevention, surveillance, and control plans.

From a methodology development perspective, by including more climate models it would be possible to better capture uncertainties associated with climate projections, which has been an important topic for climate modelers and agronomic researchers [46]. This approach is applicable to risk assessment of epidemic diseases, which could provide useful insight into underlying uncertainties.

4. CONCLUSION

We examined the relationship between HPAI H5N1 outbreaks and climate change. We used an econometric model to estimate the effects of climate on the probability of disease outbreaks. The results suggest that the risk of HPAI H5N1 outbreaks is positively related to spring temperature and winter precipitation. Using projected climate data from three GCMs, we find that countries with higherspring temperature and morewinter rainfall, or both, are likely to have more disease outbreaks during the future 20 years.

Our results indicate that climate change is a likely factor affecting HPAI H5N1 outbreaks in the past and the future. Two policy implications flow from our statistical and predicted results,

- Countries which are projected to have higherspring temperatures or more winterprecipitation might enhance their planning regarding disease prevention and surveillance so as to reduce the likelihood of future disease outbreaks.
- Climate change mitigation strategies would also have co-benefits in reducing the risks of HPAI H5N1 outbreaks in the future.

Our results also show there are other important factors that influence the risk, including history of past outbreaks, human population density and amount of international trade in live birds. These factors are related to the technology and social-economic development in a country, and are difficult to project in the future. Nonetheless, further research could examine what would happen if social-economic scenarios change together with climate change.

ACKNOWLEDGEMENTS

This study was supported in whole or in part with the funds from the DHS and EPA when the corresponding author was a Ph.D. student in the Department of Agricultural Economics at Texas A&M University, College Station, Texas. We thank the editor and two anonymous reviewers for their valuable comments and suggestions.

COMPETING INTERESTS

Authors have declared that no competing interests exist.

REFERENCES

1. Cooper RG, Tomasik C, Horbańczuk JO. Avian influenza in ostriches. Avian and Poultry Biology Reviews. 2007;18(3):87-92.
2. Centers for Disease Control and Prevention (CDC). Highly Pathogenic Avian Influenza A (H5N1) Virus; 2013. Accessed 12 December 2013. Available: http://www.cdc.gov/flu/avianflu/h5n1-virus.htm.
3. Jin Y, Mu JE. Avian Influenza in China: Consumer Perception of AI and their Willingness to Pay for Traceability Labeling. Health and Animal Agriculture in Developing Countries. 2012;36:391-399.
4. Alexander DJ. An overview of the epidemiology of avian influenza. Vaccine. 2007;25:5637-5644.
5. Sims LD. Lessons learned from Asian H5N1 outbreak control. Avian Dis. 2007;51:174-181.
6. CIRAD. Avian influenza: More is now known about the role of migratory birds. Agricultural Research for Development; 2010. Accessed 12 December 2013. Available: http://www.cirad.fr/en/news/all-news-items/articles/2010/science/avian-influenza-the-role-of-migratory-birds.
7. WHO. Cumulative number of confirmed human cases of avian influenza A (H5N1); 2012. Accessed 12 December 2013. Available:http://www.who.int/influenza/human_animal_interface/EN_GIP_20120810CumulativeNumberH5N1cases.pdf.

8. Paul M, Tavornpanich S, Abrial D, Gasqui P, Charras-Garrido M, Thanapongtharm W, Xiao X, Gilbert M, Roger F, Ducrot C. Anthropogenic factors and the risk of highly pathogenic avian influenza H5N1: prospects from a spatial-based model. Vet. Res. 2010;41:28.
9. Si YL, Skidmore AK, Wang TJ, de Boer WF, Debba P, Toxopeus AG, Li L, Prins HHT. Spatio-temporal dynamics of global H5N1 outbreaks match bird migration patterns. Geospatial Health. 2009;4:65-78.
10. Si YT, Wang AK, Skidmore WF, De Boer, Li L, Prins HHT. Environmental factors influencing the spread of the highly pathogenic avian influenza H5N1 virus in wild birds in Europe. Ecology and Society. 2010;15:26.
11. Ottaviani D, de la Rocque S, Khomenko S, Gilbert M, Newman SH, Roche B, Schwabenbauer K, Pinto J, Robinson TP, Slingenbergh J. The cold European winter of 2005-2006 assisted the spread and persistence of H5N1 influenza virus in wild birds. EcoHealth. 2010;7:226-236.
12. Lowen AC, Mubareka S, Steel J, Palese P. Influenza virus transmission is dependent on relative humidity and temperature. PLoS Pathog. 2007;3(10):e151. doi:10.1371/journal.ppat.0030151
13. Liu CM, Lin SH, Chen YC, Lin KCM, Wu TSJ, King CC. Temperature Drops and the Onset of Severe Avian Influenza A H5N1 Virus Outbreaks. PLoS ONE. 2007;2:e191. doi:10.1371/journal.pone.0000191.
14. Gilbert M, Xiao X, Pfeiffer DU, Epprecht M, Boles S, Czarnecki C, Chaitaweesub P, Kalpravidh W, Minh PQ, Otte MJ, Martin V, Slingenbergh J. Mapping H5N1 highly pathogenic avian influenza risk in Southeast Asia. ProcNatlAcadSci USA. 2008;105:4769-4812.
15. Fang LQ, de Vlas SJ, Liang S, Looman CWN, Gong P, Xu B, Yan L, Yang H, RichardusJH, Cao WC. Enviromental factors contributing to the spread of H5N1 avian influenza in mainland China. PLos ONE. 2008;3:e2268.
16. Kilpatrick AM,Chmura AA, Gibbons DW, Fleischer RC, Marra PP, Daszak P. Predicting the global spread of H5N1 avian influenza. Proceedings of the National Academy of Sciences USA. 2006;103:19368-19373.
17. Ward MP, Maftei D, Apostu C, Suru A. Association between Outbreaks of Highly Pathogenic Avian Influenza Subtype H5N1 and Migratory Waterfowl (Family Anatidae) Populations. Zoonoses and Public Health. 2009;56:1-9.
18. Chen H, Smith JD, Zhang SY, Qin K, Li J, Wang KS. H5N1 virus outbreak in migratory waterfowl. Nature. 2005;436:191-192.
19. Gilbert M,Xiao X, Domenech J, Lubroth J, Martin V, Slingenbergh J. Anatidae migration in the western Palearctic and spread of highly pathogenic avian influenza H5NI virus. Emerg Infect Dis. 2006;12:1650-1656.
20. FeareCJ.The Role of Wild Birds in the Spread of HPAI H5N1 (Papel de las ayes silvestres en la diseminación del virus de influenza aviar H5N1 de altapatogenicidad). Avian Dis. 2007;51:440-447.
21. Cui P, HouY,Xing Z, He Y,Li T, GuoS,Luo Z, Yan B, Yin Z, Lei F. Bird Migration and Risk for H5N1 Transmission into Qinghai Lake, China. Vector-Borne and Zoonotic Diseases. 2011;11:567-576.
22. Pfeiffer DU, Minh PQ, Martin V, Epprecht M, Otte MJ. An analysis of the spatial and temporal patterns of highly pathogenic avian influenza occurrence in Vietnam using national surveillance data. Vet. J. 2007;174:302–309.
23. Ward MP, Maftei D, Apostu C, Suru A. Environmental and anthropogenic risk factors for highly pathogenic avian influenza subtype H5N1 outbreaks in Romania, 2005–2006. Vet Res Commun. 2008;32:627-634.

24. Williams RAJ, Fasina FO, Peterson AT. Predictable ecology and geography of avian influenza (H5N1) transmission in Nigeria and West Africa. Transactions of the Royal Society of Tropical Medicine and Hygiene. 2008;102:471–479.
25. Williams RAJ, Peterson AT. Ecology and geography of avian influenza (HPAI H5N1) transmission in the Middle East and northeastern Africa. International Journal of Health Geographics. 2009;8:47.
26. Hogerwerf L, Wallace RG, Ottaviani D, Slingenbergh J, Prosser D, Bergmann L, Gilbert M. Persistence of highly pathogenic avian influenza H5N1 virus defined by agro-ecological Niche. EcoHealth. 2010;7:213-225.
27. Farnsworth ML, Ward MP. Identifying spatio-temporal patterns of transboundary disease spread: Examples using avian influenza H5N1 outbreaks. Veterinary Research. 2009;40:20.
28. Yupiana Y, de Vlas SJ, Adnan NM, Richardus J. Risk factors of poultry outbreaks and human cases of H5N1 avian influenza virus infection in West Java Province, Indonesia. Int J Infect Dis. 2010;14:e800-5.
29. Collins MS, Tett FB, Cooper C. The internal climate variability of HadCM3, a version of the Hadley Centre coupled model without flux adjustments. ClimDyn. 2001;17:61-81.
30. SchlenkerW, Hanemann WM, Fisher AC. The Impact of Global Warming on U.S. Agriculture: An Econometric Analysis of Optimal Growing Conditions.The Review of Economics and Statistics. 2006;88:113-125.
31. Gnanadesikan A, Dixon KW, Griffies SM, Balaji V, Barreiro M, Beesley JA, Cooke WF, Delworth TL, Gerdes R, Harrison MJ. GFDL's CM2 global coupled climate models. Part II: The baseline ocean simulation. J. Clim. 2006;19:675-697.
32. Delworth TL, Broccoli AJ, Rosati A, Stouffer RJ, Balaji V, Beesley JA, Cooke WF, Dixon KW, Dunne J, Dunne KA. GFDL's CM2 global coupled climate models. Part I: Formulation and simulation characteristics. J. Clim. 2006;19:643-674.
33. Salas-Mélia D, Chauvin FD, Douville H, Gueremy JF, Marquet P, Planton S, Royer JF, Tyteca S. Description and validation of the CNRM-CM3 global coupled model; 2005. Accessed 12 December 2013.
Available: http://www.cnrm.meteo.fr/scenario2004/paper_cm3.pdf.
34. Nakićenović N, Swart R. Special Report on Emissions Scenarios. Cambridge University Press. New York, NY; 2000.
35. Wooldridge JM. A framework for estimating dynamic, unobserved effects panel data models with possible feedback to future explanatory variables. Economics Letters. 2000;68:245-250.
36. Wooldridge JM. Econometric Analysis of Cross Section and Panel Data. MIT Press: Cambridge, MA; 2002.
37. Wooldridge JM. Simple solutions to the initial conditions problem in dynamic, nonlinear panel data models with unobserved heterogeneity. Journal of Applied Econometrics. 2005;20:39-54.
38. Chamberlain G. Analysis of covariance with qualitative data. Review of Economic Studies. 1980;47:225-238.
39. Brown JD, Swayne DE, Cooper RJ, Burns RE, Stallknecht DE. Persistence of H5 and H7 Avian Influenza Viruses in Water. Avian Diseases. 2007;51:285-289.
40. Shahid MA, Abubakar M, Hameed S, Hassan S. Avian influenza virus (H5N1): effects of physio-chemical factors on its survival. Virol J. 2009;6:38.
41. Newman SH, Siembieda J, Kock R, McCracken T, Khomenko S, Mundkur T. FAO EMPRES Wildlife Unit Fact Sheet: Wildlife and H5N1 HPAI Virus -Current Knowledge. Animal Production and Health Division, Food and Agriculture Organization of the United Nations; 2010. Accessed 12 December 2013.
Available: http://www.fao.org/avianflu/en/wildlife/index.html.

42. Sims LD, Brown IH. Multicontinental Epidemic ofH5N1 HPAI Virus (1996–2007). In Avian Influenza. D. E. Swayne, edited by Blackwell, Ames, IA. 2008;251-286.
43. IPCC. Climate Change. The Physical Science Basis. Cambridge University Press, New York, NY; 2007.
44. Kunkel KE. North American trends in extreme precipitation. Nat. Hazards. 2003;29:291-305.
45. Groisman PY, Knight RW, Easterling DR, Karl TR, Hegerl GC, Razuvaev VN. Trends in Intense Precipitation in the Climate Record. J. Clim. 2005;18:1326-1350.
46. Asseng S et al. Uncertainty in simulating wheat yields under climate change. Nature Climate Change. 2013;3:827-832.

9

Stream Flow Response to Skilled and Non-linear Bias Corrected GCM Precipitation Change in the Wami River Sub-basin, Tanzania

Frank Joseph Wambura[1*]

[1]*Department of Housing and Infrastructure Planning, Ardhi University, Box 36176, Dar es Salaam, Tanzania.*

Author's contribution

The sole author designed, analyzed and interpreted and prepared the manuscript.

ABSTRACT

The reliability of stream flow projection under changing climate cannot be guaranteed if the General Circulation Model (GCM) used for the projection of future climate does not predict well its past climate. In this study stream flows in the Wami River sub-basin were simulated under changing climate by the skilled and non-linear bias corrected GCM using a physically based and semi distributed rainfall runoff model, SWAT. The SWAT model was setup using the terrain, land use, soil, precipitation and temperature data. The baseline water uses were used to naturalise stream flows and the SWAT model was calibrated and validated using the historical stream flows. In addressing future runoff projections the domestic, livestock, irrigation and industrial water demands in the sub-basin were projected to the year 2039 using the current irrigation area growth rates, Tanzania vision 2025 and development plans for the Wami River sub-basin. The GCMs were incorporated in the hydrological model so as to factor in the effects of climate change. Precipitation was selected as the changing climatic variable for projection because runoff is very sensitive to precipitation as compared to other climatic variables like temperature. A total of twenty four GCMs from CMIP3 database representing twentieth century precipitation were interpolated into forty five sub-catchments in the sub-

**Corresponding author: Email: wafrj@yahoo.com;*

basin and evaluated for their skills. The HADCM3 model was selected due to its highest skill score in predicting past climate. Then the HADCM3 precipitation signal of scenario A2, was corrected by Non-linear Bias Correction (NBC) in the forty five sub-catchments in the sub-basin and used to simulate future stream flow. The results of stream flow simulated using skilled and non-linear corrected HADCM3 precipitation signal shows that stream flow is projected to increase for the near term climatology (2010 – 2039).

Keywords: Climate change; GCM precipitation change; Non-linear bias correction; skill score; wami.

1. INTRODUCTION

Tanzania's economy continues to be dominated by agricultural production, which accounts to more than 50% of gross domestic product [1,2]. Among the nine (9) river basins in Tanzania, Wami River sub-basin, which is within the Wami-Ruvu basin has been identified to offer a potential area for irrigation and infield rainwater harvesting agriculture [3]. Thus, changes in freshwater resource availability are likely to be one of the most important consequences of projected 21st century climate change critically affecting sustainable development of life and livelihoods [4,5] in the Wami River sub-basin. The impacts of hydrological changes resulting from projected changes in climate may be particularly severe for the Wami River system given its role as a vital area for providing food, water and livelihoods. Previous studies of the hydrological impacts of potential climate change in Tanzania have generally focused on climate forcings from selection of GCM (s), depending on the criteria of being whether most wet or dry [6]. Although the GCM simulated temperature can be relatively consistent between GCMs, the same is not true for precipitation. Indeed, projections of future precipitation from different GCMs often disagree, even in the direction of change [7]. Therefore, there is a need to address the important issue of GCM performance score in predicting past climate before it is entrusted to be used in stream flow projection. GCM performance test can be done by several methods which compare observed and historical predictions of GCMs; some of these methods are the mean and variance method and the skill score test. The mean and variance method do not capture the extremes in the process of comparing the observed data and the GCM historical predictions. Further, the use of statistics like means and standard deviations do not allow for a comparison of the entire data distribution [8]. There is at least one major advantage of evaluating a climate model based on probability density functions (PDFs) using the skill score test. If a climate model can simulate an entire probability density function (PDF), this demonstrates a capability to simulate values that are currently rare and that may become more common in the future [9]. Thus, skill score test accounts for spatial variation in comparing modelled and measuered climate parameters. However, even for the skilled GCM, its projection cannot be used without being downscaled because GCMs have coarse resolution. Most of the GCMs have a resolution of 1.3° x 2.7° latitude and longitude scale.

There are many methods available for downscaling GCM projections to a specific region or study area of interest, for discriminating between mean changes and changes in climatic variability and for ensuring consistency between climate change scenarios. The method ranges from complex procedures such as dynamical and statistical downscaling to simple approaches like bias correction methods [10,11,12]. However, Fowler et al. [12] argued that simple methods have been used for downscaling and found to be effective in simulating hydrological systems.

Dynamic downscaling is a method of extracting local scale information by developing and using limited area models (LAMs) or regional climate models (RCMs) with the coarser resolution GCM data used as boundary conditions. The basic steps are then to use the GCMs to simulate the response of the global climate. But the demerit with this method is that, the downscaling process requires computing faculties with very high computing efficiency for the computation process [10]. Therefore, this method cannot be applied in small scale or ordinary computer laboratory.

Statistical downscaling involves developing a quantitative relationship between large scale atmospheric variables (predictors) and local surface variables (predictands). From this perspective, regional or local climate information is derived by first determining a statistical model which relates large scale climatic variables to regional and local. Then the large scale output of a GCM simulation is fed into the statistical model to estimate the corresponding local and regional climate characteristics. Many statistical downscaling techniques have been developed to translate large-scale GCM output into a finer resolution. The simplest method is to apply GCM-scale projections in the form of bias correction method [12].

The most common bias correction methods for projected precipitation are simple delta method, also called Linear Bias Correction (LBC) and Non-linear Bias Correction (NBC). By LBC, the ratio of means between the future projection of GCM and its baseline is applied to the observed data to get future precipitation [11] but NBC involves the application of ratio of means between the calibrated future projection of GCM and its calibrated baseline to the observed data to get future precipitation. The main difference between LBC and NBC is that, although both approaches involve the application of GCM precipitation signal to the observed precipitation, but NBC applies the calibrated signal. To estimate the stream flow response to skilled and non-linear bias corrected GCM precipitation change, the hydrological model is the central part of linking the projected precipitation with the stream flows.

In deterministic context the hydrological models are classified as: (i) data-driven models, (ii) conceptual models; and (iii) physically based models. In this study a physically distributed rainfall runoff model known as Soil and Water Assessment Tool (SWAT) was used. This is because of the extensiveness of the sub-basin as well as the need to incorporating most of the sub-basin features in the modelling activity. SWAT is one of the common physically distributed models. It is a continuous-time distributed simulation watershed (catchment or sub-basin or basin) model [13]. SWAT is described in detail by Neitsch et al. [14], so only a brief summary of the model is provided here. SWAT was developed to predict the effects of alternative management decisions on water, sediment and chemical yields with reasonable accuracy. It uses readily available inputs, thus minimum data are required to make a run. SWAT is computationally efficient, thus simulation of very large watersheds or a variety of management strategies can be performed without excessive investment of time or money. SWAT provides an option of estimating surface runoff using the SCS curve number procedure [15]. The SCS curve number is a function of the soil's permeability, land use and antecedent soil water conditions [15]. A minimum data requirement in estimating the potential evapotranspiration is also an advantage of using SWAT model, because it incorporates the Hargreaves method which requires only air temperature in estimating potential evapotranspiration [14]. SWAT is capable of simulating a number of climate options. Climate change can be simulated with SWAT by manipulating climatic variables (precipitation, etc.) that are read into the model in each sub-catchment. It also has a less time-consuming method in setting adjustment factors for the various climatic variables. SWAT allows the adjustment terms to vary from month to month, so that the user is able to simulate seasonal changes in climatic conditions. In Nilotic countries, including Tanzania,

Kenya, Ethiopia, Rwanda, Uganda and Burundi, SWAT model has been applied in catchments for various applications [16]. Therefore, in this study the stream flow response to skilled and non-linear bias corrected GCM precipitation change, involved setting up sub-basin SWAT model, calibration and validation of the same model. Then testing of the skilled CMIP3 climate models and bias correction of the skilled GCM followed. And finally the SWAT model was customized by the projected skilled and bias corrected GCM precipitation change to simulate the stream flow response.

2. MATERIALS AND METHODS

2.1 The Wami River Sub-basin

The Wami River sub-basin is located between 5ºS to 7ºS and 36ºE to 39ºE. It extends from the semi-arid areas in central Tanzania to the humid inland swamps in east-central Tanzania to the Indian ocean (Fig. 1). It encompasses an area of approximately 41,167 km^2. The topography of the Wami river sub-basin ranges from 2 m to 2370 m above sea level. The upstream and downstream parts the sub-basin are separated by the Eastern Arc Mountains (Nguru, Ukaguru and Rubeho mountain ranges). The Wami River sub-basin has three major catchments: Kinyasungwe, Mkondoa and Wami. Each catchment has a downstream flow gauging station, namely 1GD-Godegode, 1G1-Dakawa, and 1G2-Mandera, respectively (Fig. 1). However, this study used only the 1G2 – Mandera gauge in calibration and validation of the hydrologic model due to the length of its readily available historical records as compared to other gauges. Two representative rainfall stations (9536017-Ikombo and 9636018 -Ukaguru forest) were also used to address the rainfall variability within the sub-basin.

The sub-basin has an average rainfall of 550 to 1000 mm per annum. There are two rainfall zones in the sub-basin, the western and south–west part falls within the unimodal rainfall zone (one wet period in December, January, February, March and April (DJFMA)) and the east and north–east part of the sub-basin falls within the bimodal rainfall zone (two wet periods, first rains in October, November and December (OND) and second rains in March, April and May (MAM)). Kinyasungwe catchment falls within the unimodal rainfall zone, whereas Mkondoa and Wami catchments fall within the bimodal rainfall zone. The mean annual temperature in the sub-basin ranges from 12 to 24 degree Celsius. The low lands are warm whereas the highlands are cold. The main uses of water in Wami River sub-basin are domestic, livestock, industrial and irrigation.

2.2 Data Used

In this study, data used for setting up the hydrological model were land cover (shapefile format) of scale of 1:250,000 from the Africover Land cover Classification and Natural Resource Services, FAO (1997); soil (shapefile format) of scale of 1:2,000,000 from the soil and terrain database for southern Africa developed by International Soils Reference and Information Centre, ISRIC, (2003); Digital Elevation Model (raster) of 90m resolution downloaded from the USGS/NASA STRM-DEM (2007); verified stream network of Wami River from the Ministry of Water (MoW); daily rainfall and temperature data from the Tanzania Meteorological Agency (TMA); Population data (shape file) from National Bureau of Statistics (NBS); measured stream flows at 1G2 –Mandera and water uses (irrigation and industrial water uses) from Wami-Ruvu Basin Office (WRBO) and Basin Irrigation Department (BID).

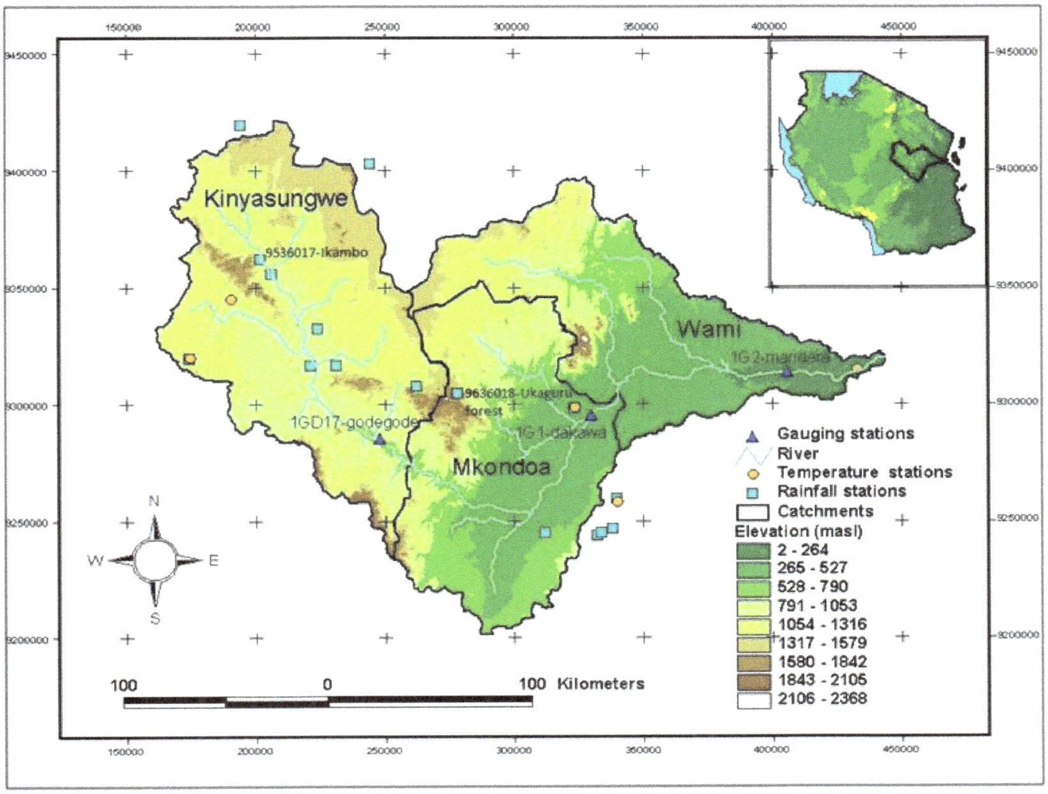

Fig. 1. Wami river sub-basin

Data used for climate customisation in the hydrological model was projected precipitation from a skilled and non-linear bias corrected GCM. For skill score test, 24 GCMs precipitation with resolution ranging from 150 to 300 km^2 were downloaded from the CMIP3 database in the NetCDF format.

2.3 Hydrologic Modelling

First, the sub-basin was divided into forty five (45) sub-catchments using the DEM which was burnt in with the verified river network dataset and the topography of each sub-catchment was derived from the same DEM, then land cover and soil were used to generate default flow parameters.

Secondly, rainfall and temperature data were used to drive the Wami River sub-basin climate. The average baseline climatological domestic, livestock, irrigation and industrial water uses were distributed in sub-catchments to naturalize stream flows. Computation of evapotranspiration was done using the Hargreaves method and the flows were routed using the Muskingum algorithm. SWAT model used water balance equation (Equation 1) as a driving force in the sub-basin [14].

$$SW_t = SW_0 + \sum_{i=1}^{t} \left(R_{day} - Q_{surf} - E_a - W_{perc} - Q_{gw}\right) \tag{1}$$

Where SW_t and SW_o are final and initial soil water content (mm/d) respectively; *t* is the time (day); R_{day} is the precipitation (mm/d); Q_{surf} is the runoff (mm/d); E_a is the evapotranspiration (mm/d); W_{perc} is the percolation (mm/d); Q_{gw} is the return flow (mm/d).

Finally, the stream flow data between 1978 and 1988 were used for calibration and between 1993 and 1996 for validation of the SWAT model. Data from the most downstream gauging station, 1G2-Mandera was used for calibration using the SWAT Calibration and Uncertainty Program (SWATCUP) because of its capability of calibrating the records with missing data. The procedure used in SWATCUP was Sequential Uncertainty Fitting (SUFI). This procedure was preferred over other methods because of its efficient optimization [17]. SUFI procedure is an inverse optimization approach that uses the Latin Hypercube Sampling (LHS) procedure along with a global search algorithm to examine the behaviour of objective functions. The procedure is sequential in nature, meaning that one more iteration can always be made before choosing the final estimates. The procedure has a Bayesian framework, indicating that the method operates within uncertainty domains (prior, posterior) associated with each parameter. It is a fitting procedure, conditioning the unknown parameter estimates on an array of observed values. Nevertheless, SUFI procedure is iterative, requiring a stopping rule which was provided by a critical value of a goal function, Nash-Sutcliffe coefficient [18].

In calibration and validation, the objective function (Equation 2) used for optimization was Nash-Sutcliffe coefficient (NSE). NSE provides a normalized estimate of the relationship between the observed and projected model values [19]. Legates and McCabe [20] argued that NSE is the most appropriate goodness-of-fit measures available because of its ability for physical interpretation owing to their meaningful comparison with zero (i.e., comparison with a long-term mean).

$$\mathrm{NSE} = 1 - \frac{\sum_{i=1}^{n}(S_i - O_i)^2}{\sum_{i=1}^{n}(O_i - \bar{O}_i)^2} \qquad (2)$$

Where n is the number of records of the time series, S_i is the simulated variable, O_i is the observed variable and $\bar{O}_i$ is the mean of the observed variable.

In model setup and simulation of the baseline conditions the 1980-2009 average water uses were used, while the 2010-2039 average water uses were used in simulation of the response of stream flows to climate change. Table 1 shows the summary of the baseline and future water uses used in this study. The future water uses were estimated under the consideration of Tanzania development goals [21], current population and irrigation area growth rates and development plans for Wami-Ruvu Basin, in which Wami River sub-basin is a part.

Table 1. Summary of water uses in Wami river sub-basin

		Population	Livestock	Irrigation	Industries
(1980-2009)	Users	1.5 mil	0.9 mil	6,722 ha	1*
	Demand (Lt/day)	38 mil	22.6 mil	1,320 mil	129.6 mil
(2010-2039)	Users	2.8 mil	1.3 mil	18,672 ha	2
	Demand (Lt/day)	72 mil	33.3 mil	2,964 mil	145 mil
	Increase	89%	47%	124%	12%

*Note: mil stands for million and Lt/day means litres per day; * number of industries (unit-less)*

2.4 Interpolation of GCMs Grids

Since most GCMs have an average resolution of 1.3° x 2.7° latitude and longitude scale, therefore interpolation of the GCM grids to the point of interest was necessary before further use of GCM data for skill score testing and bias correction. However, there are many techniques available for interpolating GCM grids to the specific region of interest. In this study, the Inverse Distance Weighting (IDW) method was chosen to be used. IDW was selected because of the assumption that things that are close to one another are more alike than those that are farther apart. Thus, IDW assumes that each known point has a local influence that diminishes with distance, and the value at the unknown point is a weighted sum of the values of the known points. Detailed explanations of IDW in interpolation of climatic variables are found in, Zhu and Jia [22] and Lin and Yu [23].

Six points of GCM grids were found covering the Wami River sub-basin; of these points, each has its own distance from each of the 45 sub-catchments in the sub-basin. The interpolation of the six (6) grid points of GCMs to each sub-catchment used the weighting factor of two (2), such that the influence of one point relative to another is a function of inverse squared distance. Weighting factor was assumed as two (2) as used by Zhu and Jia [22] and Lin and Yu [23]. Therefore, the time series of precipitation of six (6) GCM grid points were interpolated into sub-catchments' precipitation GCM time series.

2.5 Skill Score Tests

The record length between 1973 and 2000 for both observed and GCM baseline precipitation was used for skill score testing. The missing observed precipitation (mm/month) from the historical record for the 45 sub-catchments were filled by the WXGEN algorithm developed by Nicks [24]. This is the weather generator embedded in SWAT.

In skill score test, probabilities were derived for observed and the 24 GCMs precipitation for all 45 sub-catchments. Then the GCMs' skill scores were determined for each sub-catchment using Equation 3 [8]. This test compares the similarity between two (2) PDFs; it allows a comparison across the entire PDF. The skill score is the cumulative minimum value of two distributions of each binned value, thereby measuring the common area between GCM and observed precipitation. If the PDF of observed precipitation is totally different from the PDF of GCM precipitation, the skill score equals zero with negligible overlap between the observed and GCM precipitation PDFs, but if the PDF for observed and GCM precipitation is exactly the same, the skill score becomes unity [8]. Therefore the skill score ranges between zero and unity. This test is a very simple measure that provides a robust and comparable measure of the relative similarity between GCM and observed PDFs.

$$\text{Skill score} = \sum_1^n \text{minimum}\left(\mathrm{P}_{o(\mathrm{k})} \;,\; \mathrm{P}_{(\mathrm{k,r})} \right) \tag{3}$$

Where n is the fixed number of bins, because in robust comparison PDFs should be of the same class marks. $\mathrm{P}_{o(\mathrm{k})}$ is a probability of monthly precipitation in a given bin from the observed data in sub-catchment, *k*, and $\mathrm{P}_{(\mathrm{k,r})}$ is the probability of monthly precipitation in a given bin from GCM, *r,* in the sub-catchment, *k*. Probability is computed as the relative frequency of a given bin value.

The skill score tests were done for precipitation of all 24 GCMs against observed precipitation in the 45 sub-catchments and the threshold score used was 75% as the average value [8]. The performances of each GCM for all 45 sub-catchments across the sub-basin were used to evaluate the average, lower and upper dispersion bounds of skill scores for the GCMs. And GCM with the highest average score and the lowest range (uncertainty) between lower and upper scores across the sub-basin was selected as the most skilled GCM.

2.6 Non-linear Bias Correction of Skilled GCM

In bias correction of skilled GCM precipitation change, the new baseline (1980 – 2009) was chosen because it incorporates some of the strongest natural variability of climate, including the strongest El Nino Southern Oscillation (ENSO) warm event in 1997/1998 to a strong La Nina cold event in 1999/2000 [25]. And the scenario selected for prediction of near term (2010-2039) climatology was A2 because it expresses the criticality of a more divided world, independently operating, self-reliant nations, continuously increasing population and regionally oriented economic development [26].

In this study, the LBC, which is explained in detailed by Graham et al. [27], Sperna et al. [28], Eisner et al. [29] and Watabane et al. [30], was modified with the calibrated baseline and calibrated scenario GCM to obtain the NBC (Equation 4). NBC method corrects the coefficient of variation of the GCM precipitation signal [30,31,32,33,34,35]. This method is based upon transferring the calibrated monthly precipitation change between GCM baseline and GCM scenario to a daily observed time series. NBC uses only parameter, *a,* because it is only the ratio of precipitation variability signal which is required to modify the observed precipitation to simulate the future precipitation.

$$\mathrm{PCP}_{\mathrm{future}\,(i,j,k)} = \mathrm{OBS}_{(i,j,k)} \times \left(\frac{\mathrm{GCM}_{\mathrm{future}\,(j,k,r)}}{\mathrm{GCM}_{\mathrm{baseline}\,(j,k,r)}}\right)^{a} \tag{4}$$

Where $\mathrm{GCM}_{\mathrm{baseline}\,(j,k,r)}$ and $\mathrm{GCM}_{\mathrm{future}\,(j,k,r)}$ are baseline and future precipitation of GCM, *r,* in month, *j,* in the sub-catchment, *k,* respectively. $\mathrm{PCP}_{\mathrm{future}\,(i,j,k)}$ and $\mathrm{OBS}_{(i,j,k)}$ are the projected and observed precipitation on day, *i*, in month *j* in sub-catchment, *k,* respectively. Whereas *a* is the unit-less calibration parameter.

Therefore, bias correction of skilled GCM precipitation change of scenario A2 for the near term (2010-2039) climatology involved obtaining the calibration parameter, *a* in trying to match the monthly coefficient of variation of precipitation of the skilled GCM in the baseline period against observed precipitation [32,33,34,35] using equation 5. Then the same parameter, *a,* was used in equation 4 in calibration of the precipitation signal.

$$\mathrm{C_V}\left(\mathrm{OBS}_{(j,k)}\right) \cong \mathrm{C_V}\left(\left(\mathrm{GCM}_{\mathrm{baseline}\,(j,k,r)}\right)^{a}\right) \tag{5}$$

Where $\mathrm{C_V}$ is the coefficient of variation (unit-less), $\mathrm{OBS}_{(j,k)}$ is the observed precipitation in month, *j,* in sub-catchment, *k*.

2.7 Climate Customisation in SWAT

In SWAT, climate customisation was done by adjusting factors for the precipitation variable for each of the 45 sub-catchments from January to December. The adjustment factor (Equation 6) for precipitation customisation was derived from equation 4 (NBC) so that it can fit in equation 7, which is incorporated in the SWAT model [14]. Mathematically the substitution of equation 6 in equation 7 gives equation 4.

$$\mathrm{adj}_{(j,k,r)} = \left(\left(\frac{\mathrm{GCM}_{\mathrm{future}\ (j,k,r)}}{\mathrm{GCM}_{\mathrm{baseline}\ (j,k,r)}}\right)^{a} - 1\right) \times 100 \quad (6)$$

$$\mathrm{PCP}_{\mathrm{future}\ (i,j,k)} = \mathrm{OBS}_{(i,j,k)} \cdot \left(1 + \frac{\mathrm{adj}_{(j,k,r)}}{100}\right) \quad (7)$$

Where $\mathrm{PCP}_{\mathrm{future}\ (i,j,k)}$ and $\mathrm{OBS}_{(i,j,k)}$ are the projected and observed precipitation on day, *i,* in month, *j,* in sub-catchment, *k,* respectively. The adjustment factor, $\mathrm{adj}_{(j,k,r)}$ is the projected change in precipitation (%) in a given month, *j,* in sub-catchment, *k,* for GCM, *r.* $\mathrm{GCM}_{\mathrm{baseline}\ (j,k,r)}$ and $\mathrm{GCM}_{\mathrm{future}\ (j,k,r)}$ are monthly baseline and future precipitation of GCM, *r,* in month, *j,* in sub-catchment, *k,* respectively. And, *a,* is the unit-less calibration parameter.

3. RESULTS AND DISCUSSION

3.1 Calibration and Validation of the SWAT Model

The SWAT model was calibrated and validated with daily observed historical records at the most downstream gauging station (1G2 – Mandera). The period from 1978 to 1988 was considered for calibration while the model was validated with historical records from 1993 to 1996. The NSE for calibration and validation were 70% and 74%, respectively. NSEs for both calibration and validation were considered satisfactory because Santhi et al. [36] and Benaman et al. [37] suggested that the prediction efficiency of the calibrated model can be judged as satisfactory if NSE values are greater or equal to 60%. However, Moriasi et al. [38] argued that NSE greater or equal to 65% in calibration and validation, the model performance is considered good. But Saleh et al. [39] and Rossi et al. [40], argued that NSE greater or equal to 65% the model performance is very good. However, there is no consensus on what is the appropriate criterion to be used in evaluation of model performance; therefore visual inspection of time series was used in drawing conclusion on the extent and ability of the model to mimic the reality [20].

Fig. 2 shows the time series comparison of simulated and observed daily flow in the Wami river sub-basin during the calibration period. The model captures well the low flows and some peaks, although the highest flow peak of 1411 m^3/s (cms) was not well captured by the model. Looking on the other peaks before and after the highest peak, it was agreed that the highest peak might be an outlier or error resulting from flow measurements. Therefore the visual inspection of time series together with the statistical evaluation (NSE= 70%) in the calibration period was considered satisfactory. During the validation period (Fig. 3) the time series of comparison between the simulated and the observed measured stream flow at the 1G2 – Mandera gauge shows that the model captures all the peaks although there is slight over prediction of low flows in some days. The times series of comparison shows that the model is able to mimic stream flow because the simulated flow matches with the measured flow. The times series observations, together with the statistical evaluation (NSE= 74%) of the model concludes that the model is also satisfactory in the validation period.

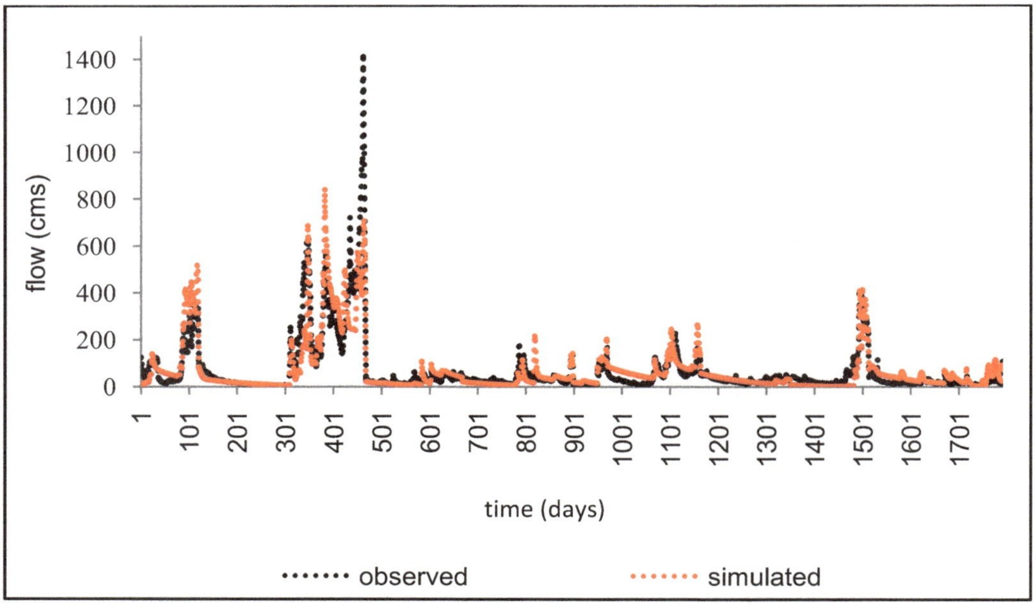

Fig. 2. Calibration curves at 1G2-Mandera, 1978 - 1988 (missing data skipped)

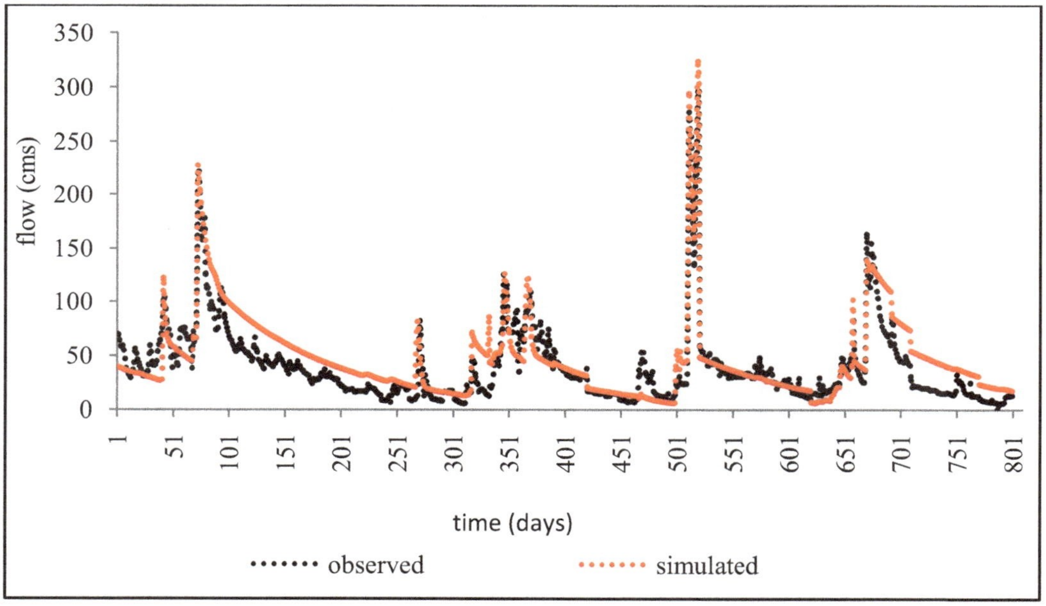

Fig. 3. Validation curves at 1G2-Mandera, 1993-1996 (missing data skipped)

3.2 Skill Score Test Results

In the skill score test, baseline/control predictions of 24 GCMs were tested against observed precipitation in each sub-catchment. And the average, lower and upper dispersion bound scores for each GCM were obtained from the sub-catchments' performances in order to represent the sub-basin skill score statistics. On average performance of GCMs across the sub-basin, it was found that GISS-EH has the lowest average sub-basin skill score, whereas

ECHAM4, ECHO, HADCM3 and HADGEM1 have average sub-basin skill scores above the threshold of 75.0% (Fig. 4). Among the four GCMs with skill scores above the threshold value, HADCM3 has the highest average sub-basin skill score of 79.0%. The other GCMs namely, ECHAM4 and ECHO both have the same average sub-basin skill score of 75.5%, but HADGEM1 has the average sub-basin skill score of 76.0%. Depending on the highest performance of GCMs across the sub-basin, it is obviously that HADCM3 was supposed to represent the future climate of the sub-basin. But nevertheless, the uncertainty of GCM performance across the sub-basin was also considered.

Uncertainty range was considered to be the interval between the upper and lower dispersion bounds of the skill scores of the 45 sub-catchments in the sub-basin. Statistically, it is twice the standard deviation of the skill scores. The uncertainty range measures the amount of variation or dispersion of the sub-catchments' skill scores from the average sub-basin skill score. A low range indicates that many sub-catchments' skill scores tend to be very close to the average sub-basin skill score; a high uncertainty range indicates that many sub-catchments' skill scores are spread out over a large range of scores. Therefore GCM with the lowest uncertainty range is the one which predicts well the natural variability of precipitation in the sub-basin, because it has the ability to model the spatial distribution of precipitation. The higher value of uncertainty ranges signifies that the GCM models only some parts of the sub-basin's precipitation patterns, whereas most parts are not well modelled. The uncertainty ranges also do identify how far the average skill scores are skewed to the few high skill scores in the sub-basin. In this study the uncertainty range of 10% was considered as the maximum threshold for selection of GCMs. This means that the sub-catchments' skill scores of prediction for a particular GCM are allowed to differ by ± 5% from the average sub-basin skill score.

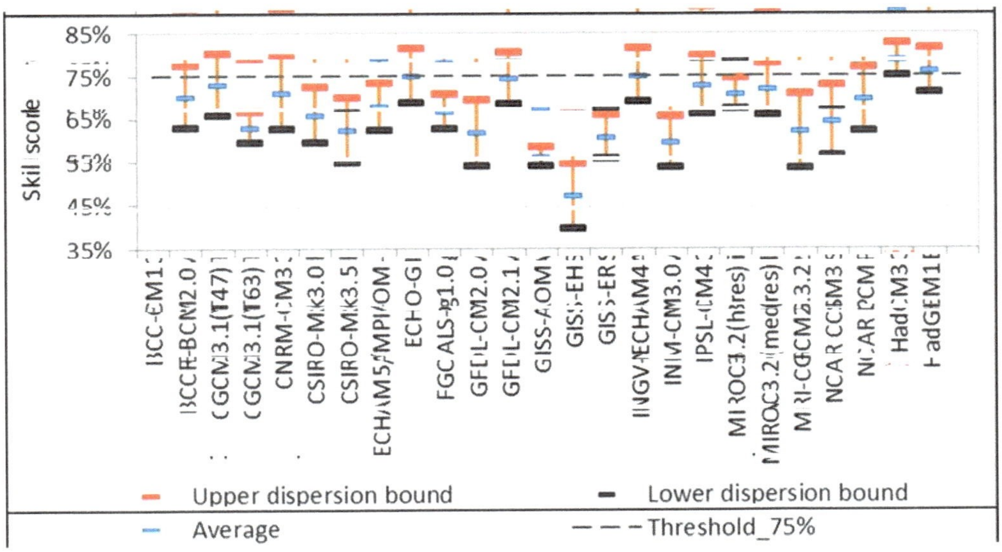

Fig. 4. Skill scores for GCMs against Observed data

Fig. 4 shows that five GCMs have an uncertainty range below 10%, these GCMs are GISS-AOM, CGCM3.1 (T63), HADCM3, MIROC3.2 (hires) and FGOALS-g1.0. GISS-AOM has the lowest uncertainty range of 4.3%, followed by CGCM3.1 (T63) and then HADCM3 with an uncertainty range of 7.3%. However the other highly skilled GCMs like ECHAM4, ECHO-G and HADGEM1 have uncertainty ranges of 12.2%, 12.8% and 10.2% respectively.

According to the uncertainty range of performance across the sub-basin, GISS-AOM, CGCM3.1 (T63), MIROC3.2 (hires) and FGOALS-g1.0 could be considered, but since their average sub-basin skill scores are far less than the threshold value, thus they were removed from consideration. Between the set of four (4) highly skilled GCMs and the set of four (4) low uncertainty range GCMs, it is only HADCM3 which is found in both sets. Therefore, HADCM3 was selected to simulate future precipitation in the Wami river sub-basin. However, the skill score test, like any other performance or efficiency measures of a model, has to be supported by visual inspection of PDFs of GCM and measured precipitation. Fig. 5 shows that, the PDFs of HADCM3 and measured precipitation matches very well from the average to high precipitation, although there is slight under prediction of low precipitation (<28 mm/month). This means HADCM3 slightly under predict "drizzle" in the Wami River sub-basin.

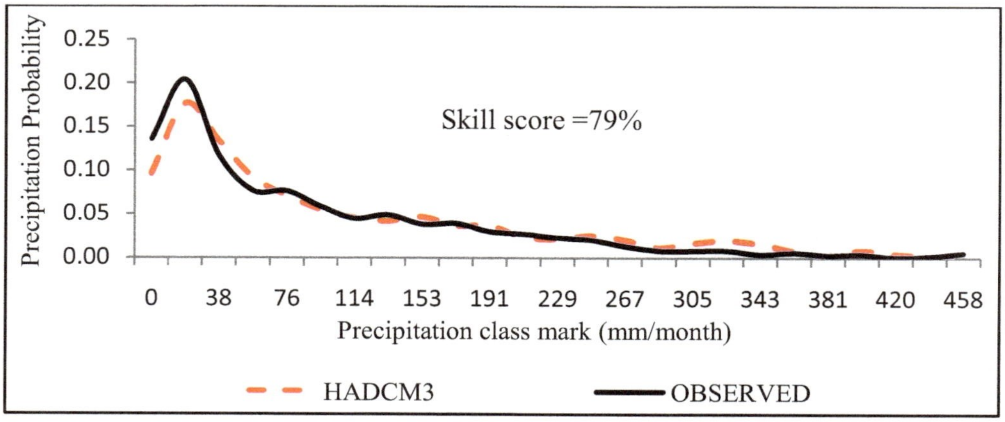

Fig. 5. GCM against Observed data

3.3 Measured and HADCM3 Scenario Precipitation

The average measured precipitation (1980–2009) in the upstream (Kinyasungwe catchment) of sub-basin is heavy from December to April, but decreases from May to October and then start to rise again from November towards December (Fig. 6a). This part of the sub-basin falls in the unimodal rainfall zone, the representative rainfall station (Fig. 6a) in the unimodal rainfall zone shows that the heaviest rainfall is experienced in DJFMA with the average measured rainfall (1980-2009) of 105 mm/month at 9536017-Ikombo station. Small amounts of rainfall are experienced from June to October with 11 mm/month in average. The temporal variability of measured rainfall (1980-2009) is in the range of ±62 mm/month. The HADCM3 scenario (2010-2039) at 9536017-Ikombo station (Fig. 6b) shows that the heaviest precipitation is projected to be in March with a magnitude of about 300mm/month. Fig. 6b shows that an average increase in precipitation is projected (2010-2039) for the DJFMA period; and it is estimated to be around 106%, whereas an average decrease from June to November is projected in the magnitude of 25%.

Fig. 6c shows that, the measured precipitation in the downstream part of the sub-basin (Mkondoa and Wami catchments) is heavy in the periods of MAM and OND. This is because this area experiences bimodal rainfall. The measured precipitation at 9636018 - Ukaguru forest station (representative rainfall station in the bimodal area of the sub-basin) has an average (1980-2009) amount of 74 mm/month in the MAM season and 73 mm/month in the OND season. The average rainfall in the little rain period of June to September is 39 mm/month and the heaviest rainfalls are found between March and April in the case of MAM season whereas in the OND period, heavy rainfalls are in December. The average temporal variability of measured precipitation (1980-2009) ranges between ±78 mm/month. The scenario predictions at 9636018 - Ukaguru forest station (Fig. 6d) shows that the heaviest precipitation (2010-2039) is projected to be in April with the magnitude of about 191 mm/month for the MAM season and in November (97 mm/month) for the case of OND season. In the MAM season the average increase in precipitation is projected to be 100% and from June to September the decrease in precipitation is projected to be 45%, but OND average precipitation is projected to decrease by 9%.

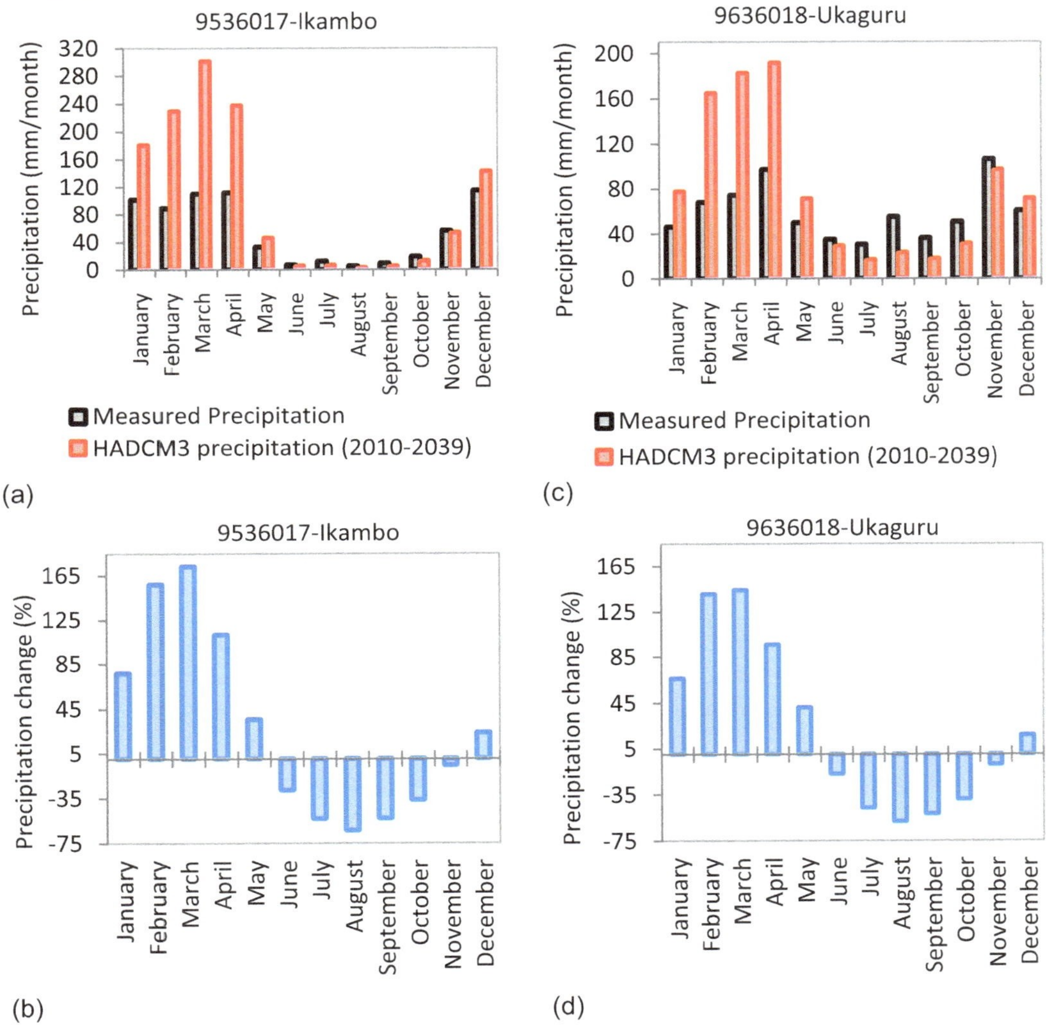

Fig. 6. Average precipitation of representative stations

In comparing precipitation of the upstream and downstream parts of the sub-basin, it is found that the Kinyasungwe catchment has one season of rainfall and the average annual measured rainfall (1980-2009) is around 670 mm but the Mkondoa and Wami catchments, which have two seasons of rainfall, have average annual rainfall amounts of 713 mm. The HADCM3 average annual increase of precipitation (2010-2039) is projected to be 81% in the upstream and 36% in the downstream. The temporal variability of measured rainfall in the Mkondoa and Wami catchments is larger than that of Kinyasungwe catchment; this is partly contributed by the presence of two seasons of rainfall in the downstream part. However, the topography shows that the upstream part of the sub-basin is on the leeward, whereas the downstream part is on the windward side of the Eastern Arc Mountains. Therefore, the influence of Indian Ocean cyclonic rains is thought to be blocked by the Eastern Arc Mountains in reaching the upstream part of the sub-basin. Hence, for this reason, more rain days are encountered on the downstream part compared to the upstream part. HADCM3 prediction (2010-2039) shows that, pattern of temporal variability is the same as the one of the measured rainfall and both upstream and downstream parts of the sub-basin exhibit decrease in rainfall between June and November but increase from December to May.

3.4 Stream Flow Responses

Upstream of the sub-basin (Fig. 7a) shows that, the patterns for average baseline flow (1980-2009) and the average HADCM3 projected flow (2010-2039) are similar. Both flow curves show that, stream flow at 1GD17- Godegode is high from December to May and recedes towards September. However, there is a slight increase of stream flow in October towards November. The maximum average baseline flow occurs in April (70 m^3/s) whereas the minimum average baseline flow occurs in August (0.6 m^3/s). For HADCM3 flow projection, the average maximum projected flow occurs in April and the minimum average projected flow occurs in October. Fig. 7a also shows that, HADCM3 projects a high increase (+392 m^3/s) in flow from February to August as the result of MAM rainfalls, but a decrease (-39%) in flow is projected from October to December. However, there is a slight increase of flow in January. The sudden deep drop of flow is projected in October, this is the most severe decrease (-99%) of flow projected by the HADCM3 in the upstream of Wami River sub-basin.

Fig. 7b shows that, at 1GD17 the baseline flow experience a slight lag between peak of precipitation and peak of flow due to the time of concentration for runoff to reach the outlet of the Kinyasungwe catchment. However, the bank storage recharge, groundwater recharge and infiltrations in the catchment are also thought to contribute to this lag in flow. This lag in flow is not experienced in the case of projected flow because HADCM3 projects a high increase in precipitation, thus the bank storage, groundwater storage and infiltrations saturate earlier than in the case of baseline flow (Fig. 7b).

At 1G1-Dakawa, in the Mkondoa catchment, the average baseline flow has the same pattern as the HADCM3 average projected flow (Fig. 8a). The highest average baseline flow occurs in March (56 m^3/s) and the lowest occurs in October (8 m^3/s). The highest average HADCM3 projected flow (2010-2039) occurs in April and the lowest occurs in November. HADCM3 projects a very high increase in stream flow from January to December in Mkondoa catchment as the results of heavy projected precipitation in the OND and MAM periods. The highest average increase in flow is projected (+581 m^3/s) in April and the lowest increase is projected in December (+75 m^3/s).

From Fig. 8b, both baseline and projected precipitation peaks in April, but the baseline flow peaks in March. Thus the lag between peak of precipitation and peak of flow is only on the baseline condition, because the HADCM3 projects a flow peak in April when the precipitation peaks. The lag in peak of baseline flow is thought to be caused by losses of water out of the streams and in the catchment as bank storage, infiltrations in the catchment as well as the groundwater recharge. However, the lag in peak of baseline flow in April also occurs in the Kinyasungwe catchment, therefore this signal of the delay is transferred to the Mkondoa catchment from further upstream (Kinyasungwe catchment). The same effect on the peak flow also causes the low flow to lag by the same duration (Fig. 8b).

Despite the Mkondoa catchment having the OND season, but the effects of these rains, lag to the end of the season for both baseline and projected flows. Fig. 8b shows that the lag is even extensive for the projected flows because in the OND period, the HADCM3 predicts decrease in precipitation from June to November (Fig. 8b). However, water losses as bank storage recharge, groundwater recharge, irrigation abstractions and infiltrations in the catchment as well as the time of concentration of runoff to reach the outlet of Mkondoa catchment are also thought to contribute to the lag in peak of flow resulting from the OND rainfall.

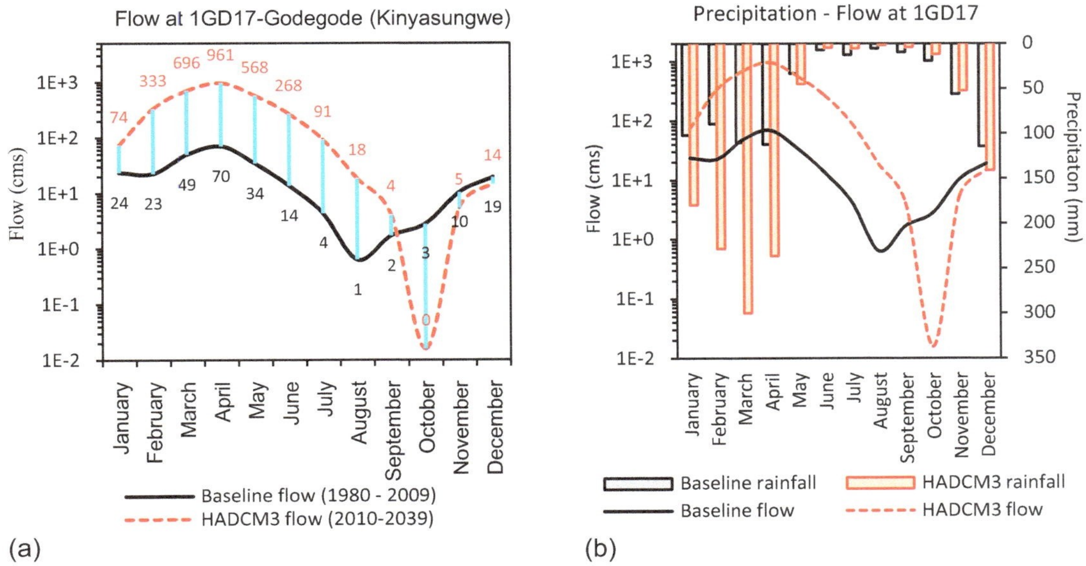

Fig. 7. Flow response at 1GD 17

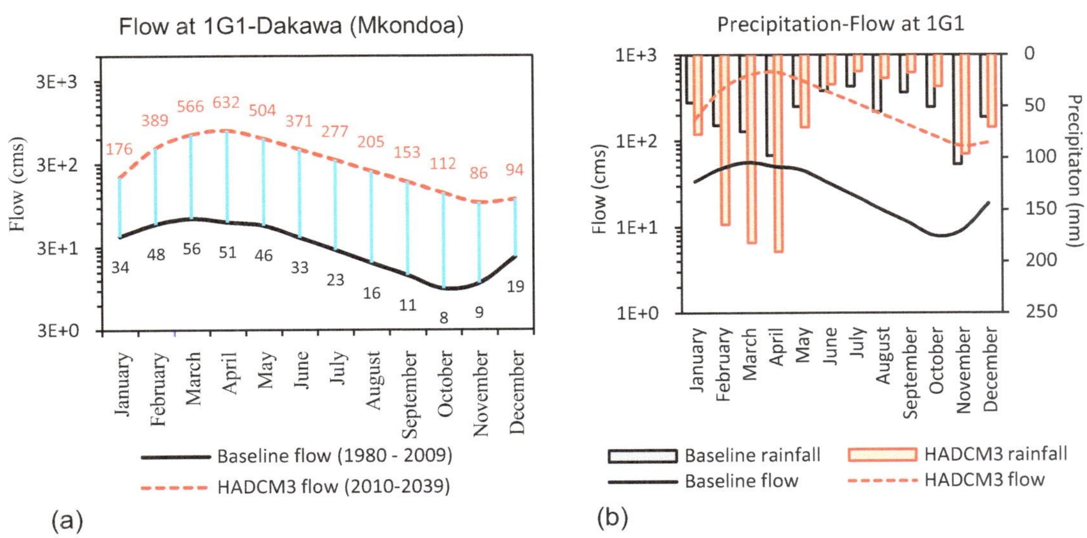

Fig. 8. Flow response at 1G1

1G2-Mandera is the most downstream flow gauge station, which collects all the stream flows from the Wami River sub-basin before they flow into the Indian Ocean. The minimum baseline flows occur in October with the magnitude of 32m^3/s, whereas the maximum baseline flows occurs in April with the magnitude of 207 m^3/s (Fig. 9a). However, Fig. 9a also shows that, HADCM3 predicts high increase in flow throughout the year from the sub-basin; this flow results from the heavy rainfall in the MAM period. The highest increase is projected to occur in April (+1168 m^3/s) and the lowest increase is projected in November (+17 m^3/s).

Fig. 9b shows that, during the MAM period, there is not lag between peak of precipitation and peak of flow for both baseline simulations and HADCM3 projections. However, like the Mkondoa catchment, the Wami catchment also experiences the lag in the peak of flows resulting from the OND rains for both baseline and projected flows (Fig. 9b). The lag in flow in the Wami catchment is also extensive in the case of the projected flow because in the OND period, the HADCM3 predicts a decrease in precipitation. However, similar water losses from streams as in the Mkondoa catchment are also thought to be happening in the Wami catchment.

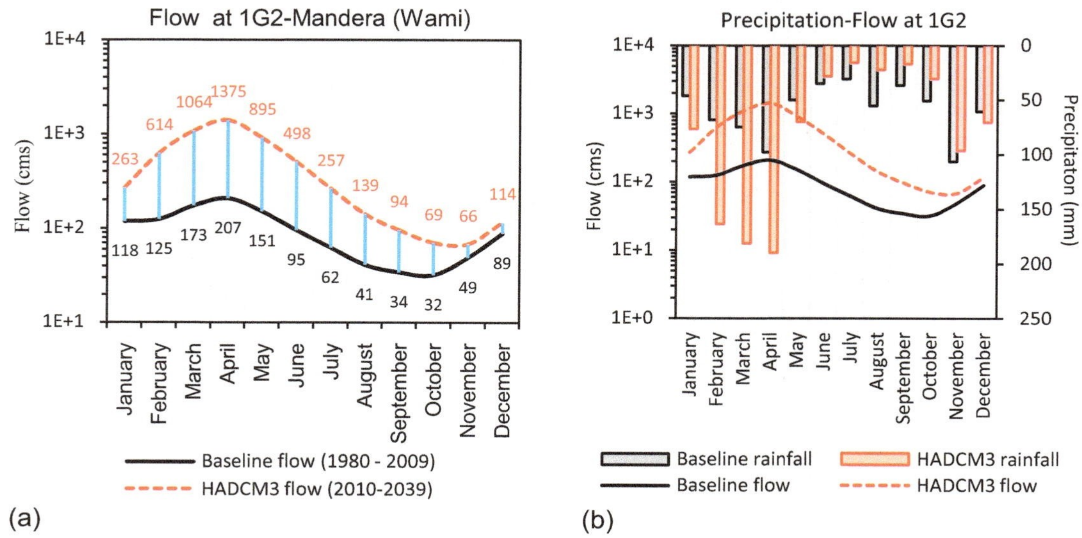

Fig. 9. Flow response at 1G2

The average annual baseline and projected stream flows from the Kinyasungwe catchment are 21m^3/s and 253m^3/s respectively (Fig. 10). Fig. 10 also shows that, the average annual stream flow from the Mkondoa catchment for baseline and projected flows are 29m^3/s and 297m^3/s respectively. From the Wami catchment, the average annual stream flow is the net flow from the whole sub-basin. The average annual flows from the sub-basin for baseline and projected cases are 98m^3/s and 454m^3/s respectively (Fig. 10). The average annual flow in the Wami River sub-basin is projected (2010-2039) to increase by 10 times in the Kinyasungwe and Mkondoa but the net outflow is projected to increase by 4 times. The baseline average annual flow shows that 72% of average annual flow from the Mkondoa catchment comes from the Kinyasungwe catchment outflow, whereas 30% of average annual flow from the Wami River sub-basin comes from the Mkondoa catchment outflow. In the case of projected flows, the average annual flow from the Kinyasungwe catchment outflow is 85% of the Mkondoa average annual flow, whereas from the whole sub-basin,

65% of average annual flow comes from the Mkondoa catchment. These quantities shows that at the baseline conditions 70% of the average annual outflow from the sub-basin results from the Wami catchment, whereas the projected average annual flow originating from the Wami catchment is 35% of the projected sub-basin outflow. This is the result of higher projected precipitation in the upstream part as compared to the downstream part of the sub-basin. Generally, a non-linear bias corrected HADCM3 (scenario A2) climate model projects (2010-2039) very high increase in flows in the Wami River sub-basin.

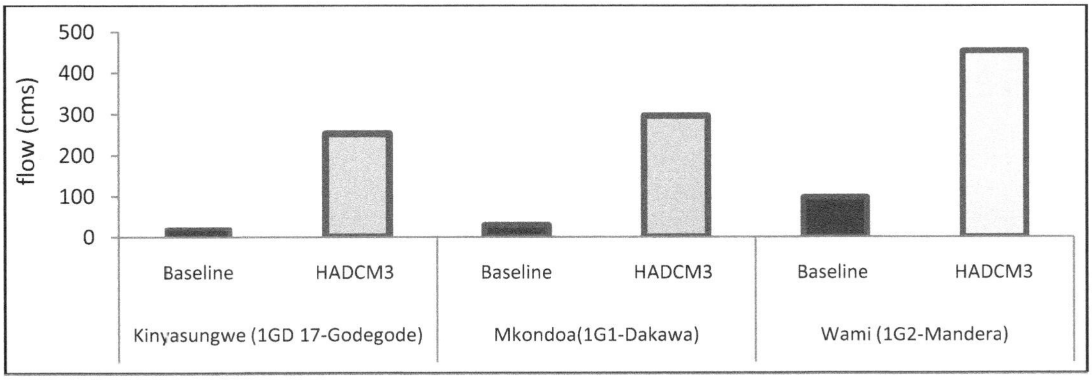

Fig. 10. Average annual stream flow in Wami river sub-basin

4. CONCLUSION

The evaluation of climate models against observed data and the use of non-linear bias correction method are important steps in building confidence in their use for evaluation of responses from hydrological systems. This study of the response of freshwater resources within the Wami River sub-basin as a result of skill score testing and non-linear bias corrected climate model has revealed a number of important findings.

Firstly, and most importantly, it has been shown that projections of stream flow in the sub-basin are highly dependent upon the direction of projected changes in precipitation because patterns drawn by the precipitation changes are the same as those of stream flows. It is likely that such changes (particularly in high and low flows) will have important implications for both the ecological and anthropogenic development of the Wami River sub-basin.

Secondly, this study provides a useful context to previous analyses of climate change impacts on stream flow, based on a single GCM or ensemble means. Skill score test in the selection of GCM gives the credibility and reliability of the use of particular climate model in future prediction, because of its strengths in predicting the control or past period. Strong performance in a skill score test provides more confidence in a climate model in question, however, a model that does well in a skill score test could still hide major limitations in (say) the frequency of no-rain days or events that are too rare to significantly contribute to the skill score.

Non-linear bias correction is one of the very strong and simple methods for bias correction of GCM precipitation. It should be noted that there is no negligible difference between bias correction and the use of the non-bias corrected GCMs. Therefore, proper bias correction is necessary to remove model bias in CMIP3 projections; otherwise it will generate considerable bias in the projected stream flows.

ACKNOWLEDGEMENTS

This study was supported by the Enhancing Climate change adaptation in Agriculture and Water (ECAW) project. The author is gratefully thanking Prof. Siza Tumbo, Dr. Preksedis Ndomba, Dr. Victor Kongo and Mr. Charles Richard; all from the ECAW project, for supporting this study.

COMPETING INTERESTS

Author has declared that no competing interests exist.

REFERENCES

1. United Republic of Tanzania. Poverty and Human Development Index. Dar es Salaam: Mkuki na Nyota Publishers; 2007.
2. Nyunza G, Mwakaje EG. Analysis of round potato marketing in Tanzania: The Case of Rungwe District, Tanzania. International Journal of Business and Social Science. 2012;3(23):86-96.
3. Wami-Ruvu Basin Water Office. Wami Basin: A situational Analysis. Morogoro, Tanzania: Wami-Ruvu Basin Water Office; 2010.
4. Bates BC, Kundzewicz ZW, Wu S, Palutikof JP, editors. Climate change and water: Technical paper of the intergovernmental panel for climate change. Geneva: IPCC Secretariat; 2008.
5. Todd MC, Taylor RG, Osborn TJ, Kingston DG, Arnell NW, Gosling SN. Uncertainty in climate change impacts on basin-scale freshwater resources – preface to the special issue: the QUEST-GSI methodology and synthesis of results. Hydrology and Earth System Sciences. 2011;15:1035–1046. DOI: 10.5194/hess-15-1035-2011.
6. Notter B, Hans H, Wiesmann U, Ngana JO. Evaluating watershed service availability under future management and climate change scenarios in the Pangani Basin. Physics and Chemistry of the Earth. 2013;61-62:1-11. DOI: 10.1016/j.pce.2012.08.017
7. Randall DA, Wood RA, Bony S, Colman R, Fichefet T, Fyfe J, et al. Climate models and their evaluation, in Climate Change: The Physical Science Basis. Contribution of Working Group 1 to the AR4: Cambridge Univ. Press; 2007.
8. Perkins SE, Pitman AJ, Holbrook NJ, Mcaneney J. Evaluation of the AR4 climate models' simulated daily maximum temperature, minimum temperature, and precipitation over Australia using probability density functions. Journal of Climate. 2007;20:4356-4376. DOI: 10.1175/JCLI4253.1.
9. Frich P, Alexander LV, Della-Marta P, Gleason B, Haylock M, Klein AMGT, et al. Observed coherent changes in climatic extremes during the second half of the twentieth century. Climate Research. 2002;19:193–212.
10. Wilby RL, Wigley TML. Downscaling general circulation mode output: a review of methods and limitations. Progress in Physical Geography. 1997;21:530–548. DOI: 10.1177/030913339702100403.
11. Prudhomme C, Reynard N, Crooks S. Downscaling of global climate models for flood frequency analysis: where are we now? Hydrological Processes. 2002;16:1137–1150. DOI: 10.1002/hyp.1054.

12. Fowler HJ, Blenkinsopa S, Tebaldib C. Review: Linking climate change modelling to impacts studies: recent advances in downscaling techniques for hydrological modelling. International Journal of Climatology. 2007;27:1547–1578. DOI: 10.1002/joc.1556.
13. Arnold JG, Srinivasan R, Muttiah RS, Williams JR. Large area hydrologic modelling and assessment. Part 1: Model development. Journal of the American Water Resources Association. 1998;34(1):73–89.
14. Neitsch SL, Arnold JG, Kiniry JR, Williams JR, King KW. Soil and water assessment tools: Theoretical documentation version. Report No: 061. 808 East Blackland Road-temple (TX): Grassland, Soil and Water Reasearch Laboratory, ARS; 2005.
15. SCS (USDA Soil Conservation Service). National Engineering Handbook. Section 4: Hydrology. 1972;4-10.
16. Ndomba PM, Birhanu BZ. Problems and prospects of SWAT model applications in NILOTIC catchments: A review. Nile Basin Water Engineering Scientific Magazine. 2008;1.
17. Yang J, Reichert P, Abbaspour KC, Jun X, Hong Y. Comparing uncertainty analysis techniques for a SWAT application to the Chaohe Basin in China. Journal of Hydrology. 2008;358:1-23. DOI: 10.1016/j.jhydrol.2008.05.012.
18. Abbaspour KC, Van Genuchten MT, Schulin R, Schla¨ppi E. A sequential uncertainty domain inverse procedure for estimating subsurface flow and transport parameters. Water Resources Research. 1997;33(8):1879–1892.
19. Nash JE, Sutcliffe JV. River flow forecasting through. Part I. A conceptual models discussion of principles. Journal of Hydrology. 1970;10:282–290.
20. Legates DR, McCabe GJ. Evaluating the use of "goodness-of-fit" measures in hydrologic and hydroclimatic model validation. Water Resources Research. 1999;35(1):233-241. DOI: 10.1029/1998WR900018.
21. United Republic of Tanzania. The Tanzania Development Vision 2025. Dar es Salaam, Tanzania; 1999.
22. Zhu HY, Jia SF. Uncertainty in the spatial interpolation of rainfall data. Progress in Geography. 2004;23(2):34–42. DOI: 10.11820/dlkxjz.2004.02.005.
23. Lin XS, Yu Q. Study on the spatial interpolation of agroclimatic resources in Chongqing. Journal of Anhui Agricultural Science. 2008;36(30):13431–13463. Chinese.
24. Nicks AD, Stochastic generation of the occurrence, pattern, and location of maximum amount of daily rainfall. In Proc. Symp. Statistical Hydrology, Aug.-Sept. 1971. Tuscon, AZ. U.S. Department of Agriculture. 1974;1275:154-171.
25. Anyamba A, Tucker CJ, Mahoney R. From El Nino to La Nina: Vegetation Response Patterns over East and Southern Africa during the 1997-2000 period. Journal of Climate. 2002;15:3096-3103. Available: http://dx.doi.org/10.1175/1520-0442(2002)015<3096:FENOTL>2.0.CO;2
26. IPCC. The scientific Basis. Contribution of working group I to the third assessment report of the intergovernmental panel on climate change. New York. Cambridge University Press; 2007.
27. Graham LP, Andréasson J, Carlsson B. Assessing climate change impacts on hydrology from an ensemble of regional climate models, model scales and linking methods-A case study on the Lule River basin. Climatic Change. 2007;81(S1):293–307. DOI: 10.1007/s10584-006-9215-2.
28. Sperna WFC, Van Beek LPH, Kwadijk JCJ, Bierkens MFP. The ability of a GCM-forced hydrological model to reproduce global discharge variability. Hydrology and Earth System Science. 2010;14(8):1595–1621. DOI: 10.5194/hess-14-1595-2010.

29. Eisner S, Voss F. Kynast E. Statistical bias correction of global climate projections – consequences for large scale modeling of flood flows. Advances in Geosciences. 2012;31:75–82. DOI: 10.5194/adgeo-31-75-2012.
30. Watanabe S, Kanae S, Seto S, Yeh PJF, Hirabayashi Y, Oki T. Intercomparison of bias-correction methods for monthly temperature and precipitation simulated by multiple climate models. Journal of Geophysical Research. 2012;117:D23114. DOI: 10.1029/2012JD018192.
31. Leander R, Buishand TA. Resampling of regional climate model output for the simulation of extreme river flows. Journal of Hydrology. 2007;332(3–4):487–496. DOI: 10.1016/j.jhydrol.2006.08.006.
32. Leander R, Buishand T, Vandenhurk B, Dewit M. Estimated changes in flood quantiles of the river Meuse from resampling of regional climate model output. Journal of Hydrology. 2008;351(3–4):331–343. DOI: 10.1016/j.jhydrol.2007.12.020.
33. Van Pelt SC, Kabat P, ter Maat HW, van den Hurk BJJM, Weerts AH. Discharge simulations performed with a hydrological model using bias corrected regional climate model input. Hydrology and Earth System Science. 2009;13(12):2387–2397. DOI: 10.5194/hess-13-2387-2009.
34. Hurkmans R, Terink W, Remko U, Torfs P. Changes in streamflow dynamics in the Rhine Basin under three high-resolution regional climate scenarios. Journal of Climate. 2010;23(3):679–699.Available: http://dx.doi.org/10.1175/2009JCLI3066.1
35. Terink W, Hurkmans RTWL, Torfs PJJF, Uijlenhoet R. Evaluation of a bias-correction method applied to downscaled precipitation and temperature reanalysis data for the Rhine basin. Hydrology and Earth System Sciences. 2010;7(1):221–267. DOI: 10.5194/hessd-7-221-2010.
36. Santhi C, Arnold JG, Williams JR, Dugas WA, Srinivasan R, Hauck LM. Validation of the SWAT model on a large river basin with point and nonpoint sources. Journal of the American Water Resources Association. 2001;37(5):1169–1188. DOI: 10.1111/j.1752-1688.2001.tb03630.x.
37. Benaman J, Christine AS, Douglas AH. Calibration and validation of soil and water assessment tool on an agricultural watershed in upstate New York. Journal of Hydrologic Engineering, ASCE. 2005;10(5):363-374. Available: http://dx.doi.org/10.1061/(ASCE)1084-0699(2005)10:5(363)
38. Moriasi DN, Arnold JG, Van Liew MW, Binger RL, Harmel RD, Veith T. Model evaluation guidelines for systematic quantification of accuracy in watershed simulations. Transactions of the ASABE. 2007;50(3):885-900.
39. Saleh A, Arnold J, Gassman PWA, Hauck L, Rosenthal W, Williams J, et al. Application of SWAT for the upper North Bosque River Watershed. Transactions of the ASAE. 2000;43:1077–1087.
40. Rossi CG, Dybala TJ, Moriasi DN, Arnold JG, Amonett C, Marek T. Hydrologic calibration and validation of the soil and water assessment tool for the Leon River watershed. Journal of Soil and Water Conservation. 2008;6:533–541. DOI: 10.2489/jswc.63.6.533.

10

Open Data Flood Mapping of Chao Phraya River Basin and Bangkok Metropolitan Region

Richard T. Cooper[1*]

[1]Southeast Asia START Regional Center, Chulalongkorn University, Bangkok, Thailand.

Author's contribution

This whole work was carried out by the author RTC.

ABSTRACT

Aims: To examine the utility of open data for flood mapping of the Bangkok Metropolitan Region and Chao Phraya River basin. The region is particularly vulnerable to flooding, having experienced recurrent major flooding events, including the some of the most extensive and prolonged in 2011.
Study Design: Novel methodologies were innovated utilising open spatial data and open source geographical software to generate flood extent/hazard maps of the Bangkok Metropolitan Region and Chao Phraya River basin. Key geospatial data were sourced from the Thai Geo-Informatics and Space Technology Development Agency and NASA's Shuttle Radar Topography Mission.
Methodology: Given limited resources for conducting detailed hydrological-hydraulic analyses, two alternative approaches were examined for flood extent/hazard mapping of the basin and city. The first method made use of publicly available historical flood data to produce an up-to-date composite flood extent/hazard map. The second approach, using the latter output as a reference source, examined the utility of a modified topographic index for delineating flood-prone areas, as integrated into the r.hazard.flood module of the open source GRASS GIS application.
Results: Compilation of multi-year historical data enabled generation of a relatively fine-scale (~100m spatial resolution) flood extent/hazard map for the basin and city. The optimal tau threshold for delineating flood exposed cells from the modified topographic index was linearly related to the sub-basin mean slope. The four most northerly sub-basins of the Chao Phraya basin, those with higher mean slopes, gave lowest total errors, ranging from 17.5 to 35.9 percent.
Conclusions: Open data in the form of multi-year spatial flood layers were effectively

**Corresponding author: Email: rcooper@start.or.th;*

combined to generate a relatively fine-scale flood extent/hazard map for the Chao Phraya River basin and Bangkok Metropolitan Region, and the modified topographic index showed promise as an alternative means for identifying flood exposed areas.

Keywords: Open data; disaster risk reduction; climate change adaptation; flood mapping; flood hazard; flood risk; digital elevation model; GRASS GIS.

1. INTRODUCTION

Flooding is an increasing hazard for low-lying coastal megacities of Southeast Asia. Climate change is projected to bring increased precipitation and frequency of intense rainfall events, increased extreme rainfall events linked to tropical cyclones and sea level rise [1]. The densely inhabited mega delta regions in the Asia-Pacific region are described as being at most risk from increased flooding from the sea and in some areas by rivers [2].

1.1 Increasing Flood Potential of Chao Phraya River Basin

The Bangkok Metropolitan Region (BMR), situated at the mouth of the Chao Phraya River, is vulnerable to both riverine and coastal flooding. The area is low-lying, with Bangkok City mostly less than 2.5 meters above mean sea level and below mean sea level in some areas [3]. In a recent OECD study, Bangkok was ranked as one of the most vulnerable to future coastal flooding with regard to population and asset exposure in the context of expected urbanisation and economic growth [4]. Coastal erosion in the upper Gulf of Thailand was described as already critical and impeding sustainable development of the BMR, with sea level rise and land subsidence due to excessive water withdrawal likely exacerbating hydrological impacts in the future [5]. Factors that potentially increase the risk of flooding in the BMR, including both human-related development activities within the Chao Phraya River catchment and climate change-related impacts [1,5-9], are shown in Fig. 1.

1.2 Bangkok Flood History

Bangkok has experienced recurrent major flooding, including from the mid-twentieth century: 1942, 1978, 1980, 1983, 1995, 1996, 2002, 2006 and 2011 [6,10-11]. The flood of 2011 was unprecedented in recent times, likely representative of more than a 100-year return period event, resulting from heavy rainfall over the entire Chao Phraya catchment [12-13]. According to DHI [12], the return period is estimated above 100 years, based on modelling conducted of the Chao Phraya River basin since the 1990s, hydrological studies and measured peak water levels, with the latter likely underestimated in 2011 given diversions of upstream flows; observed and estimated peak water levels reproduced as Table A1 in Appendix. Precipitation during the rainy season of 2011 was reported as 143 percent of the average over the 20 year period from 1982 to 2002 [14]. The flood resulted in over 800 mortalities, the displacement of thousands of city residents, an estimated 45 billion USD in damage, and economic repercussions internationally [14-17].

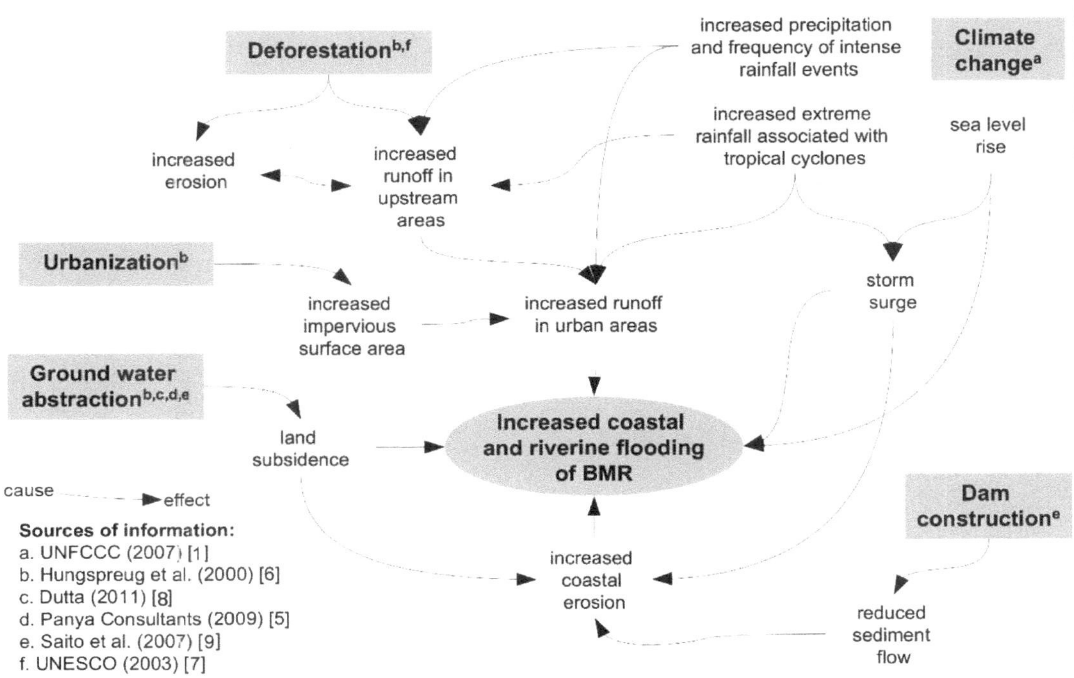

Fig. 1. The overall effect of human activities on increasing flood potential of Bangkok Metropolitan Region [1,5-9]

1.3 Flood Modelling

In order to accurately model flooding in the Bangkok Metropolitan Region, requires a detailed understanding of rainfall and run-off processes in the Chao Phraya River basin and its constituent sub-basins. The Chao Phraya River basin covers an area of approximately 160,000 km^2 (see Table 3, section 3.1) with four northern tributaries (Ping, Wang, Yom and Nan Rivers) that converge at Nakhon Sawan to form the Chao Phraya River, which subsequently branches and flows through the BMR to the Gulf of Thailand (Fig. 2) [7]. Flooding can result from a number of causes including upstream storm run-off, direct precipitation over the city, coastal storm surge and sea level rise, and the extent of flooding influenced by a range of factors such as flood management infrastructure, drainage and sewerage networks, land subsidence, sedimentation, and soil properties [18], as well as measures adopted during flood events such as water diversion (e.g., [12]). Since the mid-twentieth century, in the Chao Phraya basin, 3000 dams have been constructed, the two largest being the Bhumiphol and Sirikit dams which control 22 percent of the run-off of the whole Chao Phraya basin [7]; the latter two dams constructed in 1964 and 1974 respectively [14] (see Fig. 2).

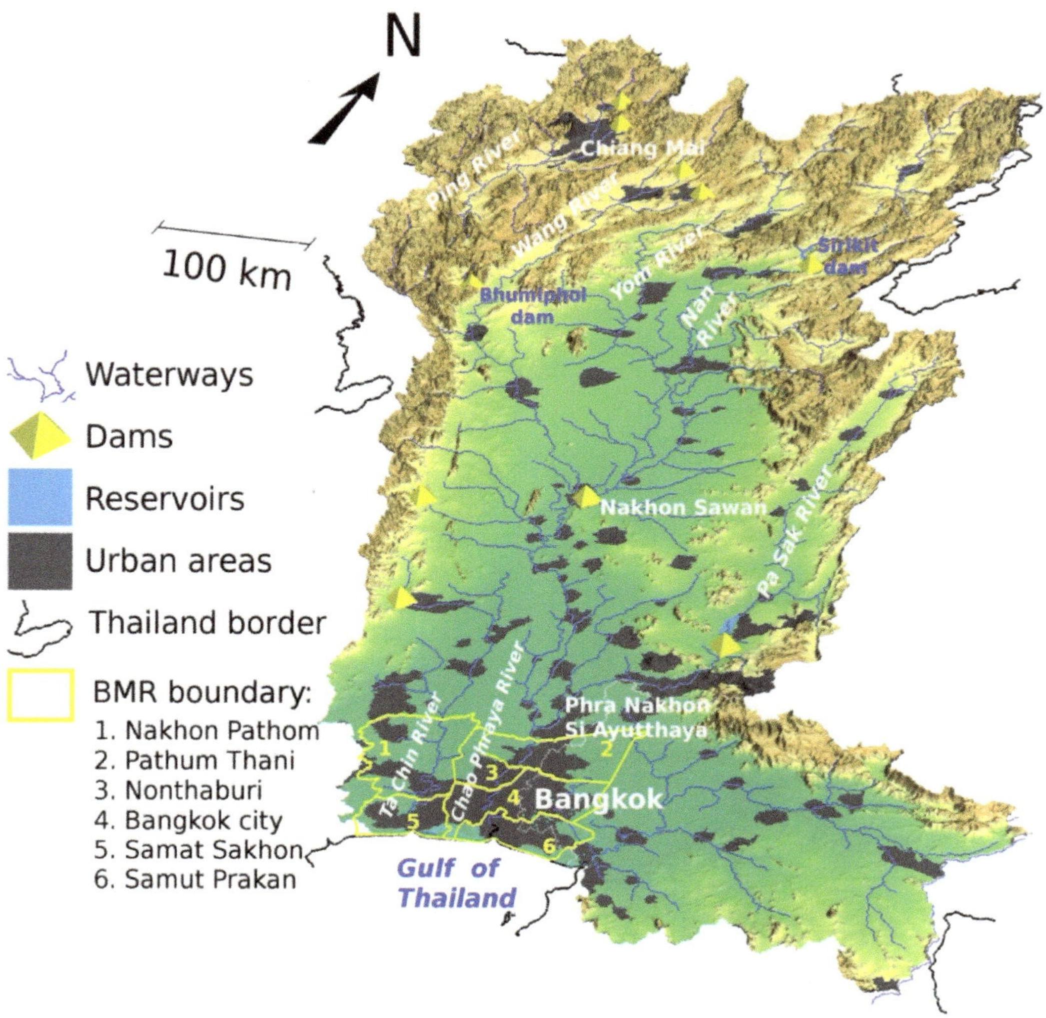

Fig. 2. Three-dimensional view of Chao Phraya River, adjacent Sa Keo sub-basin, and BMR with labelled administrative areas

Note: vertical elevation of image is exaggerated. Open data sources used in figure: digital elevation model [37]; basin/sub-basin boundaries [40]; dams and reservoirs [51]; urban areas [49]; waterways [50]; and BMR administrative areas [46]

Detailed flood inundation modelling efforts in the recent past by Panya Consultants [5] and the World Bank [18] examined the potential impact of climate change on flooding in Bangkok. However, modelling systems used in the latter studies, based on the commercial MIKE FLOOD software package developed by the Danish Hydraulic Institute (DHI), were not accessible, nor the data and financial resources available for the author to conduct an equivalent detailed investigation. Accurate and reliable flood hazard/risk maps were not available for the Chao Phraya River basin nor Bangkok Metropolitan Region. Data management concerns have previously been ascribed to multiple agencies responsible for maintaining water resources data and information [7].

1.4 Flood Extent/Hazard Mapping

This study specifically addresses creation of *flood extent/hazard* maps. In general, the term 'flood hazard map' refers to a spatial information product that illustrates geographical areas that are flood-prone or susceptible to flooding, which differs from a 'flood risk map' which indicates potential negative impacts of a flood, such as the number of people and kind of economic activities affected [19]. More specific requirements of hazard maps, as defined by the European Flood Directive (EFD) definition [19], includes identification of those areas at low, medium (return period ≥ 100 years) and high probabilities of flood, including components of extent, depth, and flow velocity where relevant. The emphasis here is on mapping the geographical extent of flooding, plus indication of flooding frequency based on historical data, with outputs to be integrated subsequently with socio-economic and health data to better understand risk and resilience as part of the Coastal Cities at Risk (CCaR) project.

1.5 International Coastal Cities at Risk Initiative

Improved knowledge of flood risk is an important focus of the ongoing five-year, 2011-2016, international Coastal Cities at Riskinitiative, being co-implemented by Chulalongkorn University, Thailand, and the University of Western Ontario, Canada [20-21]. CCaR's aim is to improve knowledge and capacity of megacities to adapt to climate change-related impacts in the context of rapid urban growth. CCaR is part of the International Research Initiative on Adaptation to Climate Change (IRIACC) of the Canadian International Development Research Centre (IDRC), and funded by the three research councils of Canada: Canadian Institutes of Health Research (CIHR), the Natural Sciences and Engineering Research Council of Canada (NSERC), and the Social Sciences and Humanities Research Council of Canada (SSHRC).

CCaR supports research in the megacities of Bangkok, Manila, Lagos and Vancouver, of which a central output is development of a spatial system dynamics simulator to model city resilience (Coastal Megacity Resilience Simulator) [22], of which developing a measure of flood risk is a key component, together with other economic, social and health inputs. However, the unavailability of flood hazard data became apparent during implementation of project in Bangkok, and thus critical that alternative approaches identified for flood mapping. Hence the focus of this paper to describe approaches taken, challenges encountered, and to share findings with wider research and practitioner communities. Moreover, with two major floods each decade since the 1980s, flooding poses a substantial climate-related risk to the livelihood of Bangkok residents. The need for accurate, accessible and up-to-date flood maps is clearly evident to facilitate decision making, whether for government disaster preparedness planning, investment by the business community, or residents wanting to better understand their potential vulnerability.

1.6 Study Objectives

The chief objective of the study was to examine the utility of open data for flood mapping of the BMR and its wider basin. While there is no single definition of 'open data' [23-24], for the purpose of this study, the broad opendefinition.org was adopted, where: 'a piece of data or content is open if anyone is free to use, reuse, and redistribute it - subject only, at most, to the requirement to attribute and/or share-alike' [25]. In this study all of the data used in the analyses were freely accessible online, apart from one set of administrative boundaries supplied by the BMA ([26-30], Table A2). However, only one dataset considered for use, NASA's Shuttle Radar Topography Mission (SRTM) data, was released and distributed without restriction; its derivative created by Consultative Group for International Agricultural Research - Consortium for Spatial Information (CGIAR-CSI) was actually used in this study, but is more restrictive with regard to data distribution (see section 2.1 and Table A2 for more details).

Key open datasets used in the study included eight years of historical inundation data, and application of a modified topographic index (TI_m) based on SRTM digital elevation model data using methodology developed by Manfreda and colleagues [31-32]. According to the Manfreda et al. [31], the TI_m-based method 'may represent a useful and rapid tool for a preliminary delineation of flooding areas in ungauged basins and in areas where expensive and time-consuming hydrological-hydraulic simulations are not affordable or economically convenient'.

Historical data and outputs from TI_m will indicate at a relatively fine scale the spatial extent of flood-prone areas. Multi-year historical data can also provide a spatio-temporal indication of flooding occurrence for hazard mapping (section 3.1). Though eight consecutive years of historical data is a relatively short time-span, the period includes two major flood episodes, including the most recent of 2011, estimated as having a return period of more than 100 years (see section 1.2); and based on discharge data from the Chao Phraya River basin, the 2006 flood had an estimated return period of 10 years [12]. Moreover, the pattern of recent historical inundation should reflect current flood management infrastructure and efforts, potentially useful given the current lack of available modelling capacity. Flood-prone areas derived from the TI_m are based on the compiled eight-year historical flood map; given that 2011 likely reflected an above 100-year flood event, outputs from the TI_m analysis were considered roughly correspondent with this return period. Data from all eight years were used in the TI_m analysis, and not 2011 coverage alone, as collectively this should capture more areas susceptible to flooding, thereby addressing spatial variability of precipitation across the basin. Findings will be directly used for examining city resilience though the Coastal Megacity Resilience Simulator (section 1.5), and should also prove beneficial for other efforts investigating and developing risk maps for the basin and city.

The study's secondary aim is to provide a quantitative historical assessment of the impact of recent flooding events in terms of area inundated, including the above 100-year event of 2011. Additionally, open source software is used in the analyses, with the aim of presenting methods that could be adopted by researchers or practitioners working outside an institutional setting with limited access to commercial GIS software, key issues which in the opinion of this author pose barriers to the uptake and application of geospatial technologies.

The *r.hazard.flood* add-on module of the open source GRASS GIS (Geographic Resources Analysis Support System Geographic Information System) [32-33] offers potential for generating flood maps based solely on the properties of a digital elevation model (DEM). The *r.hazard.flood* module is a python-based script that can be readily integrated into the main GRASS software suite, accessed through a graphical user interface window, and enables creation of flood map rasters based on a modified topographic Index (TI_m) derived from DEM physical properties. Research on applying TI_m- based flood mapping is well documented on river systems in Italy [31], and in this study, its utility further evaluated as a practical open source mapping tool by applying to the Chao Phraya River basin in Thailand.

2. MATERIALS AND METHODS

Two approaches were examined for creating flood extent/hazard maps based on the availability of open data. Firstly, multi-year spatial historical inundation data (see section 2.1) were used to create a GIS composite layer of the entire Chao Phraya basin and BMR, and the latter then used as a reference layer to assess the utility of the modified topographic index, as integrated in the GRASS GIS module *r.hazard.flood*, for delineating flood-prone areas. Open source desktop GIS and command line utilities were used, including GRASS GIS and associated add-on modules [33], and GDAL/OGR command line utilities [34]. Before examining these methods, open data sources used are highlighted in the next section.

2.1 Open Data Sources and Quality Assessment

A variety of data sources were utilised and are detailed below, together with any data quality issues encountered.

2.1.1 Historical flood inundation layers

Historical spatial flood data were obtained from the Thailand Flood Monitoring System of the Geo-Informatics and Space Technology Development Agency (GISTDA) [35]. Flood inundation area maps were downloaded as ESRI shapefiles from the latter site, with annual flood data accessible from 2005 to 2012 inclusive. The flood maps were derived from synthetic aperture radar (SAR) imagery acquired during the rainy season [GISTDA, personal communication].

2.1.2 Digital elevation data

Post-processed digital elevation data from the Shuttle Radar Topography Mission (SRTM) [36] were used in the study to create the modified topographic index (TI_m). SRTM data used were of 3 arc second spatial resolution and processed to fill no-data regions by the Consortium for Spatial Information (CSI) of the Consultative Group for International Agricultural Research (version 4.1) (CGIAR) [37-38]. ASTER Global Digital Elevation Model (ASTER GDEM) [39] data were also examined for use, but quality appeared variable in the lower Chao Phraya basin which precluded theiruse on this occasion; a linear artefact was apparent in two tiles: ASTGTM2_N14E099_dem_tif and ASTGTM2_N14E100_dem.tif.

2.1.3 Chao Phraya river basin and sub-basin boundaries

The Chao Phraya River basin and sub-basin boundaries for the study were defined using the ESRI shapefile - '*Hydrological basins in Southeast Asia (Derived from HydroSHEDS)*' - accessible from the UN FAO's GeoNetwork portal [40]. The data set was published under FAO's AQUASTAT programme, and derived from the HydroSHEDS initiative of the World Wildlife Fund (in partnership with the U.S. Geological Survey (USGS) and other organisations) [41]. While basin boundaries have been derived elsewhere (as briefly mentioned below), boundaries for both basin and sub-basins are provided, all of which are named; however, some sub-basins were incorrectly labelled in the latter resource and have been re-labelled[1] to reflect main rivers/waterways as shown in Fig. 3.

---

[1] *Sub-basins re-labelled as indicated in parentheses: Mae Nam Wang (Mae Nam Ping); Mae Nam Yom (Mae Nam Wang); Mae Nam Nan (Mae Nam Yom); Mae Nam Ping (Mae Nam Nan); Noi (Mae Nam Pa Sak); Mae Nam Pasak (Khlong Kriangkrai).*

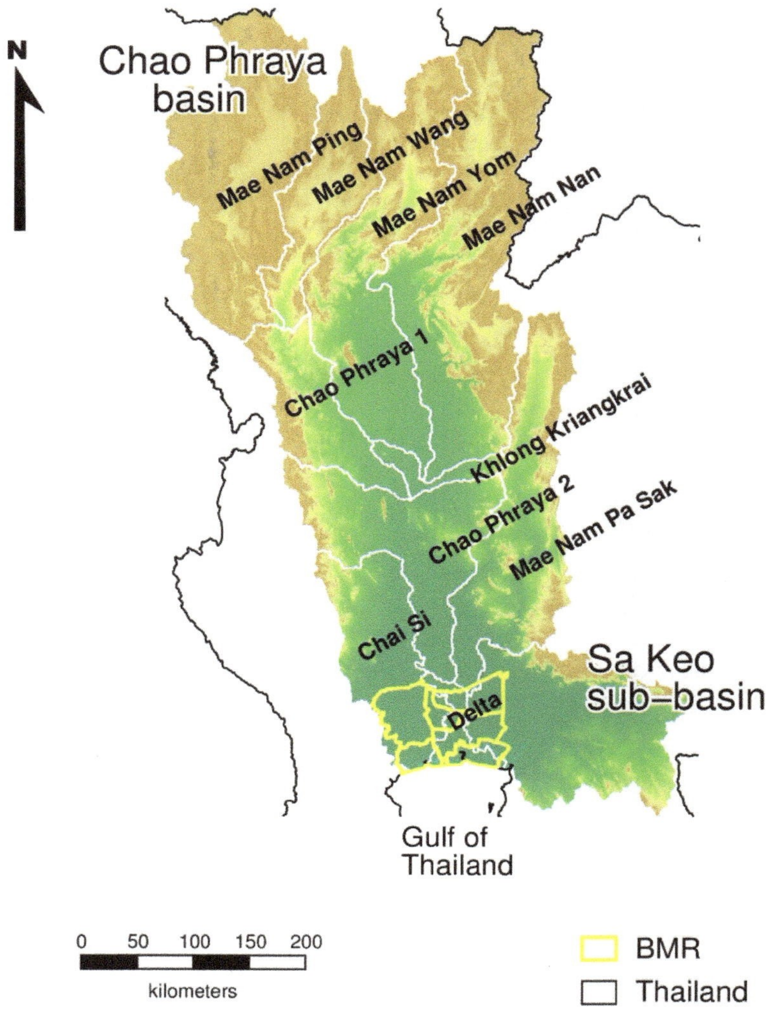

Fig. 3. Chao Phraya River basin and sub-basins and adjacent Sa Keo sub-basin

Coordinate reference system: EPSG 32647 (UTM47N WGS84). Basin boundaries (white lines) as per FAO AQUASTAT hydrological basins data set [40]. Other open data used in figure: CGIAR-CSI SRTM digital elevation model [37]

HydroSHEDS data are primarily based on elevation data acquired from NASA's Shuttle Radar Topography Mission [36] as well as the Digital Chart of the World (DCW), ArcWorld data set, and the Global Lakes and Wetlands Database (GLWD) [42]. Unlike HydroSHEDS, the basins/sub-basins are named in the FAO-amended dataset, and the boundary of the Chao Phraya basin closely matches that in HydroSHEDS, apart from some differences in the southern part of the basin, such as inclusion of the Chai Si sub-basin and small differences near the coast. The technical literature for HydroSHEDS describes the drainage basins as not yet finalised, though generally they have better accuracy than the earlier HYDRO1k elevation derivative database, ArcWorld and DCW, and there may be issues concerning the accuracy of river networks in low relief areas [42].

HYDRO1k Elevation Derivative Database is derived from the Global 30 Arc-Second Elevation (GTOPO30) data set, a 30 arc second (~1 km) digital elevation model of the world created by USGS in 1996 [43-44]. Layers created as part of the data set include derivatives for hydrological analysis including basin boundaries, streams, slope, aspect, flow direction and accumulation.

Another data set delineating the Chao Phraya basin is the Global Runoff Data Centre (GRDC) '*Major River Basins of the World*' [45], derived from HYDRO1k. The chief differences between the latter and FAO's data set concerns the inclusion of the Mae Nam Moei (Salween basin) and the Sa Keo (Gulf of Thailand Coast) sub-basins into the Chao Phraya River basin in the GRDC layer.

For the purpose of this study, FAO's delineation of the Chao Phraya basin was adopted [40], where the Bangkok Metropolitan Region (BMR) crosses two major basins - the Chao Phraya and Gulf of Thailand Coast basins. The BMR extends from the Chao Phraya basin into the Sa Keo sub-basin of the latter. The BMR and entire Chao Phraya basin are located in the UTM47N zone and all results are presented based on this projection.

2.1.4 Administrative boundaries and urban areas

The BMR, or Greater Bangkok, comprises the city of Bangkok (special administrative area) and conterminous provinces of Samut Prakan, Samut Sakhon, Nakhon Pathom, Nonthaburi, and Pathum Thani. The BMR boundary can be freely accessed from the GADM database of Global Administrative Areas [46]. The BMR covers an area of approximately 7652 km^2 (calculated from GADM) with a population of 10.5 million in 2012 [47], though the population varies by several million more in consideration of unregistered residents and commuters [48]. A BMR boundary layer supplied by the Bangkok Metropolitan Administration proved to be mostly congruent with the GADM-derived data with a total BMR area of 7689 km^2; some minor boundary differences are apparent generally with the largest differences in Samut Sakhon and Samut Prakan provinces. The BMA-supplied layer was used for determining flooded areas for each BMR administrative area, as reported in Tables 1 and 2 (Results and Discussion). Urban areas flooded in the BMR were based on open data from the Global Rural-Urban Mapping Project (GRUMP) by CIESIN et al. [49].

Other open data sets accessed include those used for displaying rivers and water bodies in Fig. 2 above [50-51].

2.2 Creation of Composite Historical Flood Map

The method adopted for creating the composite flood map is summarised diagrammatically in Fig. 4. Shapefiles were clipped and transformed from EPSG 4326 to EPSG 32647 (UTM 47N/WGS84) and rasterized to a spatial resolution of 100m using GDAL command line utilities - ogr2ogr and gdal_rasterize; the appropriate spatial resolution was determined following consultation with GISDTA. After rasterizing, a single compiled raster image was created from the series of eight flood inundation images (2005 - 2012 inclusive) by summing cell values using the GRASS *r.series* module [52]. The output layer contained cell values ranging from 1 to 8 representing areas with the lowest to highest occurrence of flood; areas unflooded during the 8 year period were assigned a value of zero. Based on the accuracy of the input inundation maps, identification of flooded areas should be approximately 80 to 90% correct, though the error might be higher in urban areas [53].

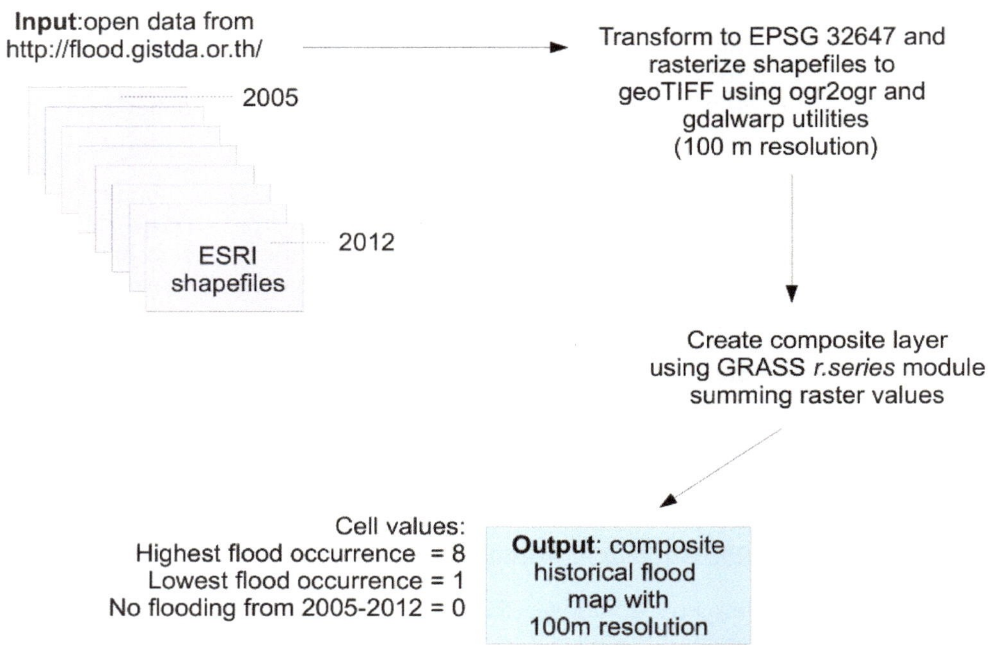

Fig. 4. Schematic diagram showing creation of composite historical flood map

2.3 Development of Flood Map Using Modified Topographic Index and Error Analysis

The second approach examined the utility of the modified topographic index (TI_m), a procedure developed by Manfreda et al. and incorporated into the GRASS *r.hazard.flood*add-on extension module [31-32,54-55]. The module delineates flood-prone areas by calculating a modified topographic index from a digital elevation model (see formula below), and categorises those cells above a given threshold value as exposed to flooding [31]. In other words, flood-prone areas are derived from topographic parameters, the supposition being that physical surface features can reveal areas exposed to flood inundation given that the land surface is shaped by hydrological processes [31]. Areas susceptible to inundation are determined by a value above a calculated threshold (*tau*, τ).

$$TI_m = \log\left[\frac{a_d^n}{\tan(\beta)}\right] \tag{1}$$

The TI_m is based on the Topographic Index (TI), also called the Topographic Wetness Index (TWI), which reflects the quantity of water flowing to a particular location and accumulation in low slope areas (e.g., [56]), but differs only by the inclusion of an exponent, n [57]. The TI_m is a function of the specific upslope contributing area (a_d) (determined by multiplying accumulation x cell size), the local slope (β), and the exponent n is a function of DEM cell size.

The threshold *tau* and exponent n values incorporated into the *r.hazard.flood* module were derived from research conducted in the Arno River basin in Italy [31], employing a calibration procedure based on error analysis of TI_m outputs compared to reference flood maps generated from an ensemble of historical flood maps, geomorphological and hydraulic studies [31,57]. The TI_m threshold value used in *r.hazard.flood* corresponds to the minimum total error with reference to the higher 500-year return period of the reference maps, where $\tau = 10.89n + 2.282$ [31]. The exponent n was found to be strongly correlated to DEM cell size, where a plot of n against DEM cell size could be expressed as $n = 0.016x^{0.46}$ [31]. Additional findings in the Italian study showed that good results were possible using DEM's with resolutions of up to approximately 100m, including reliable results from NASA's SRTM 90m DEM. For example, using this expression, a DEM of 90m cell size will give a threshold of 3.6627, with cells above this value defined as being exposed to flooding.

The method adopted in this study was to use the CGIAR-CSI SRTM 90m DEM of the Chao Phraya River basin and the composite historical flood map developed previously as a reference for assessing the utility of the modified topographic index for defining areas as exposed to flooding (see Fig. 5). Error estimation is based on the approach by Manfreda et al. [31] (see Fig. 6). While the analysis depends on the accuracy of the multi-year flood maps (~80-90% as noted previously), these currently represent the best source of information available for the Chao Phraya River basin; detailed and accurate flood maps based on hydrological-hydraulic mapping (as used in the Arno River basin study by Manfreda et al. [31]) were not available.

Additional potential constraints to error analysis of the TI_m layer include land use/land cover changes in the Chao Phraya basin, including urbanisation and flood infrastructure (e.g., construction of dikes), and water diversion during a flooding event (e.g., [12]), which could modify the extent of flooding in the reference layer. For instance, greater flood protection given to key commercial and business districts of the BMR (located in the Delta sub-basin). Furthermore, an appropriate tau threshold needs to be determined to identify flood-prone areas based on the reference historical flood map. The reference map likely represents more than a 100-year flood (see comments above concerning the 2011 flood) and thus if the default threshold in r.hazard.flood were applied, it would give an overestimation of area delineated by the module as flood-prone. The optimal tau value, corresponding to minimum total error, was interpolated from error analysis of each sub-basin, as summarised in Fig. 6.

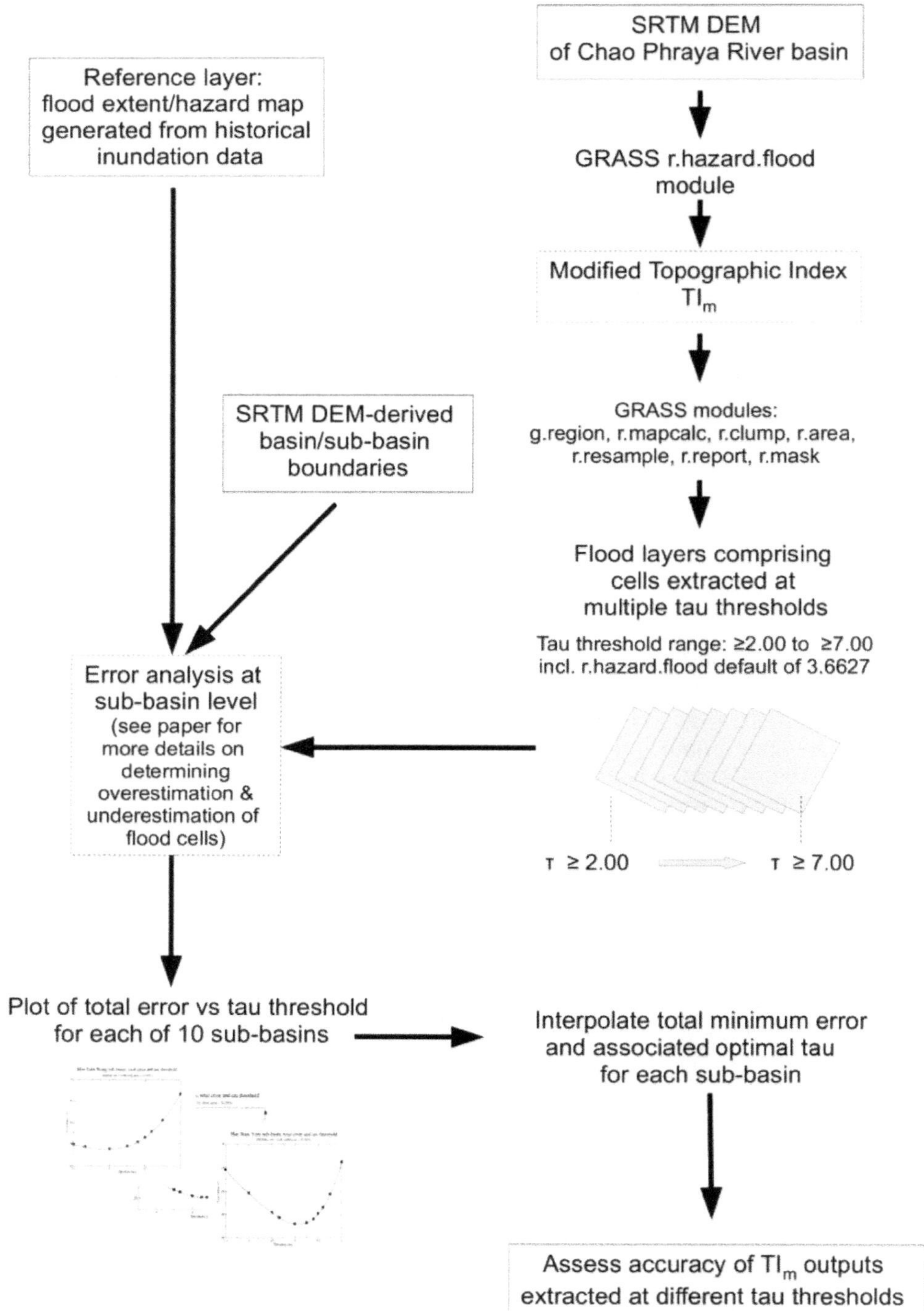

Fig. 5. Flow chart of methodology for assessing utility of modified topographic index for generating flood map

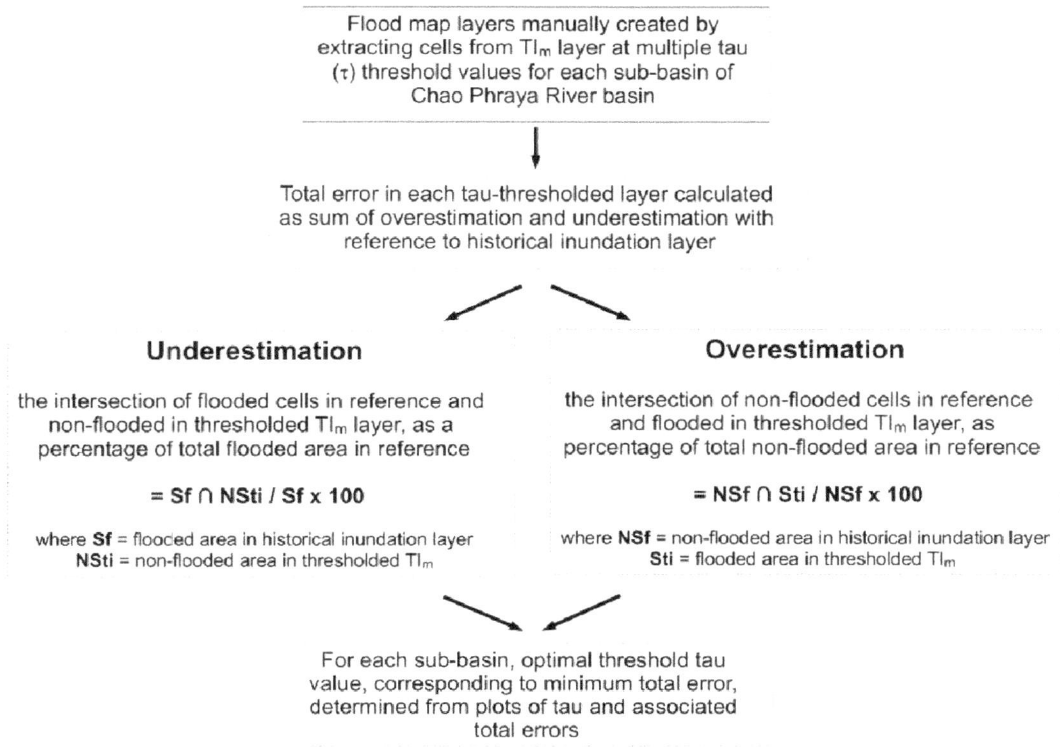

Fig. 6. Error analysis of TI_m –based flood rasters to derive optimal tau threshold values

Error estimation based on the approach by Manfreda et al. [31]

A DEM of the entire Chao Phraya River basin was created by patching together SRTM tiles downloaded from CGIAR-CSI [58] and this input to the *r.hazard.flood* add-on module to generate the TI_m layer, with flood output layers subsequently manually generated using selected tau thresholds between 2.0 and 7.0 (including the *r.hazard.flood* default of 3.6627 which corresponds to a 90m resolution DEM based on a 500-year flood return period) (Fig. 5). Selected tau thresholds and total errors were plotted for each sub-basin and the minimum threshold interpolated using a cubic spline (using GRACE 2D plotting software[2]). The mean slope of each sub-basin was also determined (output from *r.slope.aspect* [59]) and plotted against minimum total errors and tau thresholds to assess potential correlation.

3. RESULTS AND DISCUSSION

3.1 Composite Historical Flood Map

Based on historical data, flood extent/hazard maps were generated for the entire Chao Phraya River basin (Fig. 7) and Bangkok Metropolitan Region (Fig. 8). Areas of highest flood occurrence are displayed in red and least risk coloured yellow, and those areas not flooded in the eight year study period remain uncoloured. Cumulative historical inundation (i.e.,

---

[2] *GRACE 2D plotting tool: http://plasma-gate.weizmann.ac.il/Grace.*

maximum extent of flooding derived by overlaying all annual inundation layers) are presented for: the entire BMR, Bangkok City and constituent provinces (Table 1); urban areas (Table 2); and entire Chao Phraya River basin and sub-basins (Table 3). Stacked bar charts showing the annual extent of flooding for the BMR (Bangkok City and its contiguous provinces) and the entire Chao Phraya basin (all 10 sub-basins) are displayed in Figs. A1 and A2 respectively (see Appendix). Key statistics of the 2011 flood disaster indicate that approximately 4200 km^2 (56 percent) of the BMR was inundated, and at the basin level over 26,000 km^2 (17 percent) was flooded.

Table 1. Total land area flooded within BMR (data combined from 2005 to 2012)

Administrative area	Total area[1]	Land area[2]	Land area flooded[3]	
	(km^2)	(km^2)	(km^2)	(%)
BMR	7689.0	7483.1	4460.2	59.6
Bangkok city	1569.3	1519.0	829.4	54.6
Samut Sakhon	864.3	837.2	76.7	9.2
Samut Prakan	960.2	904.8	301.3	33.3
Nakhon Pathom	2141.2	2127.5	1408.6	66.2
Nonthaburi	633.4	617.4	479.4	77.6
Pathum Thani	1520.4	1477.2	1364.6	92.4

[1]*Total area: land and river/waterways/water bodies*
[2]*Land area: excluding river/waterways/water bodies;*
[3]*flood pixels overlaying river/waterways/water bodies omitted from flood area calculations*

Overall, at some time during the eight year period from 2005 to 2012, nearly 60 percent of the BMR's land area and 50 percent of its urban area was flooded. The province of Pathum Thani was the most flood-prone province with 92 percent of its land area inundated and 95 percent of its urban area. The provinces of Nonthaburi and Nakhon Pathom were also especially vulnerable with 78 and 66 percent of land inundated respectively. Bangkok City (the special administrative area), experienced flooding mainly in its eastern and western regions, while the coastal provinces of Samut Sakhon and Samut Prakan experienced flooding to a lesser extent, though not insubstantial; however, coastal flooding caused by tidal and storm surge may not be captured in the GISTDA inundation layers as they were derived from SAR data during the rainy season and focused on flooding caused by precipitation and downstream run-off (GISTDA, personal communication).

The results clearly indicate that millions of residents are vulnerable to flooding in the BMR: with a total population of over 10 million and some 50 percent of its urban area inundated on one or more occasions between 2005 and 2012 (see Table 2). It should also be noted that past historical inundations may not indicate the frequency and extent of present and future flooding events in the context of climate change and land cover/land use changes in the basin. However, given the lack of existing hydrological-hydraulic modelling data, these data do provide some indication of spatial vulnerability to flooding.

Table 2. Urban areas flooded within BMR (data combined from 2005 to 2012)

Administrative area[1]	Urban area[2] (km^2)	Population[3] Persons	Urban area flooded (km^2)	(%)
BMR	4039.1	10,455,800	1973.0	48.8
Bangkok city	1280.9	5,673,560	636.8	49.7
Samut Sakhon	511.1	508,812	64.1	12.5
Samut Prakan	616.6	1,223,302	178.4	28.9
Nakhon Pathom	681.4	874,616	300.5	44.1
Nonthaburi	373.3	1,141,673	244.0	65.4
Pathum Thani	575.8	1,033,837	549.1	95.4

[1]Administrative boundaries derived from data supplied by BMA to CCaR project

[2]Urban area calculations using data from CIESIN, IFPRI, WB, CIAT (2011) [49] and excluding river/waterway/waterbodies.

[3]2012 population data from BMA (2012) [47]

Table 3. Flooded areas within Chao Phraya River basin and sub-basins (data combined from 2005 to 2012)

Drainage basin	Total area (km^2)	Area flooded (km^2)	Area flooded (%)
Chao Phraya (entire basin)	157,633.3	31,919.3	20.2
Sub-basins			
Mae Nam Ping	26,269.8	708.4	2.7
Mae Nam Wang	10,736.7	838.3	7.8
Mae Nam Yom	24,582.1	6,099.2	24.8
Mae Nam Nan	33,293.0	4,280.2	12.9
Chao Phraya 1	9,440.8	1,600.6	17.0
Khlong Kriangkrai	2,122.3	1,143.5	53.9
Mae Nam Pasak	19,527.7	3,844.8	19.7
Chao Phraya 2	16,400.6	7,021.5	42.8
Chai Si	12,912.3	5,363.6	41.5
Delta	2,348.8	1,019.1	43.4

3.2 Modified Topographic Index (TI_m)

For each Chao Phraya sub-basin, the optimal threshold tau value, corresponding to minimum total error, was interpolated from a plot of calculated total errors and associated tau threshold values used for extracting the flood raster from the TI_m layer. Fig. A3 (Appendix) presents total error and tau threshold plots of each Chao Phraya sub-basin, and Table 4 shows the interpolated minimum total errors and associated optimal tau thresholds, as well as area and mean slope for each sub-basin.

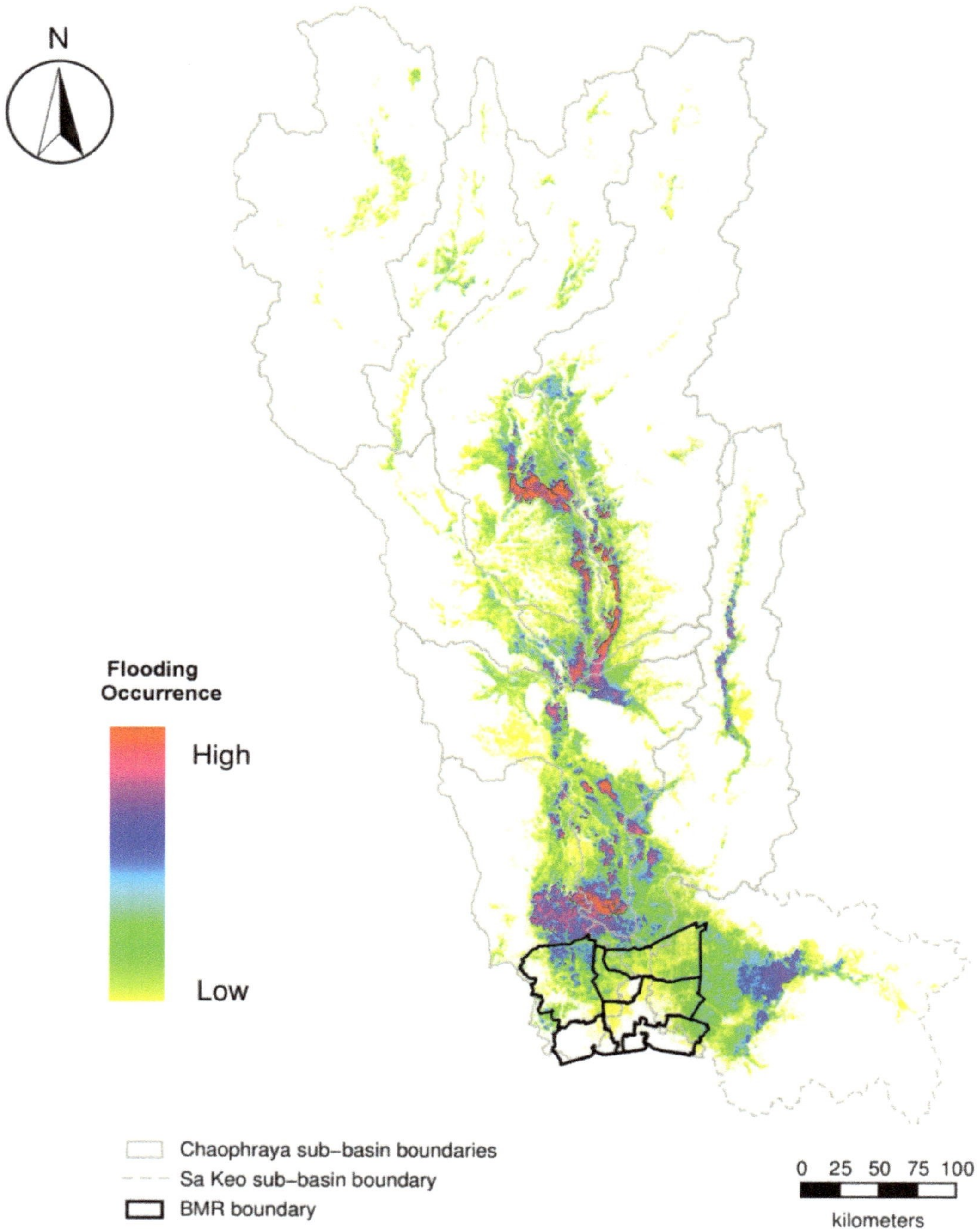

Fig. 7. Flood extent/hazard map of Chao Phraya River basin and adjacent Sa Keo sub-basin

Map shows spatial extent and frequency of past flood events from 2005 to 2012 inclusive. Coordinate reference system: EPSG 32647 (UTM47N WGS84). Spatial resolution of 100m. Areas of highest flood occurrence indicated as red

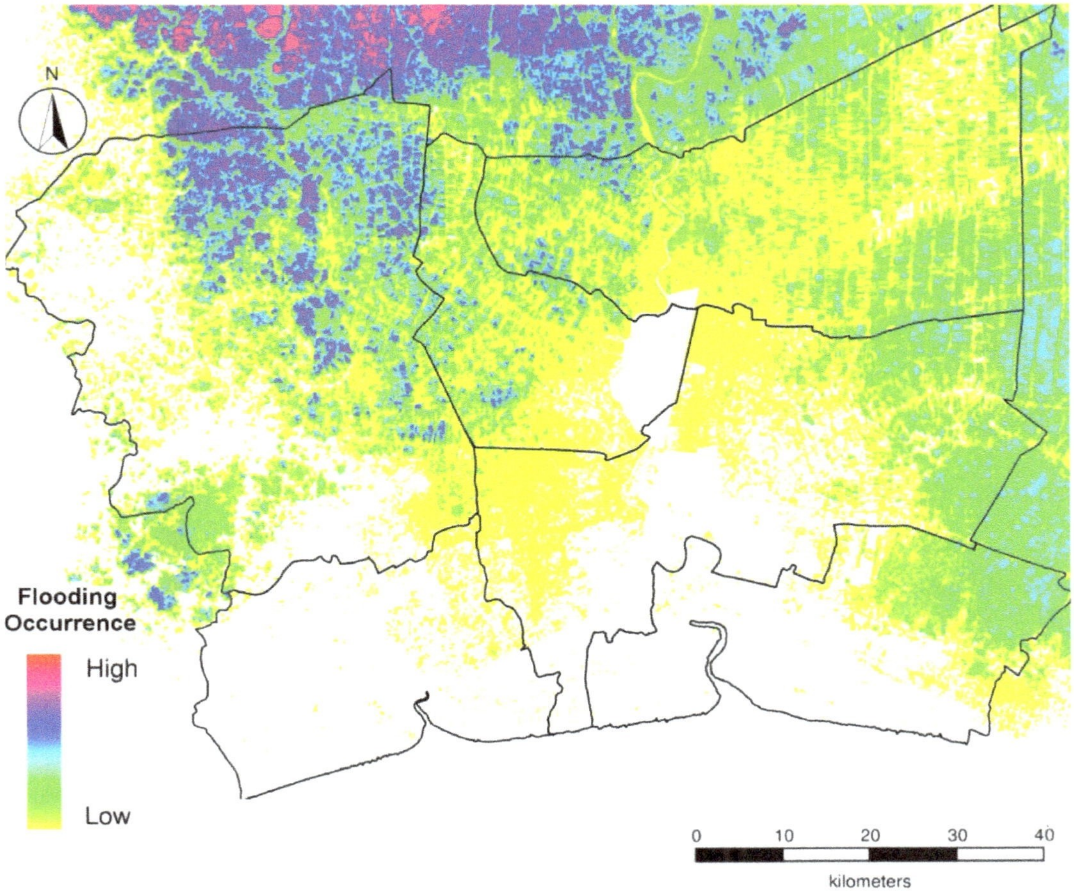

Fig. 8. Flood extent/hazard map of Bangkok Metropolitan Region

Map shows spatial extent and historical occurrence of flood events from 2005 to 2012 inclusive. Coordinate reference system: EPSG 32647 (UTM47N WGS84). Spatial resolution of 100m. Areas of highest flood occurrence indicated as red. Note: flooding is likely underestimated in region of BMR in 2005, as apparent from abrupt linear boundaries on flood polygons to north and east of BMR; perhaps related to incomplete imagery or subsequent processing

Optimal tau threshold values varied from 3.96 (Mae Nam Ping) to 5.23 (Delta). It was anticipated that the estimated tau threshold would be higher than the 3.6627 calculated using the linear function derived by Manfreda et al. [31] (see Fig. 6); at the latter threshold, much of the error in the sub-basins is attributable to overestimation of flooded areas. The reference layer used in this study was derived from eight years of inundation data, with the 2011 flood likely reflecting more than a 1-in-100 year flood, whereas the tau threshold derived by Manfreda et al. [31] was based on a considerably higher return period of 500 years.

Total minimum error decreased with increasing sub-basin mean slope. Accuracy was substantially lower when applied to areas of lower topographic relief, with minimum total error increasing linearly from approximately 10 to 70 percent for a mean slope decreasing from 20 to 2.5 percent respectively. A linear regression of minimum total error and mean sub-basin slope shows a correlation (correlation coefficient -0.9449) between increasing mean slope and reducing total error; such a linear correlation does not appear present in Manfreda et al. [31] data. The minimum total error (y) can be estimated from the linear function: $y = 78.174 - 3.3377x$, where x is the mean slope (Fig. 9). Thus, minimum total errors of 10, 25 and 50 percent correspond to mean slopes of approximately 20, 16 and 8% respectively. Significant error may be introduced into low relief areas given that the SRTM DEM comprises integer and not floating point data [31]. Unfortunately, it was not possible to further examine whether this error was related to using SRTM DEM given artefacts evident in the higher resolution ASTER DEM for the region. Overall, the four most northerly sub-basins of the Chao Phraya basin, those with higher mean slopes, gave lowest total errors, ranging from 17.5 to 35.9 percent (see Fig. 10). All the remaining sub-basins had errors ranging from 45.8 to 87.2 percent.

Table 4. Interpolated minimum total errors and associated optimal tau thresholds for each sub-basin and basin/sub-basin areas, urban extents and mean slopes

Drainage basin	Total area (km^2)	Urban area (km^2)[1]	Urban Area (% of basin/ sub-basin)	Mean slope (%)	Minimum total error (%)	Optimal threshold (tau)
Chao Phraya (entire basin)	157,633	13,652	8.7	11.70	34.07	4.50
Sub-basins						
Mae Nam Ping	26,270	1,475	5.6	20.67	17.53	3.96
Mae Nam Wang	10,737	634	5.9	14.68	35.90	4.06
Mae Nam Yom	24,582	1,284	5.2	10.78	30.50	4.50
Mae Nam Nan	33,293	1,197	3.6	15.28	20.50	4.32
Chao Phraya 1	9441	548	5.8	9.43	45.82	4.76
Khlong Kriangkrai	2,122	35	1.7	2.19	69.40	4.64
Mae Nam Pa Sak	19,528	2,750	14.1	8.40	46.65	4.67
Chao Phraya 2	16,401	1,329	8.1	4.45	58.79	5.05
Chai Si	12,912	2,846	22.0	2.82	69.30	5.18
Delta	2,348	1,553	66.1	1.24	87.16	5.23

[1]*Data derived from CIESIN, IFPRI, WB, CIAT (2011) [49]*

The optimal tau threshold representing minimum total error appears related to sub-basin mean slope. The four northernmost sub-basins, with highest mean slope values, presented optimal tau thresholds ranging from 3.96 to 4.50, with thresholds from 4.64 to 5.23 for the remaining sub-basins (Table 4). A linear relationship between optimal tau threshold and mean slope is given by the function: $y = 5.1912 - 0.06162x$, where y is the optimal threshold (Fig. 11). It is interesting to note that Manfreda et al. also noted improved performance of the TI_m method in basins of lower order [31]. Given that tau appears related to mean slope in this study, more accurate representation of flooding using TI_m may be best obtained by customising the threshold to the mean slope of each sub-basin; in this study a single tau value does not appear practicable for accurately delineating flood exposed areas.

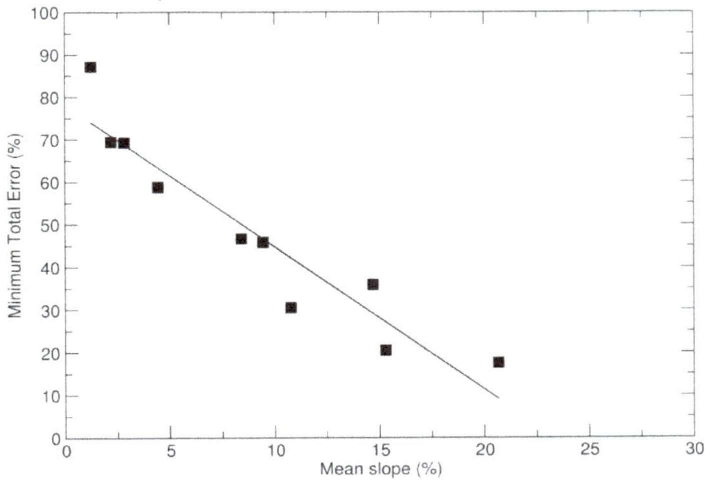

Fig. 9. Plot of minimum total errors and mean sub-basin slopes (y = 78.174 - 3.3377x; correlation coefficient = -0.9449)

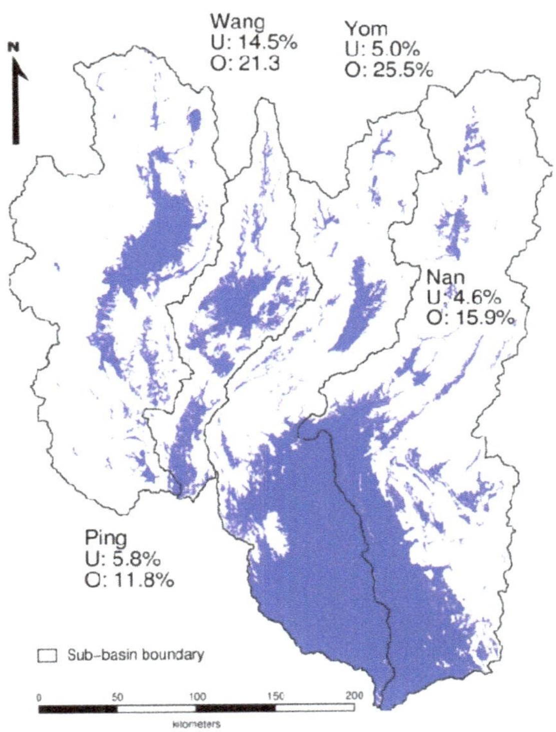

Fig. 10. Flood-prone areas (blue) of the four northern sub-basins generated from TI_m
Flood maps were created using optimal tau thresholds determined from error analysis of each sub-basin. Indicated underestimation (U) and overestimation (O) errors were calculated from the maps extracted with optimal thresholds

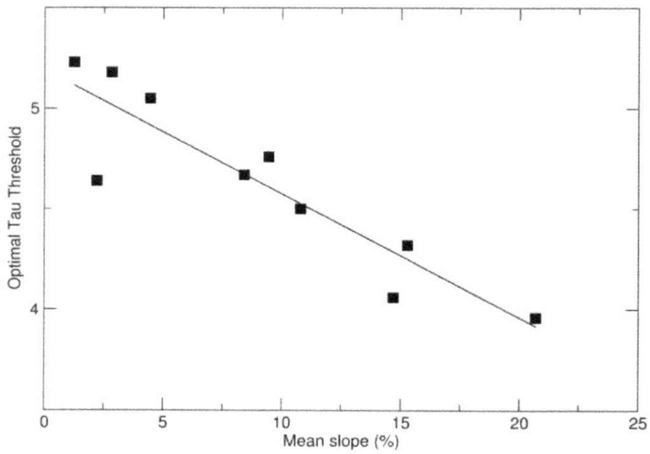

Fig. 11. Plot of optimal tau thresholds and mean slopes (y = 5.1912 - 0.06162x; correlation coefficient = -0.9061)

Potential limitations and constraints to error analysis of the modified topographic index output are depicted in Fig. 12. Limitations include the unavailability of flood data derived from hydrological-hydraulic models of the basin for strengthening the accuracy of the reference data, and more accurate DEM data for generation of the modified topographic index layer. Other potential constraints include accuracy of input historical inundation data, such as underestimation of flooded area in urban settings related to the satellite imagery, and the influence of land cover/land use changes such as urbanisation and flood defence management on flooding extent, e.g., greater flood protection of key commercial and business areas of the BMR (as located in the Delta sub-basin).

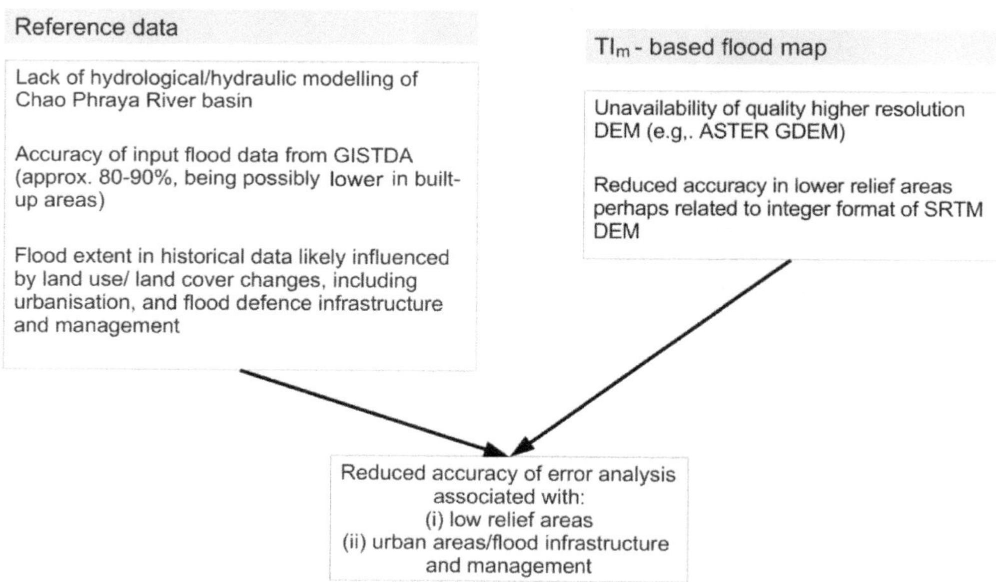

Fig. 12. Potential limitations and constraints to error analysis of TI_m-based flood map

4. CONCLUSION

Open data in the form of multi-year spatial flood layers were effectively combined to generate a relatively fine-scale (~100m spatial resolution) flood extent/hazard map for Thailand's Chao Phraya River basin and Bangkok Metropolitan Region. The maximum flooding extent is likely representative of above a 100-year return period event; however it should be noted that future flooding in the basin may not reflect the past, given new development within the basin and climate change impacts. The flood extent/hazard map offers potential to be updated with more recent data so as to better reflect land cover/land use changes and flood infrastructure developments, as well as climate change impacts. As well as being used for the CCaR project city resilience modelling, the flood maps created in this study will be made available to the Bangkok Metropolitan Administration, requesting that they are added to their existing BMA Web Map Service[4] and also to GISTDA for sharing through their flood monitoring portal [35].

The utility of using a modified topographic index from SRTM DEM for delineating flood-prone areas also offers promise for flood mapping, where other means of assessment, such as detailed hydrological-hydraulic assessments and historical inundation data are unavailable. However, as also noted by Manfreda et al. [31], the index cannot be considered a replacement for more detailed analyses, but as a preliminary assessment of flood-prone areas. For instance, the procedure could be used to identify areas especially at risk from flooding. Findings presented here indicate good correspondence between reference historical inundation data and the results of the TI_m method, at least for the four most northerly sub-basins of the Chao Phraya River basin characterised by higher mean slopes. However, selection of an appropriate tau threshold value is critical for generating an accurate flood map - the threshold value reflecting the return period to be presented, and as shown in this study, the mean slope of the sub-basin. No single tau threshold value appears practicable and might be best determined from a sub-basin's slope given the linear relationship observed above.

Performance of the TI_m-based method is reduced in areas of low topographic relief, which may relate to the limited vertical resolution of the SRTM DEM (see section 3.2). The latter observation reflects the comment by Manfreda et al. [31] that the method may be especially useful in basins with marked topography, and greater error may relate to the SRTM DEM used to generate the TI_m. Other potential errors may derive from the accuracy and nature of the reference layer compiled from historical inundation data, where flooding extent may be influenced by land cover/land use changes, such as urbanisation and flood protection measures. The two southernmost sub-basins, characterised by lowest mean slopes and highest urban extents, presented the highest total errors; however, the low mean slope and minimal urbanisation of the central Khlong Kriangkrai sub-basin suggests that mean slope might be the key factor influencing their observed error. Findings indicated that a total error of 20 percent or less required a mean slope of at least 17 percent (~10 degrees).

The modified topographic index is potentially a highly valuable tool for environmental and disaster risk management. To further validate the utility of using the modified topographic index as a tool for delineating flood-prone areas, it is recommended that future studies examine higher resolution DEMs, such as the 30m resolution ASTER GDEM, and other

---

[4]*BMA Web Map Service:* *http://203.155.220.230/wms/*

drainage basins as study sites. Moreover, it would be informative to further examine optimal tau thresholds for delineating flood exposed areas and if a linear relationship between the tau threshold and mean slope is observable elsewhere.

The methodology presented here also offers scope to create flood maps representing multiple return periods. For example, according to the report by DHI [12], the 2006 flood was estimated as a 10 year flood and spatial flood data is already accessible from GISTDA. Furthermore, given estimated return periods of 25 and 50 years for the 1996 and 1995 floods respectively [12], there should be scope to derive flood maps for these return periods.

It should also be highlighted that given the emphasis on using open data and open source GIS software, the methods adopted here offer scope for independent non-institutional researchers to conduct further investigations. Open data made generation of Chao Phraya River basin and the BMR flood hazard maps feasible in a reasonable time and with limited financial resources, apart from expertise required to conduct the GRASS-based GIS analyses.

ACKNOWLEDGEMENTS

The author would like to thank Dr. Wijitbusaba (Ann) Marome (Faculty of Architecture and Planning, Thammasat University), Team Leader of the Bangkok Coastal Cities at Risk (CCaR) Research Team, for the opportunity to contribute to ongoing research efforts of the Canadian-funded International Research Initiative on Adaptation to Climate Change (IRIACC), and also to team members for their support and encouragement. Also, thanks to Dr. Anond Snidvongs, Director, and colleagues at the Thai Geo-Informatics and Space Technology Development Agency (GISTDA) for access to national historical flood data and responding to queries regarding these data. This work was carried out with the aid of a grant from the International Development Research Centre (IDRC), the Canadian Institutes of Health Research (CIHR), the Natural Sciences and Engineering Research Council of Canada (NSERC), and the Social Sciences and Humanities Research Council of Canada (SSHRC), Ottawa, Canada. I'd also like to thank both reviewers for their valuable comments for improving this paper.

COMPETING INTERESTS

Author has declared that no competing interests exist.

REFERENCES

1. United Nations Framework Convention on Climate Change (UNFCCC). Climate Change: Impacts, Vulnerabilities and Adaptation in Developing Countries; 2007. Accessed 28 May 2014.
   Available: http://unfccc.int/resource/docs/publications/impacts.pdf.
2. Intergovernmental Panel on Climate Change (IPCC).Climate Change 2007: Synthesis Report; 2007. Accessed 19 September 2013.
   Available: http://www.ipcc.ch/pdf/assessment-report/ar4/syr/ar4_syr.pdf.
3. Royal Thai Survey Department. Map showing dikes and ground level in Bangkok districts and surrounding areas, October 2553 (Prepared by Division of Geodesy and Geophysics, Royal Thai Survey Department; 2010. Thai.

4. Nicholls RJ, Hanson S, Herweijer C, Patmore N, Hallegatte S, Corfee-Morlot J, Chateau J, Muir-Wood R. Ranking port cities with high exposure and vulnerability to climate extremes: Exposure Estimates, OECD Environment Working Papers, No. 1, OECD Publishing; 2008.
Available: http://dx.doi.org/10.1787/011766488208.
5. Panya Consultants. Climate change impact and adaptation study for bangkok metropolitan region. Final report. Panya Consultants Co. Ltd; 2009. Accessed 28 May 2014.
Available:http://siteresources.worldbank.org/INTEAPREGTOPENVIRONMENT/Resources/ClimateChangeBangkokMainReport.pdf
6. Hungspreug S, Khao-uppatum W, Thanopanuwat S. Flood management in Chao Phraya River basin. Proceedings of the International Conference on the Chao Phraya Delta: Historical Development, Dynamics and Challenges of Thailand's Rice Bowl, Kasetsart University, Bangkok Thailand; 2000. Accessed 28 May 2014.
Available:http://std.cpc.ku.ac.th/delta/conf/Acrobat/Papers_Eng/Volume%201/Wirat%20RID.pdf.
7. UN Educational Scientific and Cultural Organisation (UNESCO). The United Nations World Water Development Report 1: Water for People - Water for Life. Chao Phraya River Basin, Thailand. Chapter 16; 2003. Accessed: 29 May 2014.
Available:http://webworld.unesco.org/water/wwap/case_studies/chao_phraya/chao_phraya.pdf
8. Dutta D. An integrated tool for assessment of flood vulnerability of coastal cities to sea-level rise and potential socio-economic impacts: a case study in Bangkok, Thailand. Hydrolog Sci J. 2011;56(5):805-823.
Available: http://dx.doi.org/10.1080/02626667.2011.585611
9. Saito Y, Chaimanee N,Jarupongsakul T, Syvitski JPM. Shrinking megadeltas in Asia: Sea-level rise and sediment reduction impacts from case study of the Chao Phraya Delta. Inprint Newsletter of the IGBP/IHDP Land-Ocean Interactions in the Coastal Zone. 2007;(2):3-9. Accessed 29 May 2014.
Available: http://staff.aist.go.jp/yoshiki.saito/Saito_INPRINT_2007_2.pdf.
10. Royal Irrigation Department (RID) Flood Mitigation Management in Urban Areas. Expert Group Meeting on Innovative Strategies towards Flood Resilient Cities in Asia-Pacific; 2009. Bangkok, Thailand. Accessed 3 September 2013.
Available: http://www.unescap.org/idd/events/2009_EGM-DRR/Thailand-Flood-Mitigation-Lower-ChaoPhraya-River-Basin.pdf
11. Gemenne F, Brücker P, Ionesco D. The State of Environmental Migration 2011. Institute for Sustainable Development and International Relations (IDDRI) / International Organization for Migration (IOM); 2012. Accessed 31 August 2013.
Available:http://www.iddri.org/Publications/Collections/Analyses/SEM%202011_web.pdf
12. DHI Water & Environment(s) Pte Ltd (DHI). Thailand Floods 2011: The Need for Holistic Flood Management; 2011. Accessed 3 September 2014.
Available: http://www.globalcateye.com/media/861/Thailand_Floods_2011.pdf
13. Global Disaster Alert and Coordination System (GDACS) Orange Flood Alert in Thailand; 2011. Accessed 29 May 2014.
Available: http://www.monitor.czert.info/index.php/gdacs/floods/383-orange-flood-alert-in-thailand

14. Komori D, Nakamura S, Kiguchi M, Nishijima A, Yamazaki D, Suzuki S, Kawasaki A, Oki K, Oki T. Characteristics of the 2011 Chao Phraya River flood in Central Thailand. Hydrological Research Letters. 2012;6:41–46.
15. Bangkok Post. Japan investor post-flood wish list. Bangkok Post; 2011. Accessed 29 May 2014.
Available: http://www.bangkokpost.com/learning/learning-from-news/261352/japan-investor-post-flood-wish-list
16. East Asia Forum. Thailand's economy still recovering from devastating floods; 2012. Accessed 29 May 2014.
Available: http://www.eastasiaforum.org/2012/05/01/thailand-s-economy-still-recovering-from-devastating-floods.
17. IRIN. How to move floodwater through Bangkok. IRIN: Humanitarian news and analysis, a service of the UN Office for the Coordination of Humanitarian Affairs; 2012. Accessed 29 May 2014.
Available: http://www.irinnews.org/printreport.aspx?reportid=95680
18. World Bank (2010) Climate Risks and Adaptation in Asian Coastal Megacities: A Synthesis Report. The World Bank. Accessed 19 September 2013.
Available: http://go.worldbank.org/P8IU2IW370
19. European Exchange Circle on Flood Mapping (EXCIMAP). Handbook on good practices for flood mapping in Europe; 2007. Accessed 3 June 2014. Available: http://ec.europa.eu/environment/water/flood_risk/flood_atlas/pdf/handbook_goodpractice.pdf.
20. McBean G. Building resilience. International Innovation. 2014;126.
Available: http://www.research-europe.com/magazine/ISSUE/126/index.html. Accessed 15 May 2014.
21. Coastal Cities at Risk (CCaR). Building Adaptive Capacity for Managing Climate Change in Coastal Megacities at Risk. Project Document. Accessed 27 May 2014.
Available: http://www.start.or.th/iriacc-project/documents/project-document-for-iriacc-ccar-project/at_download/file.
22. Simonovic SP, Peck A. Dynamic Resilience to Climate Change Caused Natural Disasters in Coastal Megacities Quantification Framework. British Journal of Environment and Climate Change. 2013;3(3):378-401. Accessed 29 December 2013.
Available: http://www.sciencedomain.org/abstract.php?iid=267&id=10&aid=2034
23. Open Data Research Network. Early insight 2: Definitions of open data. The Open Data Research Network is a collaborative project, coordinated by the Web Foundation and the International Development Research Centre (IDRC); 2013. Accessed 29 May 2014.
Available:http://www.opendataresearch.org/content/2013/566/early-insight-2-definitions-open-data.
24. Opengovdata.org. The Annotated 8 Principles of Open Government Data. Accessed 29 May 2014. Available: http://opengovdata.org
25. Opendefinition.org. Open Definition. Accessed 29 May 2014.
Available: http://opendefinition.org/od/
26. NASA Jet Propulsion Laboratory. Data products, NASA/NGA Memorandum of Understanding. NASA Jet Propulsion Laboratory, 2005. Accessed 29 May 2014. Available: http://www2.jpl.nasa.gov/srtm/mou.html.

27. NASA & METI/Japan Space Systems. ASTER GDEM Policy. ASTER Global Digital Elevation Model. Accessed 10 August 2013.
Available: http://gdem.ersdac.jspacesystems.or.jp/agreement.jsp.

28. US Geological Survey (USGS). HydroSHEDS. Disclaimer; 2008. Accessed 29 May 2014. Available: http://hydrosheds.cr.usgs.gov/disclaimer.php.

29. Center for International Earth Science Information Network (CIESIN) Columbia University, International Food Policy Research Institute (IFPRI), The World Bank, and International Center of Tropical Agriculture (CIAT). Global Rural-Urban Mapping Project, Version 1 (GRUMPv1): Urban Extents Grid. Metadata. Accessed 29 May 2014.
Available: http://sedac.ciesin.columbia.edu/data/set/grump-v1-urban-extents/metadata.

30. Lehner B, Liermann CR, Revenga C, Vörösmarty C, Fekete B, Crouzet P, et al. Global Reservoir and Dam Database, Version 1 (GRanDv1): Reservoirs, Revision 01; 2011. Metadata. Accessed 29 May 2014.
Available: http://sedac.ciesin.columbia.edu/data/set/grand-v1-dams-rev01/metadata.

31. Manfreda S, Di Leo M, Sole A. Detection of flood-prone areas using digital elevation models. J Hydrol Eng. 2011;781-790. DOI: 10.1061/(ASCE)HE.1943-5584.0000367

32. Di Leo. Description: r.hazard.flood Implementation of a fast procedure to detect flood prone areas; 2014. Accessed 29 May 2014.
Available:http://svn.osgeo.org/grass/grass-addons/grass6/raster/r.hazard.flood/description.html

33. Geographic Resources Analysis Support System (GRASS) GIS. Accessed 29 May 2014. Available: http://grass.osgeo.org.

34. Geospatial Data Abstraction Library (GDAL). Accessed 29 May 2014.
Available: http://www.gdal.org

35. Geo-Informatics and Space Technology Development Agency (Public Organization) (GISTDA). Thailand Flood Monitoring System. Accessed 29 May 2014.
Available: http://flood.gistda.or.th.

36. NASA Jet Propulsion Laboratory. Shuttle Radar Topography Mission. Accessed 28 May 2014. Available: http://www2.jpl.nasa.gov/srtm/index.html.

37. Consultative Group for International Agricultural Research - Consortium for Spatial Information (CGIAR-CSI). SRTM 90m Digital Elevation Data; 2008. Accessed 28 May 2014. Available: http://srtm.csi.cgiar.org/index.asp

38. Jarvis A, Reuter HI, Nelson A, Guevara E. Hole-filled seamless SRTM data V4, available from the CGIAR-CSI SRTM 90m Database (http://srtm.csi.cgiar.org); 2008. Accessed 29 May 2014: Available: http://www.cgiar-csi.org/data/srtm-90m-digital-elevation-database-v4-1.

39. ASTER Global Digital Elevation Model (GDEM). Accessed 29 May 2014.
Available: http://gdem.ersdac.jspacesystems.or.jp/download.jsp.

40. UN Food and Agriculture Organization. Hydrological basins in Southeast Asia (Derived from HydroSHEDS). Published in the framework of the AQUASTAT programme of the Land and Water Division of the Food and Agriculture Organization of the United Nations; 2009. Accessed 29 May 2014.
Available: http://www.fao.org/geonetwork/srv/en/metadata.show?uuid=ee616dc4-3118-4d67-ba05-6e93dd3e962f&currTab=simple

41. US Geological Survey (USGS). Overview, HydroSHEDS. Hydrological data and maps based on SHuttle Elevation Derivatives at multiple Scales; 2008. Accessed 29 May 2014. Available: http://hydrosheds.cr.usgs.gov/overview.php.
42. Lehner B, Verdin K, Jarvis A. HydroSHEDS, Technical Documentation, Version 1; 2006. Accessed 29 May 2014. Available:
http://hydrosheds.cr.usgs.gov/webappcontent/HydroSHEDS_TechDoc_v10.pdf.
43. US Geological Survey (USGS). HYDRO1K; 2012. Accessed 29 May 2014.
Available: https://lta.cr.usgs.gov/HYDRO1K.
44. US Geological Survey (USGS). Global 30 Arc-Second Elevation (GTOPO30); 2012. Accessed 29 May 2014. Available: https://lta.cr.usgs.gov/gtopo30.
45. Global Runoff Data Centre (GRDC) Major River Basins of the World. Koblenz, Germany: Federal Institute of Hydrology (BfG); 2007. Accessed 29 May 2014. Available:http://www.bafg.de/GRDC/EN/02_srvcs/22_gslrs/221_MRB/riverbasins_node.html
46. GADM. GADM database of Global Administrative Areas, version 1. Accessed 26 July 2013. Available: http://www.gadm.org.
47. Bangkok Metropolitan Administration (BMA). Statistical Profile of the Bangkok Metropolitan Administration. Administrative Strategy Division, Strategy and Evaluation Department, Bangkok; 2012. Accessed 29 May 2014.
Available:http://office.bangkok.go.th/pipd/05_Stat/08Stat(En)/Stat(En)55/stat%20eng%202012.htm
48. UN Environment Programme (UNEP) Bangkok City State of Environment Report; 2003. URL: Accessed 29 May 2014.
Available: http://www.rrcap.ait.asia/pub/soe/bangkoksoe03.cfm
49. Center for International Earth Science Information Network (CIESIN) Columbia University, International Food Policy Research Institute (IFPRI), The World Bank, and International Center of Tropical Agriculture (CIAT). Global Rural-Urban Mapping Project, Version 1 (GRUMPv1): Urban Extents Grid; 2011. Accessed 29 May 2014.Available: http://sedac.ciesin.columbia.edu/data/set/grump-v1-urban-extents.
50. UN Food and Agriculture Organization. Rivers in Southeast Asia (Derived from HydroSHEDS). AQUASTAT - Programme of the Land and Water Division of the Food and Agriculture Organization of the United Nations; 2010. Accessed 24 March 2013.
Available: http://www.fao.org/geonetwork/srv/en/main.home?uuid=dc2a5121-0b32-482b-bd9b-64f7a414fa0d.
51. Lehner B, Liermann CR, Revenga C, Vörösmarty C, Fekete B, Crouzet P. et al. High-resolution mapping of the world's reservoirs and dams for sustainable river-flow management. Frontiers in Ecology and the Environment. 2011;9:494–502.
Available: http://dx.doi.org/10.1890.
52. Geographic Resources Analysis Support System (GRASS). r.series - Makes each output cell value a function of the values assigned to the corresponding cells in the input raster map layers; 2013. Accessed 29 May 2014.
Available: http://grass.osgeo.org/grass70/manuals/r.series.html
53. Geo-Informatics and Space Technology Development Agency (GISTDA). Commonly asked questions regarding GISTDA flood data. Accessed 29 May 2014.
Available:http://flood.gistda.or.th/flood/y2011/FL82_Extra/ReportQuestion.pdf. Thai

54. Di Leo M. r.hazard.flood.py; 2010. Accessed 6 June 2014. Available: http://svn.osgeo.org/grass/grass-addons/grass6/raster/r.hazard.flood/r.hazard.flood.py.

55. Manfreda S, Sole A, Fiorentino M. Can the basin morphology alone provide an insight on floodplain delineation? WIT Transactions on Ecology and the Environment. 2008;118:47-56. Accessed 6 June 2014.
Available: http://library.witpress.com/pages/PaperInfo.asp?PaperID=19286

56. Ali G, Birkel C, Tetzlaff D, Soulsby C, McDonnell JJ, Tarolli P. A comparison of wetness indices for the prediction of observed connected saturated areas under contrasting conditions. Earth Surf Process Landforms. 2014;39:399–413.

57. Di Leo M, Manfreda S, Fiorentino M. An automated procedure for the detection of flood prone areas: r.hazard.flood. Geomatics Workbooks, volume 10, FOSS4G-it, Trento; 2011. Accessed 26 June 2014.
Available:http://geomatica.como.polimi.it/workbooks/n10/GW10-FOSS4Git_2011.pdf

58. Consultative Group for International Agricultural Research - Consortium for Spatial Information (CGIAR-CSI). SRTM Data Selection Options. Accessed 29 May 2014. Available: http://srtm.csi.cgiar.org/SELECTION/inputCoord.asp.

59. Geographic Resources Analysis Support System (GRASS). r.slope.aspect - Generates raster maps of slope, aspect, curvatures and partial derivatives from an elevation raster map; 2014. Accessed 15 July 2014.
Available: http://grass.osgeo.org/grass70/manuals/r.slope.aspect

APPENDIX

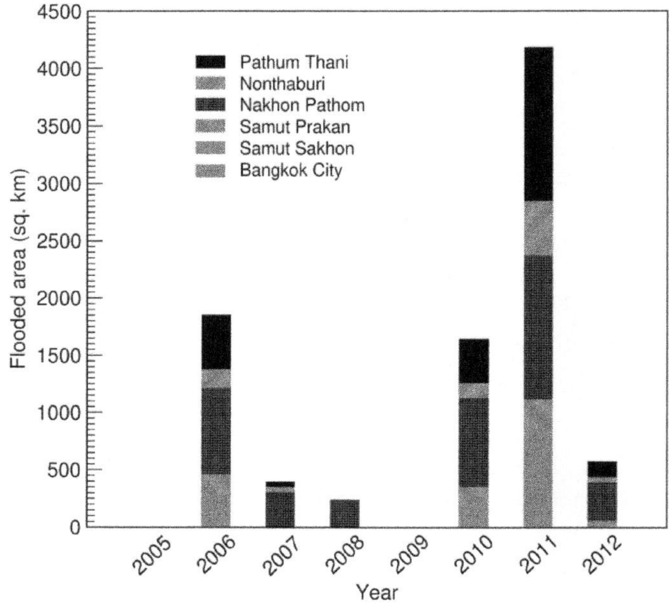

Fig. A1. Historical annual flooding in Bangkok Metropolitan Region: inundation of Bangkok City and constituent provinces

NOTE:for 2005, flooding is possibly underestimated in BMR, as apparent from abrupt linear boundaries of flood polygons to north and east of BMR; perhaps related to incomplete imagery or subsequent processing

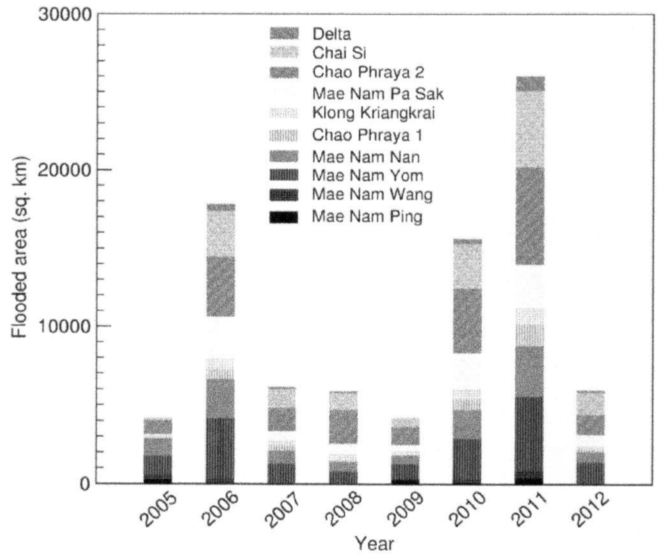

Fig. A2. Historical annual flooding of Chao Phraya River basin: inundation of each sub-basin

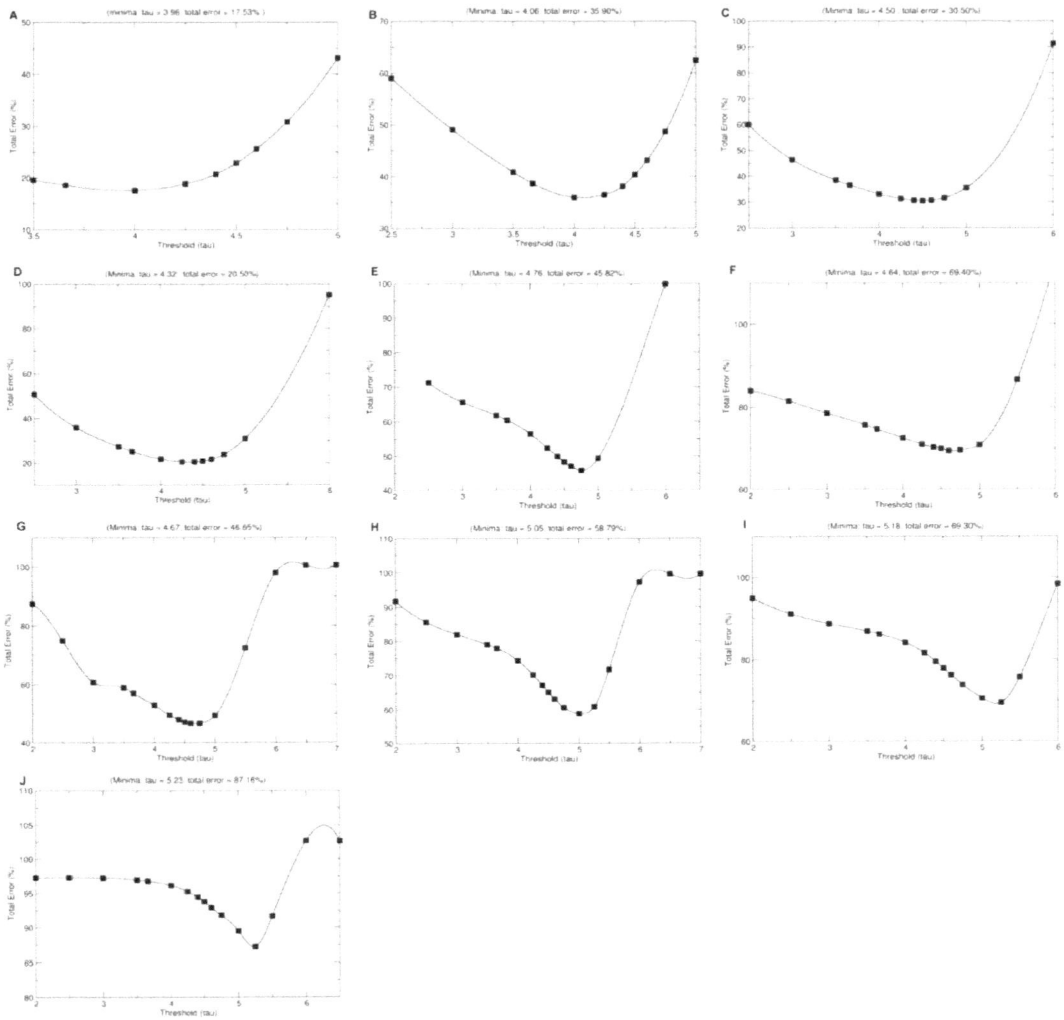

Fig. A3. Interpolation of minimum total errors and optimal tau thresholds for Chao Phraya sub-basins

Plots of total error at given tau thresholds for each sub-basin: A (Mae Nam Ping); B (Mae Nam Wang); C (Mae Nam Yom); D (Mae Nam Nan); E (Chao Phraya 1); F (KlongkriangKrai); G (Mae Nam Pa Sak); H (Chao Phraya 2); I (Chai Si); J (Delta)

Table A1. Peak water levels in Chao Phraya (metres above mean sea level) (reproduced from DHI Water & Environment(s) Pte Ltd [12])

Station	Observed Peaks			Estimated Return Period (years)					
	1983	1995	2011	2	5	10	25	50	100
Ayutthaya	4.7	5.1	5.92	3.4	4.1	4.7	5.1	5.4	5.6
Bang Sai	3.1	NA	4.21	2.6	3.1	3.4	3.7	3.8	4.0
Pakret	2.2	2.6	3.20	2.15	2.61	2.72	2.86	2.96	3.07

Table A2. Openness of data examined in study in terms of availability/access, reuse

and redistribution, and user permission

Data	Availability & access	Reuse & redistribution	User permission	Notes
CGIAR-CSI SRTM 90m Data (version 4.1)	Online [37]	Re-sale/ redistribution not permitted	Commercial use requires prior permission	Data policy bundled with data download.
NASA JPL SRTM Data (version 2-1)	Online [36]	No restriction	No restriction	The CGIAR-CSI version of the SRTM data (see this table) was used in this study. With regard to SRTM data downloaded from JPL: 'Terrain height data greater than or equal to three seconds of arc latitude and longitude (i.e. Level-1 data sets) generated from the SRTM will be released and distributed without restrictions' [26]
ASTER GDEM	Online [39]	Reuse restricted as per policy agreement (see 'Notes' column this table)	Commercial use not apparent.	Downloading data requires registration of site and selecting the policy agreement: 'I agree to redistribute the ASTER GDEM only to individuals within my organization or project of intended use or in response to disasters in support of the GEO Disaster Theme. (Required)' [27]
GADM (version 1)	Online [46]	Redistribution not allowed without prior permission	Commercial use requires prior permission.	'These data are freely available for academic and other non-commercial use. Redistribution, or commercial use, is not allowed without prior permission.' [46]
FAO Hydrological basins (derived from HydroSHEDS)	Online [40]	Re-sale and redistribution requires prior permission (for HydroSHEDS data) (see 'Notes' this table).	Non-commercial use specified (for HydroSHEDS data) (see 'Notes' this table).	Usage constraints of download not apparent on the FAO GeoNetwork metadata catalogue site nor bundled with data. From the HydroSHEDS site: 'Users may apply HydroSHEDS for non-commercial purposes.' [28]
GRDC Major	Accessible	Redistribution	Commercial use	Policy guidelines for

Data	Availability & access	Reuse & redistribution	User permission	Notes
River Basins of the World	from GRDC via email after submitting order form [45]	not allowed without prior permission	requires prior permission.	the dissemination of data accessible at http://www.bafg.de/GRDC/EN/01_GRDC/12_plcy/data_policy_node.html
CIESIN (Global Rural-Urban Mapping Project (GRUMP), v1)	Online [49]	Resale and redistribution not allowed without prior permission.	Non-commercial use only.	'Users are prohibited from any commercial, non-free resale, or redistribution without explicit written permission from CIESIN, IFPRI, The World Bank, and CIAT.' [29]
CIESIN (Global Reservoir and Dam (GRanD), v1)	Online [30]	Re-sale and redistribution not allowed without prior permission	Non-commercial use only.	'The GRanD database is freely available for non-commercial use. Users are prohibited from any commercial resale, or redistribution without explicit written permission from the authors (Lehner et al. 2011) or GWSP.' [30]
GISTDA (Flood data)	Online [35]			No data distribution policy evident at Thailand Flood Monitoring System, but shape file flood data readily accessible online.

11

Effects of Climate Change, Poverty and Macroeconomic Policies on Agricultural Trade Performance in Nigeria

Anthony O. Onoja[1*] and A. I. Achike[2]

[1]Department of Agricultural Economics and Extension, University of Port Harcourt, Nigeria.
[2]Department of Agricultural Economics, University of Nigeria, Nsukka, Nigeria.

Authors' contributions

This work was carried out in collaboration between both authors. Author AOO designed the study, performed the statistical analysis, wrote the protocol and wrote the first draft of the manuscript. Author AIA read and approved the final manuscript.

ABSTRACT

Aims: This study ascertained the joint influences of climate factors, poverty and macroeconomic environment on agricultural export performance in Nigeria.
Study Design: The study is a survey based on time series data.
Place and Duration of Study: Secondary data covering 32 years (1978-2009) obtained from Central Bank of Nigeria's Annual Report and Statistical Bulletin and National Bureau of Statistics were used for the survey.
Methodology: The sample size was 32 (years) based on data availability. Data analysis was conducted using bound testing approach of co-integration advanced by Pesaran et al. [25] otherwise known as Autoregressive Distributed Lag (ARDL).model. Test for unit roots in the series were done at their levels and first differences using Augmented Dickey Fuller and Philips Perron tests before applying the ARDL model.
Results: Preliminary results from the ARDL model indicated that climate variability (variations in mean annual rainfall), gross fixed capital formation (proxy for wealth accumulated in the economy) and macroeconomic variables including interest rate and volume of domestic credit advanced to the private sector significantly influenced the performance level of agricultural export. However, on the long-run, macroeconomic factors (interest rate and credit to the private sector) and gross fixed capital of the economy (with p values of 0.01, 0.07 and 0.03 respectively were the most significant

**Corresponding author: Email: tonyojonimi@gmail.com;*

determinants of agricultural export trade performance in the country within the review period. On the short run, it was confirmed that gross fixed capital formation (wealth) Granger caused the level of agricultural export performance while agricultural export performance level Granger caused volume of domestic credit advanced to the private sector of the economy both with p values of 0.07.

Conclusion: It was recommended that macroeconomic policies aimed at increasing farm credit and reduction of interest rate should be strengthened; while programmes to build resilience to climate variability such as irrigation facilities and capacity building in climate change adaptation should be put in place by the Nigerian government.

Keywords: Agricultural export trade; poverty; macroeconomic policy; climate change; co-integration; Autoregressive Distributed Lag Modelling (ARDL model); Nigeria.

1. INTRODUCTION

Unarguably, commodity production and trade have become a significant factor on sustainable livelihoods of the poor, as well as on the export and growth performance of the vast number of commodity-dependent developing countries as Africa [1]. Unfortunately, the World Bank [2] indicated that the systemic information that is necessary to decompose the determinants of export growth for agriculture is very limited. The information exists only for manufacturing, it noted. Therefore country level studies about determinants of agricultural export trade performance, as this study attempts to model can contribute to this knowledge gap. According to UNCTAD [1], 50 percent of all developing countries depend on non-fuel commodities for more than half of their export earnings, two thirds if fuels are included. For some countries in Africa such as Nigeria, trade in non-oil commodities especially agricultural exports, constitutes a significant share of their economic growth [3]. The CBN report indicated that Nigerian domestic economy's growth rate of 7.9 percent as at 2010 which rose from 7.0 in 2009 was largely ascribed to sustained growth in non-oil sector and improvements in oil sector. Specifically, it noted that agriculture accounted for the greatest share of GDP growth rate as it contributed 2.4 percentage followed by services with 2.1 percent while whole sale and retail trade contributed 2.0 percent. It is therefore not surprising to note that in recent years, renewed interest in agriculture as a key propeller of development and poverty reduction has been witnessed [4].

African countries face many challenges to spur diversified agricultural growth and to gain from trade, despite the implementation of the Uruguay Round Agreement on Agriculture. These impediments include problem of climate change and its attendant effects on agriculture. However, in order to take advantage of new trading opportunities Africa needs to strengthen supply-side capability especially the agricultural supply sector. Unfortunately not much study is available to give evidence on the drivers of agricultural trade performance in Africa especially in Nigeria, a major contributor to African agricultural trade. Against the foregoing problems and research gaps this study was designed to find out the major determinants of agricultural supply and trade performance in Africa using Nigeria, the most populous country in Africa as a case study while focusing especially on the probable roles of poverty, climate factors and macroeconomic policy variables (such as interest rate), external influences (such as world price index of agricultural commodities) on agricultural trade in the study area. Results are expected to provide indices for evidence based planning that will facilitate agricultural trade, economic growth, boost food production and reduce food insecurity as well as poverty at the national and regional levels. The unique aspect of this study is its bold attempt to empirically establish the linkage between climate change (physical environmental factor), macroeconomic environment and wealth of a nation with agricultural trade using relatively new econometric approach (ARDL model). Outcome of a result of this type of study will give more lessons on broader approaches to tackling improving agricultural trade's contribution to economic growth placing much emphasis on physical and economic environment and poverty reduction as possible tools of intervention to attain this noble economic goal. Such lessons are not local but can apply to any country in the world.

1.1 Literature Review

The African leaders of the New Partnership for Africa's Development (NEPAD) under the Comprehensive Actions for African Agricultural Development, CAADP framework clearly were therefore justified when they indicated that among their priorities are infrastructure and agriculture which both have an interface. However there is also the need to check factors which could influence contribution of agriculture to growth. This is why IFAD noted that globally, increasing attention is being given both to issues of adaptation to climate change in smallholder agriculture and to ways in which poor rural people can participate in, and benefit from, market opportunities linked to environmental services and climate change mitigation [4] The issue of climate change becomes very daunting when one notes Easterling et al. [5]'s report that Sub-Saharan Africa is expected to face the largest challenges regarding food security as a result of climate change and other drivers of global change. Overall, Fischer et al. [6] estimated that as a result of climate change, agricultural GDP in Africa is expected to fall by between -2 to -8 percent (HadCM3 and CGCM2) and -7 to -9 percent (CSIRO model) [6]. This will significantly influence agricultural trade performance. Many farmers in Africa are likely to experience net revenue losses as a result of climate change, particularly as a result of increased variability and extreme events, the report added. There is broad agreement that growth in agriculture usually generates the greatest improvements for the poorest people – particularly in poor, agriculture-based economies. The IFAD report recognizes that agriculture, if better suited to meeting new environmental and market risks and opportunities facing smallholders can remain a primary engine of rural growth and poverty reduction. And this is particularly true in the poorest countries such as Sub-Sahara Africa. About 70 percent of Africans and nearly 80 percent of the continent's poor live in rural areas and depends mainly on agriculture for their livelihood. The sector accounts for about 20 percent of Africa's GDP [1], 60 percent of its labour force and 20 percent of the total merchandise exports (AU-NEPAD [1]). Agriculture is the main source of income for 90 percent of rural population in Africa [1]. Unfortunately however, it was observed by NEPAD [7] that the share of Africa in world agricultural exports has dropped steadily, from 8 percent in 1971-80 to some 3.4 percent in 1991-2000 and reversing this decline will require increased efforts by the African countries, with the assistance of the international community, to surmount the hurdles, including domestic supply-side constraints [7].

NEPAD [7] rightly noted that Africa faces trade challenges at many levels: The farmer faces non-remunerative markets and loses the incentive to produce; the nation-state fails to find rewarding markets both within the region and globally; Africa as a region is marginalized as well as being often uncompetitive in the international marketplace; furthermore, Africa continues to offer mostly unprocessed produce for which prices are static or falling, the report added. Responses to these issues pose challenges that NEPAD will need to address at the appropriate level.

There are diverse schools of thoughts about the relationship between poverty and agricultural trade performance in literature. Word Bank [2] indicated that a development strategy based on agricultural commodity exports was likely to be impoverishing in the prevailing agricultural policy environment in which policymakers in many countries had mercantilist and protectionist reflexes (that, when aggregated, compromise world trade in agricultural and food products). Morrissey et al. [8] corroborated earlier literature which supports the view that low income constrains the ability of households to buy or produce the products they need. The implication of this finding is that poverty can be a cause of market performance or market performance too can also be a source of poverty alleviation to farmers who export their crops. Morrissey et al. [8] stressed that a major factor in the reduction in rural poverty was the improved incomes and earnings from marketed agricultural produce. This, according to them was in part a reflection of trade policy reforms in crops with favorable world prices. The relevance of agriculture for poverty and the poor however is not limited to export crops. Greater importance than trade per se, however is the ability of farmers to respond to opportunities, in particular to substitute crops as relative prices change. This also have implications for increases or decrease in world price of agricultural export crops. Their rise internationally can send signals to farmers to export more. While evidence shows that farmers will respond to price incentives, it is also evident that they faced major constraints in doing so.

The World Bank [9] noted that world prices of agricultural commodity and economic growth of other countries outside Africa do impact positively on market performance of agricultural exports in Africa. Specifically, it indicated that the high commodity prices of recent years, coupled with Asia's rapid economic growth, have helped Africa expand its exports to Asia, which now imports 25 percent of all African exports. The growing economies of India and China alone account for 10 percent of Africa's exports. As these countries continue to grow and demand more natural resources, African exporters may be poised to increase revenues and expand their production in concert with the Indian and Chinese economic growth.

Diouf [10] posited that climatic adversities to a certain degree, civil unrest and war; access to markets for African agricultural products and unstable world prices adversely affected agricultural development in Africa in a serious way. They do and in particular international prices of primary commodities, he added. Some African exporters of agricultural products lost up to one third of their export earnings, due to the sharp decline in world prices and the duration of drought in the Sahel during the 1970s and 1980s was longer than anywhere else outside Africa. No economy, even a developed one can sustain such circumstances for a long period without having its social balance, which is the foundation of economic growth, profoundly disrupted. NEPAD [7] noted that adequate and well-functioning infrastructure was essential for agriculture to be competitive (market driven) due to reduced costs of delivering inputs to it and of taking produce out to markets, including any storage that this may entail; energy infrastructure is essential for development of agro-industries; information infrastructure is vital for timely technological information to farmers and agro-industrialists but also between producers and markets; water infrastructure is a precondition for irrigation for Africa.

Agricultural production depends on climate variables, such as temperature, precipitation, and ligh [11]. Hussler [12] found that agricultural production growth has been sensitive to rainfall variabilities and that this sensitivity has been modified across countries.

Research has shown that Sub-Saharan African agriculture is currently facing challenges in international trade with respect to external market access conditions and competition in world markets as a result of trade liberalization efforts under the world trade organization (WTO) agreements and in particular the agreement on agriculture (AoA). It has also been observed that the performance of agricultural exports from Sub Saharan Africa, based on production and value, for the period 1990 to 2000 showed mixed trends across countries and commodities [13]. In general, there has been an increase in production and exports of non-traditional commodities but exports of traditional exports for some countries show an increasing trend while for others they have stagnated or even declined due to declining world market prices and domestic marketing problems which may be as a result of macroeconomic policies. Macroeconomic reforms, including domestic and export subsidies in developed countries was therefore advocated [13] because they reduce the competitiveness of products from Sub Saharan Africa by depressing world market prices besides making it difficult for these countries to diversify their agriculture.

It has been indicated that improved access to international markets can contribute to the expansion of the external sector at all stages of its structural development but this seems to be relatively more important at the earlier stages of structural evolution than for countries that have already achieved a high degree of structural change [1]. This suggests that the more advanced developing countries are better able to exploit market opportunities through product diversification and differentiation for example by quality upgrading and thereby also avoiding trade barriers. The less advanced countries produce more homogeneous products that are more easily targeted by trade barriers (as well as suffering from commodity price declines), so that, when barriers come down, they experience a sharper increase in performance [1]. These findings have important implications for national policies and strategies, development cooperation programmes and actions within the trading system.

FAO [14] shows how poverty can reduce market performance in tropical countries such as Sub-Sahara Africa. FAO noted that most poor countries are located in the tropics, where the higher incidence of crop and livestock diseases and pests and excessive or inadequate rainfall are further factors compromising their ability to participate in global agricultural markets.

United States Agency for International Development, USAID [15] observed that despite steady economic growth since the return to civilian rule in 1999, 2004 per capita income in Nigeria was only $500 (in current U.S. dollars). USAID noted that in Nigeria, agricultural production and export performance have been poor and showed little sign of improvement. This was attributed to a multitude of factors: the overvalued currency and poor business climate discussed above, as well as poor policies specific to the sector, such as unfavorable domestic pricing policies. In the period 1999–2003, agricultural export growth rates fluctuated but the average rate of growth was –1 percent per annum for the five-year period. The value added per agricultural worker in Nigeria averaged $807 (in constant 1995 dollars) during the five years to 2003—significantly higher than the $250 average of LI-SSA or Ghana's $346 and considerably lower than Cameroon's $1,215. The growth of added value in agriculture is in line with regional benchmarks (4.1 percent compared to the LI-SSA 4.2 percent average), but is unlikely to be sustainable. According to the World Bank as cited in USAID [15], the driving factor has been increased land use rather than improvements in technology.

Aksoy [16] showed that developing countries, especially least-developed countries, face much higher trade-related costs than other countries in getting their products into international markets. Some of these costs may reflect institutional problems within the countries themselves, such as inefficient practices and corruption and these problems require a domestic policy response. But some costs also reflect weak transportation infrastructure in many countries and firms' lack of access to standard trade-facilitating facilities such as insurance and trade finance (which may include access to credit at affordable interest rates).

Marketing can be defined as getting the right product or service in the right quantity, to the right place, at the right time and making a profit in the process [17]. Effective marketing is a result of examining every aspect of your business and how it affects the consumer's end experience. It covers everything you will need to do in order to deliver your products and services to the consumer including research, planning, pricing, packaging, promotion, selling and distribution. Marketing involves supply of goods and services. Agricultural export is a component of aggregate supply in an economy. Supply theory holds that ceteris paribus, as price changes, quantity supplied will change (i.e. movement along the same supply curve) such that when factors other than price changes, supply curve will shift. The theory holds that determinants of the supply curve (shifters) include production cost, technology (e.g. use of tractors, irrigation technology and adoption of climate resilient technologies such as fertilizer use, crop rotation, reforestation, etc.), number of sellers (level of competitiveness), environmental factors (weather and climate), government policies, population and expectation for future prices [18].

1.2 Analytical Framework

According to Stocks and Watson [19] the estimation of a long-run relationship involving co-integrated variables has been the focus of a lot of recent papers. Many studies have reported alternative co-integrating vector estimators and their asymptotic properties [20,21]. Stocks and Watson [19] noted that the asymptotic properties are not affected by endogeneity or serial correlation if the estimators are properly corrected. However, the applied researcher does not usually have enough data to justify the application of asymptotic theory. For this reason it is important to consider the small-sample performance of alternative co-integrating vector estimators. On the one hand, the general result points to a large bias in small samples for any estimator that ignores short-run dynamics [22]. The error correction mechanism (ECM) estimator, which considers explicitly knowledge of the short-run dynamics has problems in terms of t-statistics far from their theoretical distributions. Past literature identified the following models as efficient estimations methods for estimating single equation co-integrated regression: Fully Modified OLS [23], Canonical Co-integration Regression developed by Park [24] and Dynamic OLS postulated by Saikkonem in 1992 and Stock and Watson in 1993 [24]. However, these models have been flawed on various grounds which are beyond the scope of our immediate discussion. Interested readers should consult [19,20,21,25]. The most recent co-integration approach for single equation modelling gaining popularity now is the bound co-integration approach proposed by Pesaran et al. [25]. This present study applied this model. More explanation of this model can be found at the methods section.

2. METHODOLOGY

2.1 Area of Study

Nigeria is the most populous country in Africa with a population estimated at 162,265,000 by mid-2011, population living below poverty line of $2 per day of 84 percent, Gross National Income per capita of US $2,070 and CO_2 emission rate of 0.4 metric tonnes per capita [26]. According to CBN [3] the GDP of Nigeria stood at about US$4.9 billion (N775.4 billion) at

1990 basic prices. There are credible reports of climate change effects in Nigeria [27,28,29,30].

For instance IFAD [31] 2003/04 Nigeria living standards survey indicated that states in the Sudan/Sahel region (which are more arid) than all other regions of Nigeria recorded the highest incidence of poverty, with about 80 per cent of the population described as poor. The Niger Delta region has also witnessed growing poverty over the last 20 years, owing mainly to negligence by Government at all three levels. In addition, environmental degradation, perpetrated by oil companies in particular has eroded the resource base, notably land and water, thereby constraining agriculture and fishery development. This situation is further compounded by incessant socio-economic conflicts and restiveness among young people, which have hampered development efforts and severely compromised the security of life and property [31]. Average rainfall hovers around 1282.2 mm varying from 500 - 1800mm. In 2007 agriculture contributed 42.08% to Nigerians GDP. Out of this figure, crops, livestock, forestry and fishing contributed 37.54%, 2.64%, 0.53% and 1.37% to the country's economy respectively. Agricultural *Products* include cocoa, palm oil, yams, cassava, sorghum, millet, corn, rice, livestock, groundnuts, cotton. Crude oil export is the major source of revenue. Industry types include textiles, cement, food products, footwear, metal products, lumber, beer, detergents and car assembly [32].

2.2 Data Source and Methodological Framework

Secondary data, mainly time series data from Central Bank of Nigeria's Annual Report and Statistical Bulletin containing data from National Population. Secondary data from World Bank Meta-files [33] were equally used. We purposively selected 1978 to 2009 i.e. a sample of 32 years for ease of accessing data that will cover all the variables or series within the period of study.

The series were tested for unit roots at their levels and first differences using Augmented Dickey Fuller and Philips Perron tests [34,35]. The results are presented in Appendix 1. When it was noted that some of the series were not I(0) but became stable at I(1) the researchers decided to apply the bound co-integration testing approach following Pesaran, et al. [25]. Standard econometric diagnosis were performed including Breusch- Godfrey serial correlation LM test, Jacque-Bera normality test and Ramsey RESET specification test following [37,36,35,34].

Granger pair wise causality tests were performed to ascertain the short run bi-directional causality of the statistically significant independent variables in relationship with the dependent variable. From the test of dynamic short-run causality presence among the relevant variables of the model, If the null hypothesis of no causality is rejected, then we conclude that a relevant variable Granger-caused agricultural export performance.

The use of the bounds technique is based on three validations. First, Pesaran et al. [25] advocated the use of the ARDL model for the estimation of level relationships because the model suggests that once the order of the ARDL has been recognised, the relationship can be estimated by OLS. Second, the bounds test allows a mixture of I(1) and I(0) variables as regressors, that is the order of integration of appropriate variables may not necessarily be the same. Therefore, the ARDL technique has the advantage of not requiring a specific identification of the order of the underlying data. Third, this technique is suitable for small or finite sample size [25].

Moreover, the bounds testing procedure [25] employed in this study is robust for small sample study. Furthermore, the bound testing approach is possible even when the explanatory variables are endogenous. The ARDL co-integration test, assumed that only one long run relationship exists between the dependent variable and the exogenous variables [25] (assumption 3). The bound test is basically computed based on an estimated unrestricted error-correction models (UECM) or error correction version of autoregressive distributed lag (ARDL) model, by Ordinary Least Square (OLS) estimator [25]. Basically, the bound test is the Wald test (F-statistic version of the bound testing approaches) for the lagged level variables in the right-hand side of UECM. That, we test the null hypothesis of non-co-integrating relation (Ho: $\delta_1 = \delta_2 = \delta_3 = \ldots = \delta_n = 0$) against the alternative hypothesis (H_A: $\delta_1 \neq \delta_2 \neq \delta_3 \neq \ldots \neq \delta_n \neq 0$) (a long-run relationship exists). The computed *F*-statistic value will be evaluated with the critical values tabulated in table CI (iii) of Pesaran et al. [25]. According to these authors, the lower bound critical values assumed that the explanatory variables x_t are integrated of order zero or I(0), while the upper bound critical values assumed that x_t are integrated of order one or I(1). Therefore, if the computed *F*-statistic is smaller than the lower bound value, then the null hypothesis is not rejected and we conclude that there is no long-run relationship between agricultural export performance and its determinants. Conversely, if the computed *F*-statistic is greater than the upper bound value, then agricultural export performance and its determinants share a long-run level relationship. On the other hand, if the computed *F*-statistic falls between the lower and upper bound values, then the results are inconclusive. The researcher must conduct test to determine the order of integration before making a conclusive inference in this case.

Following Pesaran et al. [25,38] we assembled the vector autoregression (VAR) of order *p*, denoted VAR (*p*), for the following function:

$$Z_t = \mu + \sum_{i=1}^{p} \beta_i z_{t-i} + \varepsilon_t \tag{1}$$

where z_t is the vector of both x_t and y_t, where y_t is the dependent variable defined as agricultural export (agexport), x_t is the vector matrix which represents a set of explanatory variables i.e., trade openness (TOP), financial development (M2) and *t* is a time or trend variable. According to Pesaran et al [25], y_t must be I(1) variable but the regressor x_t can be either I(0) or I(1). We further developed a vector error correction model (VECM) as follows:

$$\Delta z_t = \mu + \alpha t + \lambda z_{t-1} + \sum_{i=1}^{p-i} \gamma_t \Delta y_{t-i} + \sum_{i=1}^{p-1} \gamma_t \Delta x_{t-i} + \varepsilon_t \tag{2}$$

where Δ is the first-difference operator. The long-run multiplier matrix λ as:

$$\lambda = \begin{bmatrix} \lambda_{YY} \lambda_{YX} \\ \lambda_{XY} \lambda_{XX} \end{bmatrix}$$

The diagonal elements of the matrix are unrestricted, so the selected series can be either I(0) or I(1). If $\lambda_{YY} = 0$, then *Y* is I(1). In contrast, if $\lambda_{YY} < 0$, then *Y* is I(0).

The vector error correction mechanism (VECM) procedures described above are compulsory in testing of at most one cointegrating vector between dependent variable y_t and a set of regressors x_t. To derive the model, postulations made by Pesaran et al. [25] in Case III, that is, unrestricted intercepts and no trends were followed.

The general co-integration hypothesis model can be stated as the following unrestricted error correction model (UECM):

$$\Delta\ (AGEXPORT)_t = \beta_0 + \beta_1(AGEXPORT)_{t-1} + \beta_2\ (lnRAINF)_{t-1} + \beta_3\ (lnWPIAGR) + \beta_4(lnINTR) + \beta_5\ (lnDOMCRPRIV) + \beta_6(lnGFC) + \beta_7\sum_{i=1}^{p}\Delta\ (AGEXPORT)_{t-1} + \beta_8\sum_{i=0}^{q}\Delta\ (lnRAINF)_{t-1} + \beta_9\sum_{i=0}^{r}\Delta\ (lnWPIAGR)_{t-1} + \beta_{10}\sum_{i=0}^{s}\Delta\ (lnINTR)_{t-1} + \beta_{11}\sum_{i=0}^{t}\Delta\ (lnDOMCRPRIV)_{t-1} + \beta_{12}\sum_{i=0}^{u}\Delta\ (lnGFC) + \varepsilon_t \quad (3)$$

Where Δ is the first-difference operator and ε_t is a white-noise disturbance term. Y = agricultural export performance also represented by AGEXPORT (market performance of agriculture measured in terms of agriculture trade as percentage of total trade), while hypothesized drivers of agricultural export trade performance include: X_1 = GFC or asset ownership (or poverty proxied by Gross fixed capital, formation in the economy in millions of naira per year, DOMCREPRIV (X_2) = domestic credit to the private sector of the economy (in millions of naira per annum), RAINF (X_3)= climate change (mean annual precipitation in Nigeria in mm), INTR (X_4) = interest rate (average ending rate per annum in percentage, WPIAGR (X_7) = world price of index agricultural commodities, while ε_t = stochastic error term. It refers to natural log of the respective variables raised to power e. β_0 ... β_{11} are co-efficients of the respective variables.

Equation (3) also can be viewed as an ARDL of order (p, *q, r, s, t, u*). Equation (3) indicates that agricultural export performance tends to be influenced and explained by its past values. The structural lags are validated by using minimum Akaike's information criteria (AIC). Judging from the estimation of unrestricted error correction mechanisms (UECMs), the long-run elasticities are the coefficient of one lagged explanatory variable (multiplied by a negative sign) divided by the coefficient of one lagged dependent variable [39]. For example in equation (3), the long-run rainfall and world price of index agricultural commodities elasticities are (β_2 / β_1) and (β_3 / β_1) respectively. The short-run effects are explained by the coefficients of the first-differenced variables in equation (3).

3. RESULTS AND DISCUSSION

3.1 Determinants of Agricultural Export Trade in Nigeria

The starting point of our analysis was the consideration of the results of tests of stability of the models' series using ADF and PP tests. The details of these analyses are presented in Appendix 1 and 2. In summary, going by the ADF test statistics, it was found that only two series, rainfall level, was stable at levels (i.e. I(0)). Other variables or series became stable after the first differences (i.e. I(1). The performance of the series with respect to unit root tests was not too different when PP test was conducted. It was found that none of the series was stable at levels using the PP criterion. They all became stable in their first differences thus indicating that they were integrated in the order 1 (i.e. I(1).). Obviously, the mixture of both I(0) and I(1) variables would not be possible under the Johansen procedure. This gives a good justification for using the bounds test approach or ARDL model.

The estimation of Equation (3) using the ARDL model is reported in Table 1. Using Hendry's general-to-specific method, the goodness of fit of the specification, that is, R-squared and adjusted R-squared, is 0.76 and 0.60 respectively. The robustness of the model has been validated by several diagnostic tests carried out, such as Breusch- Godfrey serial correlation LM test, ARCH test, Jacque-Bera normality test and Ramsey RESET specification test. All the tests indicated that the model has the desirable econometric properties, it has a correct functional form and the model's residuals are serially uncorrelated, normally distributed and homoskedastic. Therefore, the results reported are serially uncorrelated, normally distributed and homoskedastic. The results reported are valid for reliable interpretations.

Table 1. Estimated model based on equation (3) dependent variable: AGEXPORT white heteroskedasticity-consistent standard errors & covariance

Variable	Coefficient	Standard errors	t-statistics
Intercept	10.429	39.946	0.261
AGEXPORT(-1)	-0.561	0.268	-2.092*
D(AGEXPORT(-1))	-0.202	0.18	-1.127
lnRAINF(-1)	-5.375	6.199	-0.867
D(lnRAINF(-1))	8.466	3.113	2.720**
lnWPIAGR(-1)	-0.354	0.306	-1.156
D(lnWPIAGR(-1))	-0.27	1.144	-0.236
lnINTR(-1)	-2.723	1.028	2.649**
D(lnINTR(-1))	-1.58	0.764	-2.067**
lnDOMCREPRIV(-1)	2.464	1.298	1.899*
D(lnDOMCREPRIV(-1))	1.445	1.371	1.054
lnGFC(-1)	1.024	0.424	2.413**
D(lnGFC(-1))	0.001	0.446	0.002
R-squared	0.7	F-statistic	4..93***
Adjusted R-squared	0.507	Prob(F-statistic)	0.01

Table 2. Econometric diagnostic checks

Type of diagnositic test	*Coeff.*	*p Value*	*Remark*
Breusch-Godfrey Serial Correlation LM Test	2.073	0.16	Accept the null hypothesis of no presence of 2nd order serial correlation in the model
Jarque-Bera test	2.514	0.28	Accept the null hypothesis of residual coming from a normally distibiuted sample
Breusch-Pagan-Godfrey test for heteroscedasticity	1.322	0.28	Accept the null hypothesis of no heteroscedasticity in the model
Ramsey RESET Test (F test estimated at 1, 16 d.f.)	0.1188	0.73	Accept the null hypothesis that the correct specification is linear

Source: Analysis of CBN [32] and National Bureau of Statistics Data using EViews Package by Author

The estimation of Equation (3) using the ARDL model is reported in Table 1. Using Hendry's general-to-specific method, the goodness of fit of the specification, that is R-squared and adjusted R-squared is 0.70 and 0.50 respectively. The result implies that the model explained correctly 50 percent of the variation in the agricultural export trade performance model, which is a fairly good fitting, using the adjusted R-square criterion. The robustness of the model has been validated by several diagnostic tests such as Breusch- Godfrey serial correlation LM test, ARCH test, Jacque-Bera normality test and Ramsey RESET specification test (Table 2). All the tests disclosed that the model has the aspiration econometric properties, it has a correct functional form and the model's residuals are serially uncorrelated, normally distributed and homoskedastic. Therefore, the outcomes reported are valid for reliable interpretation and policy analysis.

In Table 3 the results of the bounds co-integration test demonstrate that the null hypothesis of no co-integration against its alternative of presence of co-integration is hereby rejected at the 10% significance level. The computed *F*-statistic of 4.93 is greater than the upper critical bound value of 4.78, thus indicating the existence of a steady-state long-run relationship among agricultural export trade performance indicator, mean annual rainfall, interest rate (lending rate), domestic credit volume advanced to the private sector of the economy, world price index of agricultural commodities and poverty level (gross fixed capital formation).

Table 4 indicates that lending rate (or interest rate), volume of loans advanced to the private sector, and poverty level (GFC) have a positive impact on agricultural export performance in Nigeria over the period in review. If there is one percent increase in volume of loans advanced to the private sector, agricultural export performance rate grows by approximately 4.40 percent; while a unit increase in gross fixed capital formation (or wealth increase) is accompanied by increase in trade performance in the economy by 1.82 percent. In terms of lending rate, a unit increase in this factor will be accompanied by a reduction in the rate of agricultural export performance by approximately 4.86 percent. These imply that macroeconomic variables are significant determinants of long run increase in agricultural export trade performance in Nigerian economy.

Table 3. Bounds test for co-integration analysis

Estimated F value	Lag	Bound critical values (unrestricted intercept no trend)*		
		Critical value	Lower Bound Value I(0)	Upper bound value I(1)
4.93	1	1%	7.41	8.37
		5%	4.98	5.73
		10%	4.04	4.78

**Based on Critical F values for conducting bound test for cointegration as cited in Pesaran et al. [25], in "Table C1.iii Case III, unrestricted intercept no trend"*

Table 4. Long-run estimated coefficient

Variable	Coefficient	P value
RAINFALL	-9.589NS	0.30
WPIAGR	-0.631NS	0.26
INTR	-4.858**	0.01
DOMCRPRIV	4.396*	0.07
GFC	1.826**	0.03
Intercept	18.605NS	0.80

*NS = "Not Significant", ****, ** and * indicate significance at 0.01, 0.05 and 0.10 level respectively*

The dynamic short-run causality among the relevant variables is shown in Table 5. From this test, we infer that gross fixed capital formation (or level of wealth of the nation) is statistically significant in Granger-causing agricultural trade performance in Nigeria at 10 percent significance level while volume of domestic credit advanced to the private sector in the economy over the period in review was significantly Granger caused by level of Nigerian agricultural export trade performance.

Table 5. Short-run causality test (Granger Pair wise causality Test F-statistic)

Null Hypothesis	Obs	F-statistic	P value
LNINTR does not Granger Cause AGEXPORT	31	0.334	0.56
AGEXPORT does not Granger Cause LNINTR		0.107	0.74
LNDOMCREPRIV does not Granger Cause AGEXPORT	31		4.17
AGEXPORT does not Granger Cause LNDOMCREPRIV			0.07*
LNGFC does not Granger Cause AGEXPORT	31	3.370	0.07*
AGEXPORT does not Granger Cause LNGFC		0.217	0.64

NB: (***) = figures are significant at p =<0.01; (**) = figures are significant at p =<0.05, and (= figures are statistically significant at p =<0. 10.*

4. CONCLUSION

This research has established, through appropriate time series econometric approach (co-integration analysis) that macroeconomic factors, wealth stock of the nation, climate and other environmental variables (especially rainfall level and arable land expansion) have significant influences on export trade performance of Nigerian agricultural sector. It would therefore suggest that proper policy mix that will reduce poverty for instance (economic empowerment policies such as provision of agricultural credit), proactive policies to adapt to deleterious effects of extreme weather or climate events (e.g. flood management, irrigation based farming etc) and proper management of the macroeconomic regulations can improve agricultural trade. Based on the above findings, it is hereby recommended that macroeconomic policies aimed at increasing farm credit and reduction of interest rate should be strengthened while programmes to build resilience to climate variability such as irrigation facilities and capacity building in climate change adaptation should be put in place by the Nigerian government.

COMPETING INTERESTS

Authors have declared that no competing interests exist.

REFERENCES

1. United Nations Conference on Trade and Development UNCTAD. Determinants of export trade. In UNCTAD, Developing countries in international trade. Trade and development index; 2005.
   Available: http://www.unctad.org/en/docs/ditctab20051ch2_en.pdf

2. World Bank. Global agricultural trade and developing countries. World Bank; 2005. Washington D.C. Retrieved on 14th July 2012. Available:http://siteresources.worldbank.org/INTPROSPECTS/Resources/GATfulltext.pdf
3. Central Bank of Nigeria, CBN. Annual report. Abuja: CBN; 2010.
4. IFAD. Federal Republic of Nigeria Country strategic opportunities programme; 2010. Rome: IFAD EB/2010/99/R.11.Easterling WE, Aggarwal PK, Batima PK, Brander M, Lin ED, Howden SM, Kirilenko AP, Morton J, Soussana J, Schmidhuber J, Tubiello F. Food, fibre and forest products. In Climate Change 2007: Impacts, Adaptation and Vulnerability. Contribution of Working Group II to the Fourth Assessment Report of the Intergovernmental Panel on Climate Change. Ed. Parry M., Canziani OF., Palutik of J, van der Linden PJ, and Hanson, CE, Cambridge: Cambridge University Press; 2007.
5. Easterling WE, Aggarwal PK, Batima PK, Brander M, Lin ED, Howden SM, Kirilenko AP, Morton J, Soussana J, Schmidhuber J, Tubiello F. Food, fibre and forest products. In Climate Change 2007: Impacts, Adaptation and Vulnerability. Contribution of Working Group II to the Fourth Assessment Report of the Intergovernmental Panel on Climate Change. Ed. Parry M., Canziani OF., Palutikof J, van der Linden PJ, Hanson, CE,. Cambridge: Cambridge University Press; 2007.
6. Fischer GM, Shah F, Tubiello van Velthuizen HT. Socio-economic and climate change impacts on agriculture: an integrated assessment, 1990-2080. Phil. Trans. R. Soc. B. 2005;360:2067-2083.
7. New Partnership for Africa's Development (NEPAD). Comprehensive Africa Agriculture Development Programme (CAADP); 2002. Retrieved on 18th February, 2012 from available:http://www.fao.org/docrep/005/y6831e/y6831e00.htm#TopOfPage
8. Morrisey O, Rudaheranwa N, Moller L. Uganda trade and poverty project (UTPP): trade policies, performance and poverty in Uganda; 2003. Retrieved on 14th July, 2012 Available:http://citeseerx.ist.psu.edu/viewdoc/download?doi=10.1.1.199.1013&rep=rep1&type=pdf
9. World Bank, Building competitiveness in Africa's agriculture a guide to value chain concepts and applications. Weber CM & Labaste P(Eds). The International Bank for Reconstruction and Development/The World Bank; 2010.
10. Diouf J. The challenge of agricultural development in Africa. A paper presented at Sir John Crawford Memorial Lecture. Washington,DC; 1989. Accessed 18/2/12 Available: http://www.worldbank.org/html/cgiar/publications/crawford/craw5.pdf
11. Moula EL. Climate variability, vulnerability and effectiveness of farm-level adaptation options: the challenges and implications for food security in Southwestern Cameroon. Environment and Development Economics. 2002;7:529–545. Cambridge University Press DOI:10.1017/S1355770X02000311 Printed in the United Kingdom. Retrieved on 14th July 2012. Available: http://nersp.nerdc.ufl.edu/~vecy/LitSurvey/Molua.pdf
12. Hussler S. Econometric study on the impact of rainfall variability on security in the Sahel region. 2010 OECD's Sahel and West Africa Club Secretariat, French Ministry of Foreign and European Affairs and the UK Foreign and Commonwealth Office. Retrieved on 14th July, 2012. Available: http://www.oecd.org/dataoecd/9/31/44245104.pdf
13. Nyangito HO. Performance of African agricultural exports and external market access conditions under international trade reforms, 2004, No 9518, 2004 Inaugural Symposium; 2004. Nairobi, Kenya from African Association of Agricultural Economists (AAAE). Retrieved on 10th July, 2012. Available: http://econpapers.repec.org/paper/agsaaaeke/9518.htm

14. Food and Agricultural Organization (FAO). Food and Agriculture in National and International Settings. In World agriculture: towards 2015/2030 Summary report; 2005. FAO: Rome. Retrieved on 14th July, 2012.
Available: http://www.fao.org/docrep/004/y3557e/y3557e07.htm
15. United States Agency for International Development, (USAID). Nigeria Economic Performance Assessment. Nathan Associates for United States Agency for International Development; 2006. Retrieved on 14th July, 2012.
Available: http://pdf.usaid.gov/pdf_docs/PNADF350.pdf
16. Aksoy MA, Beghin JC. (Eds). World Bank Global agricultural trade and developing countries. World Bank. Washington D.C; 2005. Retrieved on 14th July 2012. Available: http://siteresources.worldbank.org/INTPROSPECTS/Resources/GATfulltext.pdf
17. Small Business Development Corporation. Marketing; 2012. Retrieved on 10th July 2012. Available: http://www.smallbusiness.wa.gov.au/marketing/#what
18. Samuelson PA, Nordhaus, WA. Economics. Tata Mc-Graw-Hill: New Delhi; 2005.
19. Stock J, Watson M. A simple estimator of cointegrating vectors in higher order integrated systems, Econometrica.1967;61(4):783-820.
20. Phillips P, Loretan M. Estimating long-run economic equilibria, Review of Economic Studies. 1991;58:407-436.
21. Phillips PBC, Hansen BE. Statistical inferences in instrumental variables regression with I(1) processes. Econometric theory, Cambridge University Press. 2003;7(02):213-221.
22. Hansen BE. Testing for parameter instability in linear models. Journal of policy modeling 1992;14:517 -533.
23. Phillips P. Fully Modified Least Squares and Vector Autoregression, Econometrica, Econometric Society. 1995;63(5):1023-78, September. Retrieved on 25th February, 2012. Available: http://ideas.repec.org/p/cwl/cwldpp/1047.html
24. EViews 7. Estimating a cointegrating regression. Eviews 7, Enterprise Edition; 2009
25. Pesaran MH, Shin Y, Smith R. Bounds testing approaches to the analysis of level relationships, Journal of Applied Econometrics. 2001;16:289-326.
26. Population Reference Bureau, PRB. Nigeria; 2011. Retrieved on 18th February. Available: http://www.prb.org/DataFinder/Geography/Data.aspx?loc=269
27. Nigerian Environmental Study/action Team (NEST), Climate change in Nigeria: A communication guide for reporters and educators, NEST, Ibadan; 2003.
28. Odjugo PAO. An analysis of rainfall pattern in Nigeria, Global Journal of Environmental. Sciences. 2005;4(2):139-145.
29. Adefolalu DO, Pam JL. Habbi HM. Climate change and safety of air transportation – A Nigerian perspective. Proceedings of the International Conference on the impacts of extreme weather and climate on socio-economic development in Africa held at the Federal University of Technology, Akure, Nigeria. 2007;1-15.
30. Ikhile CI. Impacts of climate variability and change on the hydrology and water resources of the Benin-Owena River Basin, Ph D thesis submitted to the Department of Geography and Regional Planning. University of Benin, Benin City, Nigeria; 2007.
31. International Fund for Agricultural Development (IFAD) Rural poverty report 2011: New realities, new challenges: new opportunities for tomorrow's generation; 2011. Rome: International Fund for Agricultural Development (IFAD).

32. Central Bank of Nigeria, CBN. Annual report and Statistical Bulletin; 2007. Abuja: CBN.
33. World Bank. Nigeria. Meta files; 2012. Retrieved on 14th February 2012. Available: http://data.worldbank.org/country/nigeria
34. Gujarati DN. Essentials of econometrics. 3rd Edition. New York: Mc-Graw Hill; 2006.
35. Greene WH. Econometric analysis, 5th edition. New Delhi: McGraw Hill; 2008.
36. Studenmund AH. Using econometrics - a practical guide. Addison-Wesley-Longman. 2001;193-95.
37. Patterson K. An introduction to applied econometrics. Macmillan. 2000;184-5
38. Atif RA, Jadoon A, Zaman K, Ismail A, Seemab R. Trade liberalisation, financial development and economic growth: evidence from Pakistan (1980-2009). Journal of International Academic Research. 2010;10:2.
39. Bardsen G. Estimation of long-run coefficients in error correction models, Oxford Bulletin of Economics and Statistics. 1989;51:345-50.

APPENDIX 1

Results of stationary tests using Augmented Dickey Fuller (ADF) Statistic

Variable	ADF Value Before Differencing	ADF Value after differencing	Critical Value	Level of Integration
agexport	-2.128862NS	-7.146375***	-2.963972	I(1)
Intrrt	-2.558349NS	-8.544426***	-2.960411	I(1)
Indomcred	1.005332NS	-3.943077***	-2.960411	I(1)
Inforex	-1.514146NS	-5.058507***	-2.960411	I(1)
Inpgfc	-2.112403NS	-5.115874***	-2.960411	I(1)
Inrainf	-3.507717***	n.a.	-2.960411	I (0)
Inwpiagr	-0.814151NS	-4.701954***	-2.960411	I(1)

Source : Analysis of CBN (2011) and National Bureau of Statistics Data using EViews Package by Author

Appendix 2

Results of stationary tests using Philips Perron (PP) Statistic

Variable	PP Value Before differencing	PP Value after differencing	Critical Value	Level of integration
export	-2.163646NS	-7.080327***	-2.960411	I(1)
intrrt	-2.468201NS	-8.557680***	-2.960411	I(1)
Indomcred	1.005332NS	-3.921001***	-2.960411	I(1)
Inpovgfc	-2.161171NS	-5.125802***	-2.960411	I(1)
Inrainf	-3.472086NS	-9.085967***	-2.960411	I (0)
Inwpiagr	-0.822765NS	-4.674503***	-2.960411	I(1)

Source: Analysis of CBN (2011) and National Bureau of Statistics Data using EViews Package by Author

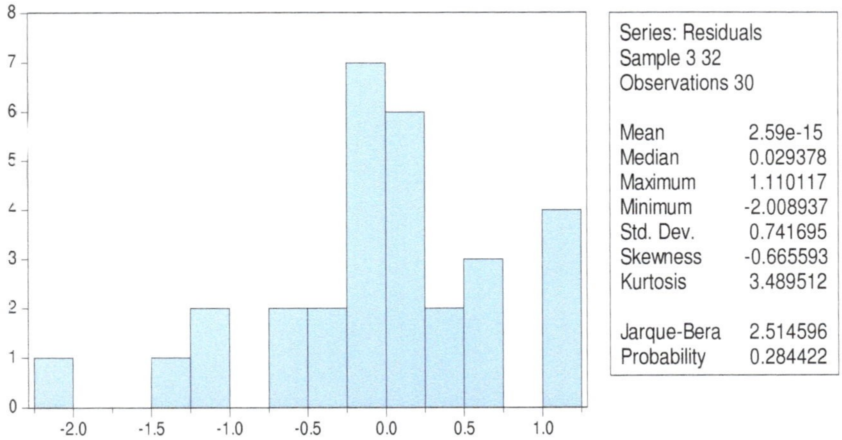

Fig. 1. Jarque-Bera test results

Heteroskedasticity Test: Breusch-Pagan-Godfrey

F-statistic	1.322244	Prob. F(12,17)	0.2915
Obs*R-squared	14.48288	Prob. Chi-Square(12)	0.2709
Scaled explained SS	5.788881	Prob. Chi-Square(12)	0.9263
Breusch-Godfrey Serial Correlation LM Test:			
F-statistic	2.073554	Prob. F(2,15)	0.1603
Obs*R-squared	6.497757	Prob. Chi-Square(2)	0.0388

Breusch-godfrey serial correlation LM Test 2.073 (p = 0.160)

Ramsey RESET Test			
Equation: UNTITLED			
Specification: D(AGEXPORT) C AGEXPORT(-1)D(AGEXPORT(-1)) LNRAINF(-1) D(LNRAINF(-1)) LNWPIAGR(-1) D(LNWPIAGR(-1)) LNINTR(-1) D(LNINTR(-1)) LNDOMCREPRIV(-1) D(LNDOMCREPRIV(-1)) LNGFC(-1) D(LNGFC(-1))			
Omitted Variables: Squares of fitted values			
	Value	**df**	**Probability**
t-statistic	0.344769	16	0.7348
F-statistic	0.118866	(1, 16)	0.7348
Likelihood ratio	0.222049	1	0.6375
F-test summary:			
	Sum of Sq.	df	Mean Squares
Test SSR	0.117644	1	0.117644
Restricted SSR	15.95323	17	0.938425
Unrestricted SSR	15.83559	16	0.989724
Unrestricted SSR	15.83559	16	0.989724
LR test summary:			
	Value	df	
Restricted LogL	-33.09512	17	
Unrestricted LogL	-32.98409	16	

Permissions

All chapters in this book were first published in BJECC, by SCIENCE DOMAIN International; hereby published with permission under the Creative Commons Attribution License or equivalent. Every chapter published in this book has been scrutinized by our experts. Their significance has been extensively debated. The topics covered herein carry significant findings which will fuel the growth of the discipline. They may even be implemented as practical applications or may be referred to as a beginning point for another development.

The contributors of this book come from diverse backgrounds, making this book a truly international effort. This book will bring forth new frontiers with its revolutionizing research information and detailed analysis of the nascent developments around the world.

We would like to thank all the contributing authors for lending their expertise to make the book truly unique. They have played a crucial role in the development of this book. Without their invaluable contributions this book wouldn't have been possible. They have made vital efforts to compile up to date information on the varied aspects of this subject to make this book a valuable addition to the collection of many professionals and students.

This book was conceptualized with the vision of imparting up-to-date information and advanced data in this field. To ensure the same, a matchless editorial board was set up. Every individual on the board went through rigorous rounds of assessment to prove their worth. After which they invested a large part of their time researching and compiling the most relevant data for our readers.

The editorial board has been involved in producing this book since its inception. They have spent rigorous hours researching and exploring the diverse topics which have resulted in the successful publishing of this book. They have passed on their knowledge of decades through this book. To expedite this challenging task, the publisher supported the team at every step. A small team of assistant editors was also appointed to further simplify the editing procedure and attain best results for the readers.

Apart from the editorial board, the designing team has also invested a significant amount of their time in understanding the subject and creating the most relevant covers. They scrutinized every image to scout for the most suitable representation of the subject and create an appropriate cover for the book.

The publishing team has been an ardent support to the editorial, designing and production team. Their endless efforts to recruit the best for this project, has resulted in the accomplishment of this book. They are a veteran in the field of academics and their pool of knowledge is as vast as their experience in printing. Their expertise and guidance has proved useful at every step. Their uncompromising quality standards have made this book an exceptional effort. Their encouragement from time to time has been an inspiration for everyone.

The publisher and the editorial board hope that this book will prove to be a valuable piece of knowledge for researchers, students, practitioners and scholars across the globe.

List of Contributors

Tariq Usman Saeed
Department of Transport and Mass Transit, Government of Khyber Pakhtunkhwa, Peshawar, Pakistan

Haleema Attaullah
Department of Civil Engineering, University of Engineering and Technology, Peshawar, Pakistan

Ching-Pin Tung
Department of Bioenvironmental Systems Engineering, National Taiwan University, Taiwan

Tzu-Ming Liu
Department of Bioenvironmental Systems Engineering, National Taiwan University, Taiwan

Szu-Wei Chen
Department of Bioenvironmental Systems Engineering, National Taiwan University, Taiwan

Kai-Yuan Ke
Department of Bioenvironmental Systems Engineering, National Taiwan University, Taiwan

Ming-Hsu Li
Institute of Hydrological and Oceanic Sciences, National Central University, Taiwan

Xiaodong Qu
State Key Laboratory of Environmental Criteria and Risk Assessment, Chinese Research Academy of Environmental Sciences, Beijing, China
Laboratory of Riverine Ecological Conservation and Technology, Chinese Research Academy of Environmental Sciences, Beijing, China

Ying Zhou
Laboratory of Riverine Ecological Conservation and Technology, Chinese Research Academy of Environmental Sciences, Beijing, China
Liaoning Provincial Key Laboratory for Hydrobiology, College of Fisheries and Life Science, Dalian Ocean University, Dalian, China

Rui Zhao
State Key Laboratory of Environmental Criteria and Risk Assessment, Chinese Research Academy of Environmental Sciences, Beijing, China
Laboratory of Riverine Ecological Conservation and Technology, Chinese Research Academy of Environmental Sciences, Beijing, China

Catherine Bentsen
State Key Laboratory of Environmental Criteria and Risk Assessment, Chinese Research Academy of Environmental Sciences, Beijing, China
Laboratory of Riverine Ecological Conservation and Technology, Chinese Research Academy of Environmental Sciences, Beijing, China

Xuwang Yin
Liaoning Provincial Key Laboratory for Hydrobiology, College of Fisheries and Life Science, Dalian Ocean University, Dalian, China

Yuan Zhang
State Key Laboratory of Environmental Criteria and Risk Assessment, Chinese Research Academy of Environmental Sciences, Beijing, China
Laboratory of Riverine Ecological Conservation and Technology, Chinese Research Academy of Environmental Sciences, Beijing, China

Aisha Bello-Dambatta
Centre for Water Systems, College of Engineering, Mathematics and Physical Sciences; University of Exeter; North Park Road; Exeter EX4 4QF, United Kingdom

Zoran Kapelan
Centre for Water Systems, College of Engineering, Mathematics and Physical Sciences; University of Exeter; North Park Road; Exeter EX4 4QF, United Kingdom

David Butler
Centre for Water Systems, College of Engineering, Mathematics and Physical Sciences; University of Exeter; North Park Road; Exeter EX4 4QF, United Kingdom

Chew-Hung Chang
National Institute of Education, Nanyang Technological University, 1 Nanyang Walk, Singapore

Kim N. Irvine
National Institute of Education, Nanyang Technological University, 1 Nanyang Walk, Singapore

Sushban Shrestha
Trimble Navigation Limited, Auburn, Alabama 36830, USA

Xing Fang
Department of Civil Engineering, Auburn University, Alabama 36849, USA

Rajesh Sawant
School of Forestry and Wildlife Sciences, Auburn University, Alabama 36849, USA

Luke J. Marzen
Department of Geology and Geography, Auburn University, Alabama 36849, USA

Sophia Barinova
Institute of Evolution, University of Haifa, 199 Aba Khoushy Ave., Mount Carmel, Haifa, 3498838, Israel

Viktor Gabyshev
Institute for Biological Problems of Cryolithozone Sb Ras, 41 Lenin Avenue, Yakutsk 677980, Russia

Olga Gabysheva
Institute for Biological Problems of Cryolithozone Sb Ras, 41 Lenin Avenue, Yakutsk 677980, Russia

Jianhong E. Mu
Department of Applied Economics, Oregon State University, Corvallis, OR, 97330, USA

Bruce A. McCarl
Department of Agricultural Economics, Texas A&M University, College Station, TX, 77843, USA

Ximing Wu
Department of Agricultural Economics, Texas A&M University, College Station, TX, 77843, USA

Michael P. Ward
Faculty of Veterinary Science, University of Sydney, Camden, NSW, 2570, Australia

Frank Joseph Wambura
Department of Housing and Infrastructure Planning, Ardhi University, Box 36176, Dar es Salaam, Tanzania

Richard T. Cooper
Southeast Asia START Regional Center, Chulalongkorn University, Bangkok, Thailand

Anthony O. Onoja
Department of Agricultural Economics and Extension, University of Port Harcourt, Nigeria

A. I. Achike
Department of Agricultural Economics, University of Nigeria, Nsukka, Nigeria